Molecular Biology of RNA

Molecular Biology of RNA

Editor: Anna Robinson

Callisto Reference
www.callistoreference.com

Callisto Reference,
118-35 Queens Blvd., Suite 400,
Forest Hills, NY 11375, USA

Visit us on the World Wide Web at:
www.callistoreference.com

ISBN: 978-1-64116-760-4 (Hardback)

Cataloging-in-Publication Data

Molecular biology of RNA / edited by Anna Robinson.
p. cm.
Includes bibliographical references and index.
ISBN 978-1-64116-760-4
1. RNA. 2. Molecular biology. 3. Molecular genetics. I. Robinson, Anna.
QP623 .M65 2023
572.88--dc23

Table of Contents

Preface

This book has been an outcome of determined endeavour from a group of educationists in the field. The primary objective was to involve a broad spectrum of professionals from diverse cultural background involved in the field for developing new researches. The book not only targets students but also scholars pursuing higher research for further enhancement of the theoretical and practical applications of the subject.

Ribonucleic acid (RNA) is a nucleic acid which plays an integral role in coding, decoding, regulation and expression of genes. Some of the different types of RNA are messenger RNA, non-coding RNA, small chain RNA and long chain RNA. Messenger RNA transmits the information from the DNA to the ribosome. Non-coding RNAs (ncRNAs) regulate the expression of genes at the transcriptional and post-transcriptional level. There are two major subtypes of regulatory non-coding RNAs, namely, long non-coding RNA (lncRNA) and micro RNA (miRNA). They both interact not only with each other but also with other intracellular components to modulate the inter-related steps and mediators of regulated forms of cell death, such as necrosis and apoptosis. As a result, they could be in a key position for regulating cell death. This book provides significant information to help develop a good understanding of the molecular biology of RNA. Those with an interest in this field would find it helpful.

It was an honour to edit such a profound book and also a challenging task to compile and examine all the relevant data for accuracy and originality. I wish to acknowledge the efforts of the contributors for submitting such brilliant and diverse chapters in the field and for endlessly working for the completion of the book. Last, but not the least; I thank my family for being a constant source of support in all my research endeavours.

Editor

1

miR-125b Regulates the Early Steps of ESC Differentiation through Dies1 in a TGF-Independent Manner

Marica Battista [1,2,3], Anna Musto [1,2], Angelica Navarra [1,2], Giuseppina Minopoli [1,2], Tommaso Russo [1,2] and Silvia Parisi [1,2,*]

[1] Department of Molecular Medicine and Medical Biotechnology, University of Naples, "Federico II", Via Sergio Pansini 5, 80131 Naples, Italy;
E-Mails: battistam@ceinge.unina.it (M.B.); anna.musto@unina.it (A.M); angelica.navarra@unina.it (A.N.); minopoli@dbbm.unina.it (G.M.); tommaso.russo@unina.it (T.R.)

[2] Ceinge Biotecnologie Avanzate, Via Gaetano Salvatore 486, 80145 Naples, Italy

[3] European School of Molecular Medicine (SEMM), 80145 Naples, Italy

* Author to whom correspondence should be addressed; E-Mail: silvia.parisi@unina.it

---

Abstract: Over the past few years, it has become evident that the distinctive pattern of miRNA expression seen in embryonic stem cells (ESCs) contributes to important signals in the choice of the cell fate. Thus, the identification of miRNAs and their targets, whose expression is linked to a specific step of differentiation, as well as the modulation of these miRNAs, may prove useful in the learning of how ESC potential is regulated. In this context, we have studied the expression profile of miRNAs during neural differentiation of ESCs. We have found that miR-125b is upregulated in the first steps of neural differentiation of ESCs. This miRNA targets the BMP4 co-receptor, Dies1, and, in turn, regulates the balance between BMP4 and Nodal/Activin signaling. The ectopic expression of miR-125b blocks ESC differentiation at the epiblast stage, and this arrest is rescued by restoring the expression of Dies1. Finally, opposite to miR-125a, whose expression is under the control of the BMP4, miR-125b is not directly regulated by Transforming Growth Factor beta (TGFβ) signals. These results highlight a new important role of miR-125b in the regulation of the transition from ESCs to the epiblast stage and add a new level of control on TGFβ signaling in ESCs.

Keywords: miRNAs; TGFβ; embryonic stem cells; differentiation; epiblast stem cells; signaling

1. Introduction

Embryonic stem cells (ESCs) have the unique characteristics of self-renewing in culture and generating a wide range of specialized cell types upon differentiation cues. For their characteristics, they represent an appealing system to study embryonic development and, more importantly, to obtain functional cells for replacement therapy. Of course, a precise and detailed knowledge of the gene programs regulating ESC fate is necessary to allow the use of these cells. Among the signals that regulate ESC fate, there are extrinsic signals, such as Leukemia Inhibitory Factor (LIF) and Transforming Growth Factor beta (TGFβ). The LIF pathway is already well characterized, and it is required to maintain the undifferentiated state of ESCs [1]. TGFβ pathways, *i.e.*, Bone Morphogenetic Protein 4 (BMP4) and Nodal/Activin, have multiple functions, both in stemness maintenance and during the differentiation [2–6], and they seem to work in a perfectly balanced way, so that a misregulation of one of these pathways has relevant consequences on the other ones [7,8]. Another important and widely studied class of regulators of ESC gene programs are the transcription factors (TFs), which have the important characteristic that one TF may control the expression of numerous genes to execute whole differentiation programs [9,10]. Another emerging cohort of molecules holding the same ability is that of miRNAs, the non-coding RNA assigned to regulate post-transcription gene expression. Like TFs, the temporal expression of miRNAs is highly regulated and responsive to changes, depending on the stem cell status. With the aim of deeply understanding the biology of ESCs and to learn more in the control of their potential, these cells have been profiled using different methods to identify miRNAs that have potential roles in stemness and differentiation [11,12]. These studies have revealed several miRNA families that are highly expressed, specifically, in undifferentiated cells [13], as well as the miRNAs that are specifically expressed when differentiation occurs, such as the well-known family of let-7 [14]. Over the past few years, it has become apparent that the distinctive pattern of miRNA expression seen in ESCs contributes to many of the unique phenotypic properties of these cells. Indeed, the switch from pluripotent to lineage-specific cells is marked by the downregulation of pluripotency markers and the activation of lineage-specific gene expression, which are accompanied by changes in the expression of many miRNAs. Whereas some miRNAs function in promoting exit from the pluripotent state by targeting pluripotency factors, other miRNAs stabilize the pluripotent state [15–18]. The introduction or depletion of miRNAs involved in regulating ESC fate may be useful in inducing differentiation along a particular lineage. Recent discoveries have revealed a model in which miRNA regulatory events are linked with transcription factor and signaling networks that control cell fate and differentiation, modulating their activity through positive and negative feedback loops to modulate stem cell fate decisions [19–21]. We have performed a miRNA profiling in ESCs undergoing neural differentiation, and we have demonstrated that some miRNAs are able to control ESC differentiation by targeting important regulators of chromatin remodeling [22]. Furthermore, we have recently illustrated a model in which miR-125a is linked to the signaling of BMP4 and Nodal/Activin, modulating their activity through negative feedback loops to regulate ESC fate decisions [8]. Another miRNA belonging to the same family of miR-125a is miR-125b. We have analyzed the function of this miRNA, which shows an interesting expression profile during ESC differentiation. In this paper, we show that the modulation of this miRNA results in significant changes of the ESC state. These effects are due to the modulation of BMP4 and Nodal signaling during the first step of ESC differentiation. Moreover, we found that miR-125b, opposite to that observed for miR-125a, does not convey the control from TGFβ signaling.

2. Results and Discussion

2.1. miR-125b Overexpression Blocks ESC at the Epiblast Stage

We screened for miRNAs differentially regulated during ESC neural differentiation and found many miRNAs that are specifically expressed in undifferentiated or differentiated cells [22]. Among these miRNAs, we found that miR-125b is expressed at low levels in ESCs, but its expression increases during the first steps of differentiation. At later time points, miR-125b expression reaches a higher level in differentiated cells (Figure 1A), and in agreement, it is highly expressed in many adult mouse tissues (Figure 1B). We have previously demonstrated that miR-125b, as well as miR-125a, are able to regulate the expression of the BMP4 co-receptor, Dies1 [8], in ESCs, but we still don't know whether miR-125b may have a role in the early phases of ESC differentiation, when BMP4 regulates the differentiation fate of these cells. The analysis of undifferentiated markers clearly indicated that miR-125b overexpression (Figure S1) does not impair the undifferentiated state of ESCs (Figure 2A); thus, we explored whether miR-125b affects the differentiation program. ESCs transfected with pre-miR-125b were induced to differentiate as serum-free embryoid bodies (SFEBs), which mainly give rise to neuroectoderm derivatives at four days of differentiation. We found that the overexpression of miR-125b blocks ESC differentiation. Indeed, we found a decrease of neuroectodermal markers, whereas the expression of stemness markers is maintained at a high level (Figure 2B,C and Figure S2). These effects are accompanied by an impairment of ERK activation that suggests the failure of proper differentiation (Figure 2D). Considering that SFEB differentiation favors the transition through the epiblast stage that then leads to the formation of neuroectoderm, we performed an analysis to see if the block of differentiation observed upon miR-125b overexpression occurs before or after the epiblast transition. We found that the epiblast markers, Fgf5, Cerberus and Dnmt3b (Figure 2E), were significantly high at four days of differentiation, indicating that the cells are blocked in the epiblast stage. To verify that miR-125b overexpression is able to maintain the epiblast stem cell (EpiSC) phenotype, rather than simply slowing down the differentiation, we analyzed the methylation state of epiblast marker genes. We found that differentiated cells at day 4 upon miR-125b overexpression had epigenetic markers similar to those of EpiSCs (Figure 2F), thus suggesting, again, that these cells may still be pluripotent. To verify this hypothesis, we decided to test the pluripotency of this epiblast cell obtained upon miR-125b overexpression *in vivo*. Thus, we injected, into immunodeficient mice, cells transfected with miR-125b or with a control miR and pre-differentiated *in vitro* for three days. We found that miR-125b overexpressing cells differentiated for three days are still able to form an extensive differentiated teratoma (Figure 2G) in four out of the five mice injected with the cells. The control cells induced the formation of a small, not completely differentiated tumor (data not shown) only in one mouse over five injected. These results demonstrated that miR-125b overexpression is able to sustain the undifferentiated phenotype, even three days after the induction of differentiation and that a large fraction of these cells maintains the pluripotency.

Figure 1. miR-125b expression in embryonic stem cells (ESCs) and mouse tissues. (**A**) miR-125b expression levels were analyzed by qPCR in undifferentiated ESCs and during neural differentiation through serum-free embryoid bodies (SFEBs) formation; (**B**) Analysis of miR-125b expression in mouse adult tissues. The data were normalized to the U6 internal control (* $p < 0.05$).

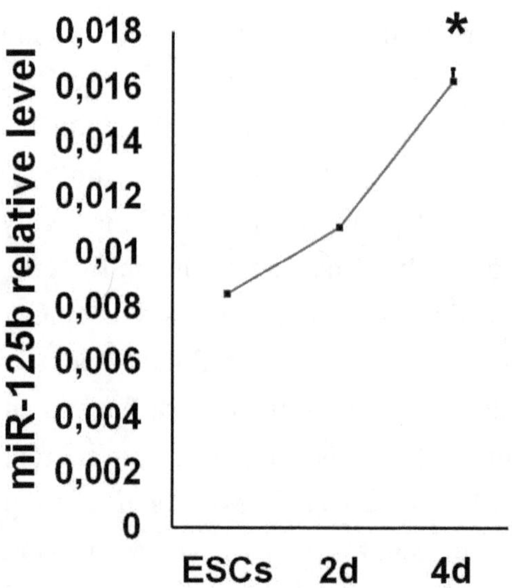

(A)

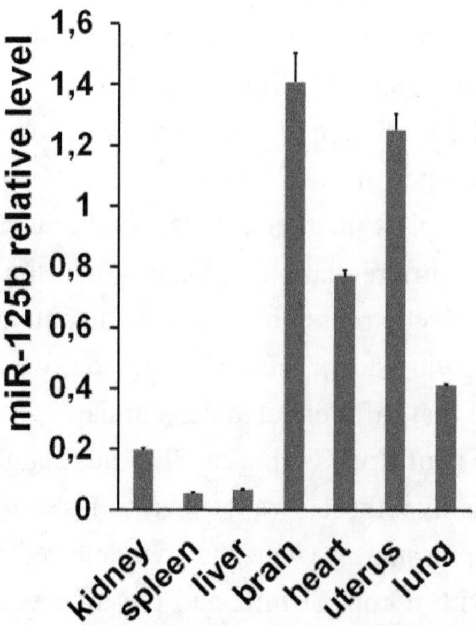

(B)

Figure 2. Effects of miR-125 ectopic expression *in vitro* and *in vivo*. (**A**) Analysis of stemness markers (Oct3/4, Nanog, Klf2 and Klf5) in undifferentiated ESCs transfected with pre-miR-125b or with control pre-miR (pre-miR-ctrl). The fold change is calculated by assigning the arbitrary value, one, to the amount found in cells transfected with control pre-miRNA; (**B**) Immunofluorescence analysis of four-day differentiated SFEBs upon miR-125b overexpression. Markers of pluripotency (Oct3/4, Nanog) and neuroectoderm (Sox1) are shown. Scale bar: 50 μm; (**C**) qPCR analysis of stemness (Oct3/4 and Nanog) and neuroectodermal (Pax6) markers in differentiating ESCs upon miR-125b overexpression. The fold change is calculated by assigning the arbitrary value, one, to the time point showing the highest amount of the indicated mRNA; (**D**) The level of active ERK (P-ERK) were analyzed by Western blot in cells transfected with pre-miR-125b or the control pre-miR after four days of differentiation—the experiment shown in the Figure is representative of two independent experiments; (**E**) The level of the epiblast marker, Fgf5, was measured by qPCR in undifferentiated ESCs and during differentiation upon pre-miR transfection. The epiblast markers, Cerberus and Dnmt3b, were measured at four days of SFEB differentiation in the cells transfected with the indicated pre-miR. The fold change is calculated as indicated in (**C**); (**F**) ChIP-qPCR analysis was performed on chromatin from ESCs transfected with the indicated pre-miR and induced differentiation for four days as SFEBs. The graphs show the methylation state of histone H3 on the promoters of pluripotency (Nanog and Klf2) and epiblast (Fgf5) markers. Data are expressed as fold enrichment relative to the control; (**G**) Immunodeficient mice were injected with ESCs transfected with the pre-miR-125b (right side) and ctrl pre-miR (left side) after three days of differentiation *in vitro* (left panel). Teratomas generated by ESCs overexpressing miR-125b were explanted after one month, and the tissues were analyzed after eosin-hematoxylin staining (right panels) (* $p < 0.05$).

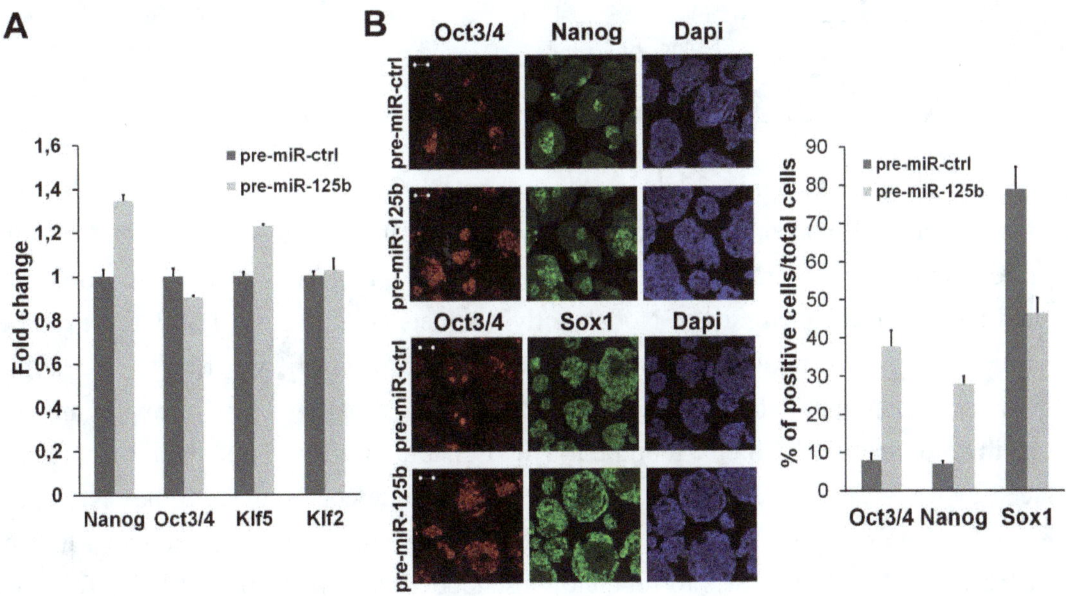

Figure 2. *Cont.*

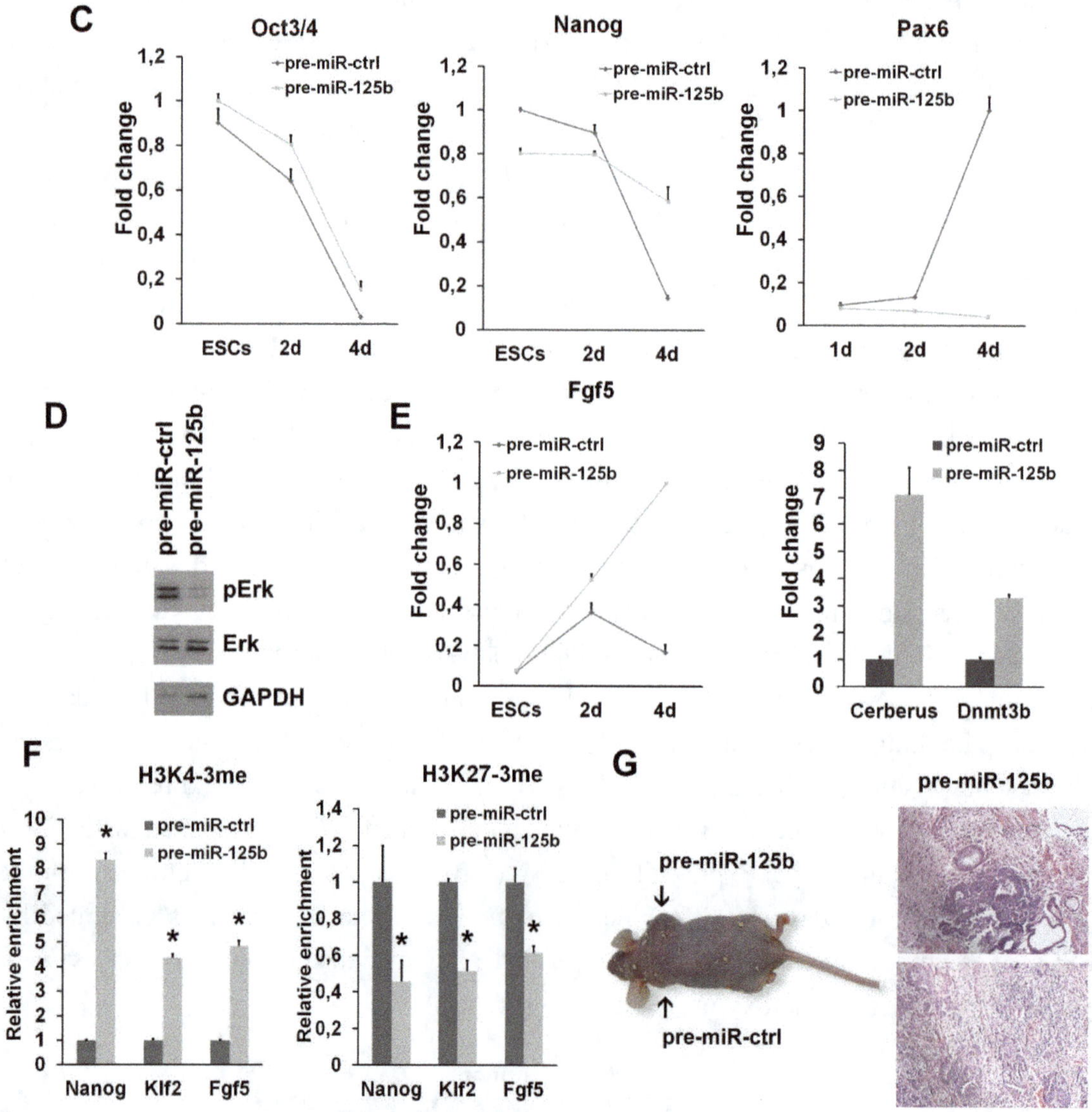

2.2. miR-125b Effects on the ESC-Epiblast Transition Are Due to Dies1

Based on the evidence that miR-125b is able to control the expression of Dies1 [8], the BMP4 co-receptor, we analyzed whether miR-125b overexpression alters the signaling of BMP4. miR-125b overexpression induced a significant decrease of BMP4 targets during ESC differentiation, accompanied by an evident increases of Nodal/Activin targets (Figure 3A). To address whether these effects of miR-125b overexpression are due to the suppression of Dies1, we tried to rescue the proper differentiation and the proper balance between BMP4 and Nodal pathways by re-expressing a form of Dies1 insensitive to the miR-125b. We found that Dies1 is able to fully rescue the block at the epiblast stage induced by miR-125b (Figure 3B,C). Moreover, this rescue corresponds to the restoration of the proper expression levels of BMP4 and Nodal/Activin targets (Figure 3D). A recent paper has indicated that miR-125b targets Lin28 in ESCs to regulate mesendodermal differentiation [23]. To explore this point, we analyzed the mRNA and protein levels of Lin28 upon miR-125b overexpression during SFEB differentiation. Interestingly, we found that Lin28 expression is not impaired in this context (Figure 3E), indicating that the block at the epiblast stage induced by miR-125b overexpression is not due to the repression of Lin28. Instead, we found that miR-125a overexpression induces a slight decrease of Lin28 protein level, suggesting that these two miRNAs can act during the early phases of ESC differentiation by regulating different subsets of targets.

Figure 3. The effects of miR-125b overexpression on TGFβ signaling are mediated by the BMP4 co-receptor, Dies1. (**A**) qPCR analysis of the expression levels of BMP4 (Id1, Id3) and Nodal/Activin (Nodal, Cripto, Lefty1, Lefty2) target genes upon miR-125b overexpression; (**B**) Analysis of the phenotype of ESCs co-transfected with the indicated pre-miR and with the vector expressing Dies1 lacking its 3'UTR or with the empty vector (mock). The expression of stemness (Oct3/4) and neuroectodermal (Sox1) markers was analyzed by immunostaining in cells differentiated as SFEBs for four days. Scale bar: 20 μm; (**C**) q-PCR analysis of the effects of Dies1 re-expression in ESCs transfected with the indicated pre-miR. After four days of differentiation, the expression of stemness (Oct3/4, Nanog) and epiblast (Fgf5) markers was analyzed; (**D**) q-PCR analysis of the expression of BMP4 (Id1) and Nodal/Activin (Lefty1 and Lefty2) targets in four-day differentiated ESCs re-expressing, or not, Dies1 upon miR-125b overexpression. Data in (**C**) and (**D**) are shown as fold changes relative to cognate controls; (**E**) The effects of miR-125a and b overexpression on Lin28 level were analyzed in four-day differentiated SFEBs by means of q-PCR (left panel) and Western blot (right panel). Data are expressed as fold change relative to the control (* $p < 0.05$).

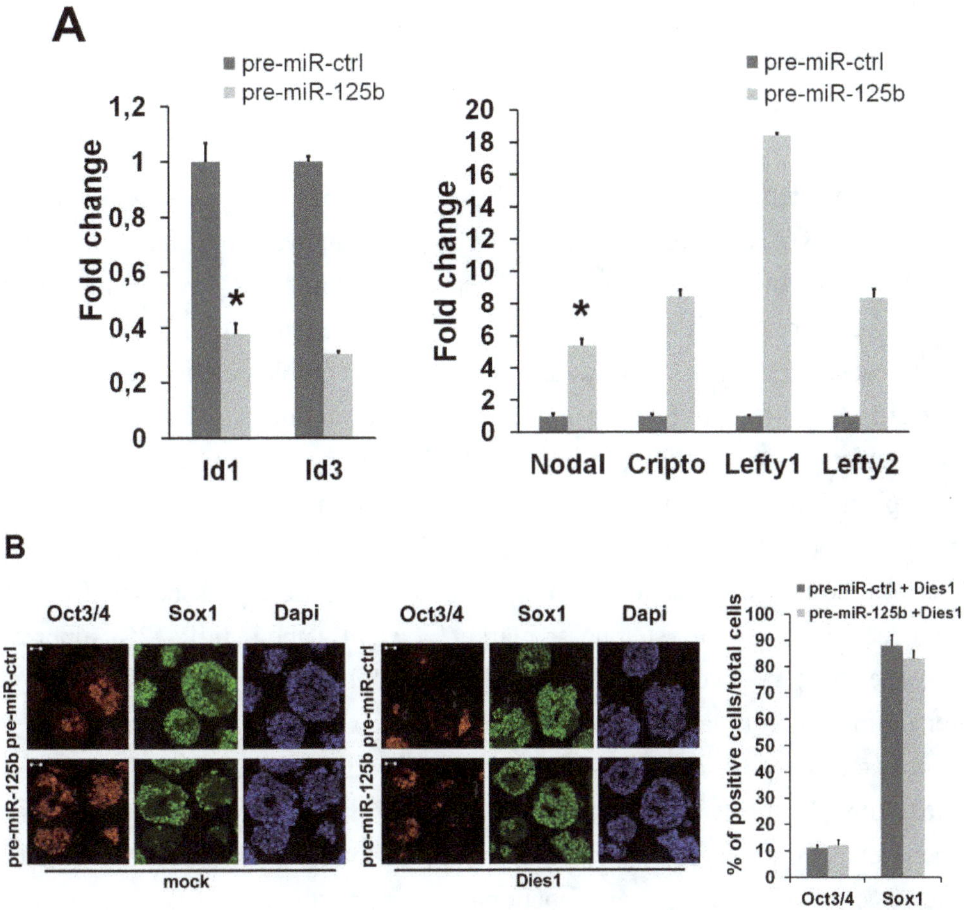

Figure 3. *Cont.*

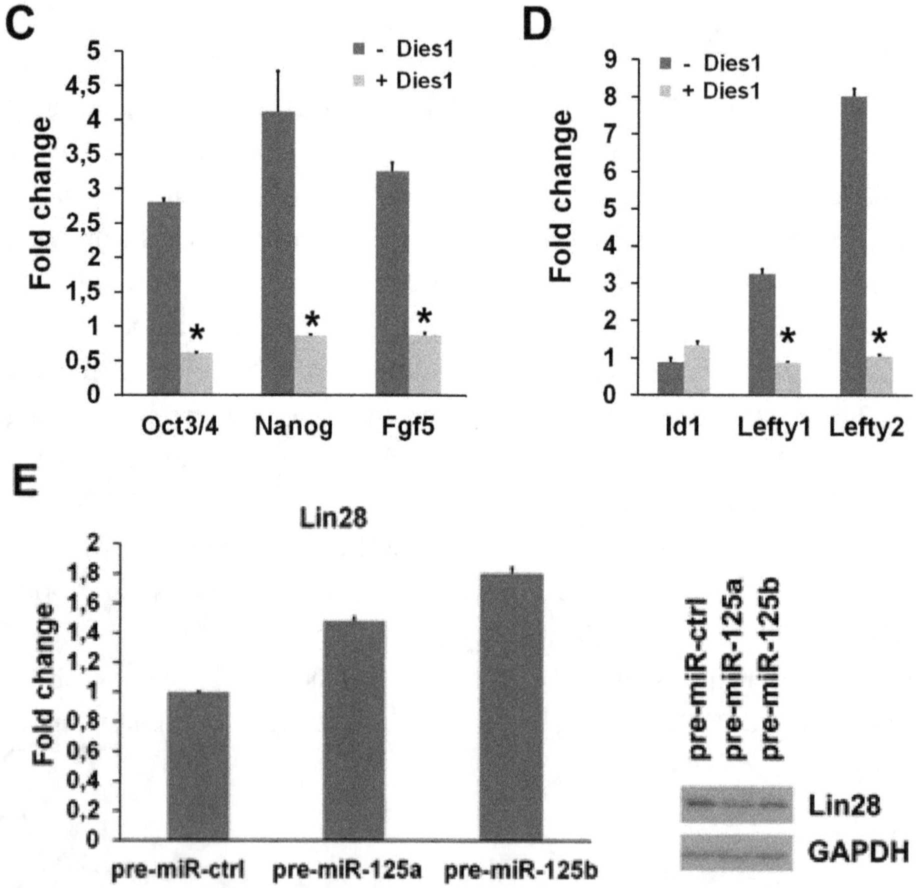

2.3. miR-125 Suppression Promotes ESC Differentiation

Considering the evident phenotype induced by miR-125b overexpression, we asked if the suppression of this miR may have an effect on ESCs. To analyze the effects of the depletion of miR-125b in ESC differentiation, we suppressed the endogenous miRNAs by transfecting a mix of anti-miR-125a and anti-miR-125b to avoid the levels of endogenous miR-125a from being able to be substituted for the absence of miR-125b. An anti-miR with no complementarity to any known miRNAs was used as negative control. After anti-miR transfection, we cultured ESCs at low density, and after seven days, we performed an alkaline phosphatase (AP) staining to assay their ability to maintain an undifferentiated phenotype. As shown in Figure 4A, the cells transfected with anti-miR-125 showed a reduction in the number of AP-positive colonies compared to the control, indicating that miRNA suppression causes the loss of the undifferentiated state, even in the presence of LIF. Moreover, we found a rapid decrease of the stemness marker, Oct3/4, upon miR-125 suppression during differentiation, already at day two of SFEB formation, which became more evident at day four (Figure 4B). To better define the correlation between miR-125b and Dies1 in this context, we analyzed the effects of Dies1 ectopic expression in ESCs in the same conditions used for miRNA suppression. We found that Dies1 forced expression induces the decrease of the number of AP-positive colonies, thus resembling the phenotype observed upon miR-125 suppression (Figure 4C). These results support the idea that the levels of the two miR-125 are functionally correlated to that of Dies1 in ESCs.

Figure 4. mir-125 suppression induces ESC differentiation. (**A**) Alkaline phosphatase (AP) staining was performed on cells transfected with the mix of anti-miR-125a and b (anti-miR-mix) and with the control anti-miR (anti-miR-ctrl) and cultured for seven days at clonal density in the presence of Leukemia Inhibitory Factor (LIF). The histogram represents the number of AP-positive and -negative colonies; (**B**) q-PCR analysis showing the expression of Oct3/4 in differentiating ESCs upon suppression of miR-125a and b. Data are expressed as fold change relative to the control; (**C**) AP staining of ESCs overexpressing Dies1 after seven days of culture at clonal density in the presence of LIF. The number of AP-positive and -negative colonies is reported in the graph (* $p < 0.05$).

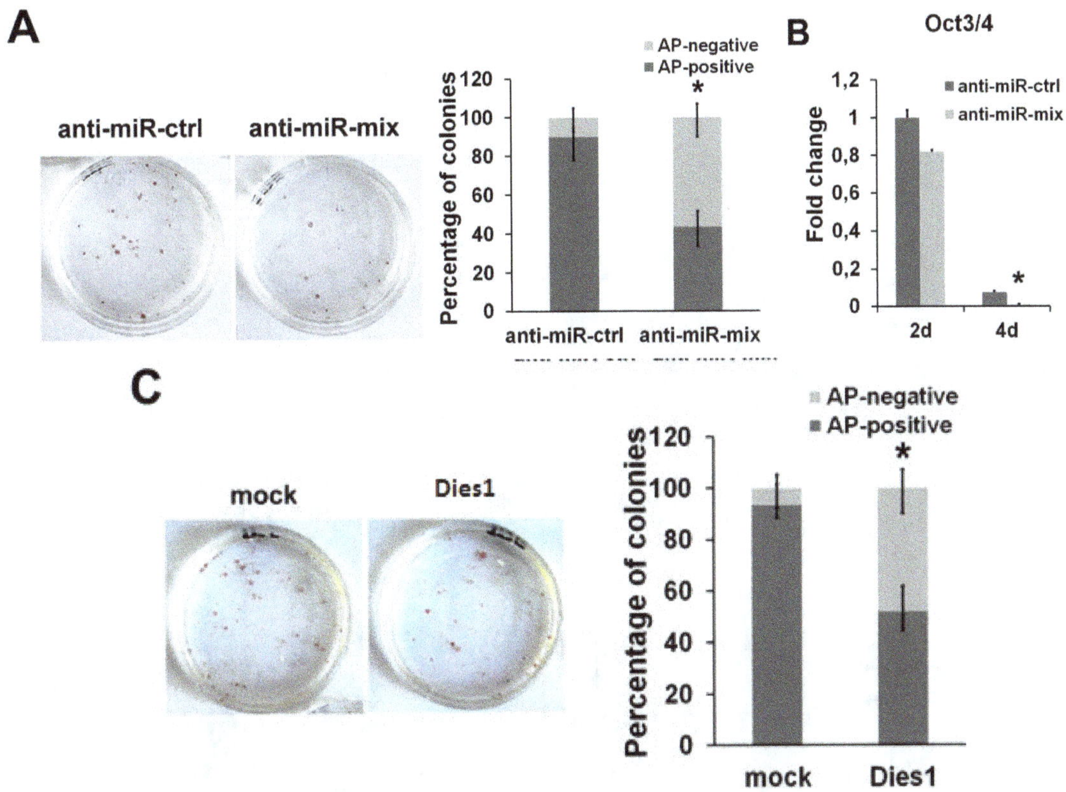

2.4. miR-125b Is Not Regulated by a TGFβ Signaling

We have shown that miR-125b overexpression impairs both BMP4 and Nodal/Activin pathways. To understand if the miR-125b can be directly regulated by one of these signalings, we exposed ESCs to BMP4 or Activin and analyzed the level of the two pri-miR-125b transcribed from the two miR-125b genes and the mature miR-125b. As shown in Figure 5A, we did not find any significant changes in the expression of miR-125b upon BMP4 or Activin treatment, indicating that this miR is not directly regulated by these two pathways. To verify that this independence of miR-125b from TGFβ is not context-dependent, we analyzed the changes in the possible expression of miR-125b in a different experimental setting. To this aim, we used C2C12, a system in which both miR-125b and BMP4 are able to regulate the differentiation [24,25] and in which miR-125b is highly expressed (Figure 5B). After the exposure of these cells to BMP4, we didn't find any changes in the expression level of miR-125b (Figure 5C). All these data suggest that miR-125b is able to impair the TGFβ pathway, but is not directly controlled by these molecules.

Figure 5. miR-125b expression is independent from TGFβ. (**A**) Analysis of the level of miR-125b after the exposure of ESCs to BMP4 or Activin. The expression of pri-miRs (left panel) was measured after 1 h of treatment with the indicated molecules by using specific primers that distinguish between the transcripts deriving from the two miR-125b genes (pri-miR-125b-1 and pri-miR-125b-2). The level of mature miR-125b was measured after 24 h of BMP4 or Activin treatment; (**B**) Expression level of miR-125b in undifferentiated and four-day differentiated C2C12 cells; (**C**) q-PCR analysis of the expression of miR-125b in C2C12 cells after 24 h of treatment with BMP4. All the data are expressed as fold change relative to the control.

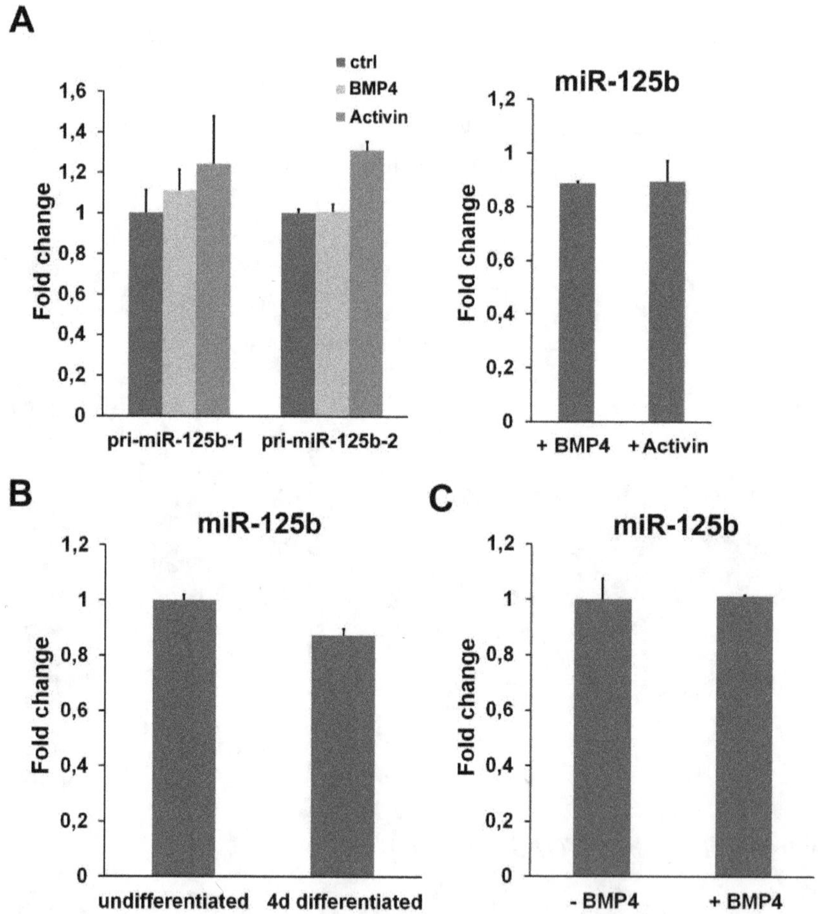

3. Experimental Section

3.1. Cell Culture, Transfection and Differentiation

E14Tg2a mouse ESCs (BayGenomics, San Francisco, CA, USA) were maintained on feeder-free, gelatin-coated plates in the following medium: Glasgow Minimum Essential Medium (GMEM, Sigma, St. Louis, MO, USA) supplemented with 2 mM glutamine (Invitrogen, Carlsbad, CA, USA), 1 mM sodium pyruvate (Invitrogen, Carlsbad, CA, USA), 1× nonessential amino acids (Invitrogen, Carlsbad, CA, USA), 0.1 mM β-mercaptoethanol (Sigma, St. Louis, MO, USA), 10% FBS (Hyclone,Watham, MA, USA) and 10^3 U/mL Leukemia Inhibitory Factor (LIF, Millipore, Billerica, MA, USA). Serum-free embryoid body (SFEB) differentiation was performed by plating 1×10^6 ESCs in 100 mm Petri dishes (BD Biosciences, San Jose, CA, USA) in the following differentiation medium: GMEM supplemented with 2 mM glutamine, 1 mM sodium pyruvate, 1× nonessential amino acids, 0.1 mM β-mercaptoethanol and 10% Knock-out Serum Replacement (KSR, Invitrogen, Carlsbad, CA, USA). In this condition, after 2 days, the cells express a high level of epiblast markers that strongly decrease at 4 days of SFEB differentiation. EpiSCs can be derived with high yield at 2 days of SFEB differentiation, whereas at 3 and 4 days of differentiation, the yield of EpiSC derivation strongly decreases [5].

C2C12 myoblasts were maintained in Dulbecco's Modified Eagle Medium (Sigma, St. Louis, MO, USA) supplemented with 2 mM glutamine (Invitrogen, Carlsbad, CA, USA), 100 U/mL penicillin/streptomycin (Invitrogen, Carlsbad, CA, USA) and 10% FBS (GIBCO, Life Technologies Italia, Monza, Italy). For differentiation to myotubes, 20×10^3 cells/cm^2 were plated, and the following day, the medium was substituted with the following differentiation medium: DMEM with 2 mM glutamine (Invitrogen, Carlsbad, CA, USA) and 2% horse serum (Sigma, St. Louis, MO, USA). After 4 days of differentiation, the cells were collected.

Transfection of pre-miRs, anti-miRs (both from Ambion, Austin, TX, USA) and plasmid for Dies1 expression was performed using Lipofectamine 2000 (Invitrogen, Carlsbad, CA, USA), following the manufacturer's instructions.

3.2. Cell Treatment and Alkaline Phosphatase Staining

For cell treatment, ESCs and C2C12 were grown overnight in KSR containing medium with LIF or DMEM plus 1% FBS, respectively and, then, treated for the indicated time with 20 ng/mL of BMP4 or Activin (both from R&D, Minneapolis, MN, USA).

Alkaline phosphatase staining was performed culturing ESCs at clonal density (20–50 cells/cm^2). After 7 days, the cells were fixed in 10% cold Neutral Formalin Buffer (10% formalin, 110 mM Na_2HPO_4, 30 mM NaH_2PO_4 in H_2O) for 15 min and, then, rinsed in distilled water for 15 min. The staining was obtained by incubation for 45 min at room temperature with the following staining solution: 0.1 M Tris-HCl, 0.01% naphthol AS MX-PO4 (Sigma, St. Louis, MO, USA), 0.4% *N*,*N*-dimethylformamide (Sigma, St. Louis, MO, USA) and 0.06% Red Violet LB salt (Sigma, St. Louis, MO, USA).

3.3. RNA Isolation, q-PCR and TaqMan Analysis

Total RNA from ESCs and C2C12 was extracted by using TRI-Reagent (Sigma, St. Louis, MO, USA). For the first-strand cDNA synthesis, the manufacturer's instructions (M-MLV RT, New England BioLabs, Ipswich, MA, USA) were followed. Q-PCR was carried out on an ABI PRISM 7900HT Sequence Detection System (Applied Biosystems, Foster City, CA, USA) using Power SYBR Green PCR Master mix (Applied Biosystems, Foster City, CA, USA). The housekeeping GAPDH mRNA was used to normalize the samples, using the $2^{-\Delta\Delta Ct}$ method. The gene-specific primers used are listed in Table S1.

For the measurement of mature miRNA, RNA was extracted with a mirVana microRNA Isolation kit (Ambion, Austin, TX, USA), according to the manufacturer's instructions. Single-stranded cDNA was synthetized from 10 ng of total RNA combined with the specific primer for miR-125a or miR-125b or U6 as internal control by using a TaqMan MicroRNA reverse transcription kit (Applied Biosystems, Foster City, CA, USA). miRNA levels were measured by using a TaqMan MicroRNA detection kit (Applied Biosystems, Foster City, CA, USA) with the 7500 Real Time PCR System instrument and the Sequence Detection Systems (SDS) software version 1.4 (Applied Biosystems, Foster City, CA, USA) [26].

3.4. Northern Blot Analysis

For Northern blot analysis total RNA was isolated by using the TRI Reagent (Sigma, St. Louis, MO, USA), according to the manufacturer's instructions. For each sample, 20 μg of total RNA were fractionated on 15% TBE-Urea gel (Criterion precast Gel, Bio-Rad, Segrate, Italy), stained with EtBr (Sigma, St. Louis, MO, USA) for loading control, then transferred to a Hybond membrane (Amersham Pharmacia, Milan, Italy) and, finally, fixed by UV cross-linking in a Stratalinker (Stratagene,La Jolla, CA), according to the manufacturer's instruction. The membrane was hybridized with 10 pmol of miRCURY LNA Detection Probe, digoxigenin labeled (Exiqon, Vedbaek, Denmark), according to the manufacturer's instruction. The signal detection was obtained by an anti-digoxigenin-alkaline phosphatase antibody (Roche, Basel, Switzerland) and a chemiluminescent substrate (CDP-Star, Roche, Basel, Switzerland), according to the manufacturer's instruction.

3.5. Protein Extraction and Western Blot Analysis

For protein extracts, cells were lysed in a buffer containing 20 mM Tris-HCl (pH 7.5), 150 mM NaCl, 1 mM EDTA, 1% Triton, 1% sodium deoxycholate and protease inhibitor cocktail (Sigma, St. Louis, MO, USA). The antibodies used for blotting were: Anti-Lin28 (1:700, Abcam,Cambridge, UK); anti-phospho-Erk1 (1:1000, Cell Signaling, Danvers, MA, USA); anti-Erk1 (1:1000, Santa Cruz, Santa Cruz, CA, USA); anti Oct3/4 (1:1000, Santa Cruz, Santa Cruz, CA, USA); and anti-GAPDH (1:1000, Santa Cruz, Santa Cruz, CA, USA).

3.6. Immunostaining

For immunostaining, 4-day differentiated SFEBs were fixed in 4% paraformaldehyde and processed, as described in Parisi *et al.*, 2012 [12]. The following primary antibodies were used: anti-Oct3/4 (1:200, Santa Cruz, Santa Cruz, CA, USA), anti-Nanog (1:500, Calbiochem, San Diego, CA, USA) and anti-Sox1 (1:100, Santa Cruz, Santa Cruz, CA, USA). The appropriate secondary antibodies were used (1:400, Alexa Molecular Probes, Invitrogen, Carlsbad, CA, USA). Confocal microscopy was performed with an LSM 510 Meta microscope (Zeiss, Milan, Italy) using LSM 510 Meta software [27] and the LSM Image Browser (Zeiss, Milan, Italy). The brightness, contrast and color balance of the images were adjusted in Photoshop CS2 (Adobe Systems, Agrate Brianza, Italy).

3.7. Teratoma Formation

ESCs transfected with pre-miR-125b or pre-miR-ctrl were differentiated as SFEBs for 3 days. Then, SFEBs were dissociated, and 2×10^6 cells were used for subcutaneous injection in nude mice. Four weeks after the injection, tumors were surgically dissected from the mice. Samples were fixed in 4% paraformaldehyde and embedded in paraffin. Sections were stained with hematoxylin and eosin.

3.8. Chromatin Immunoprecipitation (ChIP)-qPCR Analysis

Chromatin immunoprecipitation was performed as described previously [28]. To immunoprecipitate soluble chromatin extracts, anti-H3K4-3me (Millipore, Billerica, MA, USA) and anti-H3K27-3me (Millipore, Billerica, MA, USA) antibodies were used. Appropriate IgGs were used as negative control. Supernatant obtained without antibody was used as input control. After q-PCR, the amount of precipitated DNA was calculated relative to the total input chromatin and expressed as the percentage of total chromatin, according to the formula $2^{\Delta Ct}$, where Ct represents the cycle threshold and $\Delta Ct = Ct$ (input) − Ct (immunoprecipitation). The gene-specific primers used are listed in Table S1.

3.9. Statistics

Data are presented as the means ± SD of at least three independent experiments. Whenever necessary, the statistical significance of the data was analyzed using the Student's *t*-test (* $p < 0.05$).

4. Conclusions

We demonstrated that miR-125b is essential for the proper differentiation of ESCs, based on the relevant effects of its overexpression and suppression. We have previously demonstrated that miR-125a had similar effects on ESC differentiation [8], thus suggesting that these two miRNAs work cooperatively through the suppression of Dies1, the co-receptor of BMP4 [29]. This relationship is an elegant example of how multiple miRNAs can converge on a single pathway to promote a common outcome. Interestingly, while miR-125a is directly regulated by BMP4, miR-125b seems to not be regulated by TGFβ signaling. This observation suggests that the BMP4 signaling in the first steps of ESC differentiation undergoes different regulations that are dependent (miR-125a) or independent (miR-125b) by itself. This can probably be due to the relevance that the balance between BMP4 and Nodal/Activin pathways have in the control of the transition from ESC to the epiblast stage, thus indicating that ESCs modulate in different ways such important pathways.

Acknowledgments

The authors thank Danila Scarpa for precious assistance. This work was supported by grants from Associazione Italiana Ricerca sul Cancro and Italian Ministry of Research (PON 01_02782).

References

1. Niwa, H.; Ogawa, K.; Shimosato, D.; Adachi, K. A parallel circuit of LIF signalling pathways maintains pluripotency of mouse ES cells. *Nature* **2009**, *460*, 118–122.
2. Parisi, S.; D'Andrea, D.; Lago, C.T.; Adamson, E.D.; Persico, M.G.; Minchiotti, G. Nodal-dependent Cripto signaling promotes cardiomyogenesis and redirects the neural fate of embryonic stem cells. *J. Cell Biol.* **2003**, *163*, 303–314.

3. Seuntjens, E.; Umans, L.; Zwijsen, A.; Sampaolesi, M.; Verfaillie, C.M.; Huylebroeck, D. Transforming growth factor type beta and Smad family signaling in stem cell function. *Cytokine Growth Factor Rev.* **2009**, *20*, 449–458.
4. Watabe, T.; Miyazono, K. Roles of TGF-beta family signaling in stem cell renewal and differentiation. *Cell Res.* **2009**, *19*,103–115.
5. Zhang, K.; Li, L.; Huang, C.; Shen, C.; Tan, F.; Xia, C.; Liu, P.; Rossant, J.; Jing, N. Distinct functions of BMP4 during different stages of mouse ES cell neural commitment. *Development* **2010**, *137*, 2095–2105.
6. Lee, K.L.; Lim, S.K.; Orlov, Y.L.; Yit le, Y.; Yang, H.; Ang, L.T.; Poellinger, L.; Lim, B. Graded Nodal/Activin signaling titrates conversion of quantitative phospho-Smad2 levels into qualitative embryonic stem cell fate decisions. *PLoS Genet.* **2011**, doi:10.1371/journal.pgen.1002130.
7. Galvin, K.E.; Travis, E.D.; Yee, D.; Magnuson, T.; Vivian, J.L. Nodal signaling regulates the bone morphogenic protein pluripotency pathway in mouse embryonic stem cells. *J. Biol. Chem.* **2010**, *285*, 19747–19756.
8. Parisi, S.; Battista, M.; Musto, A.; Navarra, A.; Tarantino, C.; Russo, T. A regulatory loop involving Dies1 and miR-125a controls BMP4 signaling in mouse embryonic stem cells. *FASEB J.* **2012**, *26*, 3957–3968.
9. Chambers, I.; Tomlinson, S.R. The transcriptional foundation of pluripotency. *Development* **2009**, *136*, 2311–2322.
10. Parisi, S.; Russo, T. Regulatory role of Klf5 in early mouse development and in embryonic stem cells. *Vitam Horm.* **2011**, *87*, 381–397.
11. Ivey, K.N.; Srivastava, D. MicroRNAs as regulators of differentiation and cell fate decisions. *Cell Stem Cell* **2010**, *7*, 36–41.
12. Martinez, N.J.; Gregory, R.I. MicroRNA gene regulatory pathways in the establishment and maintenance of ESC identity. *Cell Stem Cell* **2010**, *7*, 31–35.
13. Houbaviy, H.B.; Murray, M.F.; Sharp, P.A. Embryonic stem cell specific microRNAs. *Dev. Cell* **2003**, *5*, 351–358.
14. Viswanathan, S.R.; Daley, G.Q.; Gregory, R.I. Selective blockade of microRNA processing by Lin28. *Science* **2008**, *320*, 97–100.
15. Sinkkonen, L.; Hugenschmidt, T.; Berninger, P.; Gaidatzis, D.; Mohn, F.; Artus-Revel, C.G.; Zavolan, M.; Svoboda, P.; Filipowicz, W. MicroRNAs control de novo DNA methylation through regulation of transcriptional repressors in mouse embryonic stem cells. *Nat. Struct. Mol. Biol.* **2008**, *15*, 259–267.
16. Zovoilis, A.; Smorag, L.; Pantazi, A.; Engel, W. Members of the miR-290 cluster modulate in vitro differentiation of mouse embryonic stem cells. *Differentiation* **2009**, *78*, 69–78.
17. Wang, Y.; Baskerville, S.; Shenoy, A.; Babiarz, J.E.; Baehner, L.; Blelloch, R. Embryonic stem cell-specific microRNAs regulate the G1-S transition and promote rapid proliferation. *Nat. Genet.* **2008**, *40*, 1478–1483.
18. Tay, Y.; Zhang, J.; Thomson, A.M.; Lim, B.; Rigoutsos, I. MicroRNAs to Nanog, Oct4 and Sox2 coding regions modulate embryonic stem cell differentiation. *Nature* **2008**, *455*, 1124–1128.

19. Juan, A.H.; Kumar, R.M.; Marx, J.G.; Young, R.A.; Sartorelli, V. Mir-214-dependent regulation of the polycomb protein Ezh2 in skeletal muscle and embryonic stem cells. *Mol. Cell* **2009**, *36*, 61–74.
20. Delaloy, C.; Liu, L.; Lee, J.A.; Su, H.; Shen, F.; Yang, G.Y.; Young, W.L.; Ivey, K.N.; Gao, F.B. MicroRNA-9 coordinates proliferation and migration of human embryonic stem cell-derived neural progenitors. *Cell Stem Cell* 2010, *6*, 323–335.
21. Krichevsky, A.M.; Sonntag, K.C.; Isacson, O.; Kosik, K.S. Specific microRNAs modulate embryonic stem cell-derived neurogenesis. *Stem Cells* 2006, *24*, 857–864.
22. Tarantino, C.; Paolella, G.; Cozzuto, L.; Minopoli, G.; Pastore, L.; Parisi, S.; Russo, T. miRNA-34a, 100, and 137 modulate differentiation of mouse embryonic stem cells. *FASEB J.* **2010**, *24*, 3255–3263.
23. Wang, J.; Cao, N.; Yuan, M.; Cui, H.; Tang, Y.; Qin, L.; Huang, X.; Shen, N.; Yang, H.T. MicroRNA-125b/Lin28 pathway contributes to the mesendodermal fate decision of embryonic stem cells. *Stem Cells Dev.* **2012**, *21*, 1524–1537.
24. Ge, Y.; Sun, Y.; Chen, J. IGF-II is regulated by microRNA-125b in skeletal myogenesis. *J. Cell Biol.* **2011**, *192*, 69–81.
25. Dahlqvist, C.; Blokzijl, A.; Chapman, G.; Falk, A.; Dannaeus, K.; Ibâñez, C.F.; Lendahl, U. Functional Notch signaling is required for BMP4-induced inhibition of myogenic differentiation. *Development* **2003**, *130*, 6089–6099.
26. *Sequence Detection Systems (SDA)*, version 1.4; software for detecting nucleic acid sequence Available online: http://www.lifetechnologies.com/global/en/home/technical-resources/software-downloads/applied-biosystems-7500-fast-real-time-pcr-system.html (accessed on 31 January, 2008).
27. *LSM 510 Meta software*; Available online: http://microscopy.zeiss.com/microscopy/n_us/downloads/lsm-5-series.html (accessed on 27 February 2007).
28. Parisi, S.; Cozzuto, L.; Tarantino, C.; Passaro, F.; Ciriello, S.; Aloia, L.; Antonini, D.; de Simone, V.; Pastore, L.; Russo, T. Direct targets of Klf5 transcription factor contribute to the maintenance of mouse embryonic stem cell undifferentiated state. *BMC Biol.* **2010**, doi:10.1186/1741-7007-8-128.
29. Aloia, L.; Parisi, S.; Fusco, L.; Pastore, L.; Russo, T. Differentiation of embryonic stem cells 1 (Dies1) is a component of bone morphogenetic protein 4 (BMP4) signaling pathway required for proper differentiation of mouse embryonic stem cell. *J. Biol. Chem.* **2010**, *285*, 7776–7783.

2

Long and Short Non-Coding RNAs as Regulators of Hematopoietic Differentiation

Franck Morceau [1,†], Sébastien Chateauvieux [1,†], Anthoula Gaigneaux [1], Mario Dicato [1] and Marc Diederich [2,*]

[1] Laboratoire de Biologie Moléculaire et Cellulaire du Cancer, Kirchberg Hospital, 9 Rue Edward Steichen, 2540, Luxembourg; E-Mails: franck.morceau@lbmcc.lu (F.M.); sebastien.chateauvieux@lbmcc.lu (S.C.); anthoula.gaigneaux@lbmcc.lu (A.G.); mdicato@gmail.com (M.D.)

[2] Department of Pharmacy, College of Pharmacy, Seoul National University, Seoul 151-742, Korea

† *These authors contributed equally to this work.*

* Author to whom correspondence should be addressed; E-Mail: marc.diederich@lbmcc.lu

---

Abstract: Genomic analyses estimated that the proportion of the genome encoding proteins corresponds to approximately 1.5%, while at least 66% are transcribed, suggesting that many non-coding DNA-regions generate non-coding RNAs (ncRNAs). The relevance of these ncRNAs in biological, physiological as well as in pathological processes increased over the last two decades with the understanding of their implication in complex regulatory networks. This review particularly focuses on the involvement of two large families of ncRNAs, namely microRNAs (miRNAs) and long non-coding RNAs (lncRNAs) in the regulation of hematopoiesis. To date, miRNAs have been widely studied, leading to a wealth of data about processing, regulation and mechanisms of action and more specifically, their involvement in hematopoietic differentiation. Notably, the interaction of miRNAs with the regulatory network of transcription factors is well documented whereas roles, regulation and mechanisms of lncRNAs remain largely unexplored in hematopoiesis; this review gathers current data about lncRNAs as well as both potential and confirmed roles in normal and pathological hematopoiesis.

Keywords: ncRNA; miRNA; lncRNA; transcription factors; regulatory network; erythropoiesis; leukemia; lymphoma; differentiation

1. Introduction

Hematopoiesis is the physiological process leading to the production of all circulating blood cells. Hematopoietic stem cells (HSCs) are pluripotent cells with high self-renewal capacity. Asymmetric division is the main approved model for self-renewal and commitment towards a specific differentiation pathway [1], which allows maintaining a steady state HSCs population by preserving hematopoietic homeostasis. As shown in Figure 1, at least two models have been proposed for the hematopoietic hierarchy.

Figure 1. Two models of the hierarchical hematopoiesis process. Hematopoietic cell differentiation proceeds by successive hierarchical maturation steps. (**A**) Pluripotent hematopoietic stem cells (HSC) give rise to multipotential progenitors (MPP) leading to common lymphocyte progenitors (CLP) and common myeloid progenitors (CMP). CLPs directly generate cells of the immune system. CMPs give rise to megakaryocyte-erythroid progenitors (MEP) and granulocyte-macrophage progenitors (GMP); (**B**) The alternative model differs by the involvement of an intermediate lymphoid-primed multipotential progenitor (LMPP) to generate GMP and CMP. Both models lead to the production of differentiated hematopoietic cells (M, monocyte; G, granulocytes; E, Erythrocyte; MK, megakaryocyte; T and B, lymphocytes; NK, natural killers DC, dendritic cells).

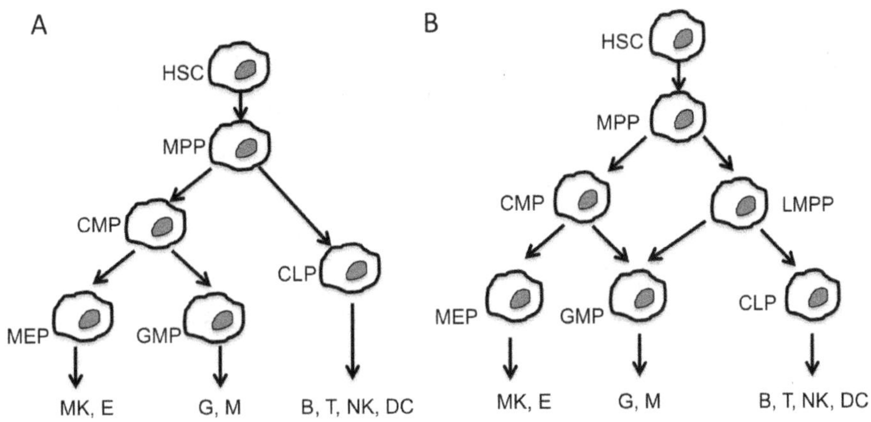

Cell fate decision is regulated by a complex network of extra- and intracellular regulatory factors to ensure HSCs commitment, survival, differentiation as well as maturation depending on physiological requirements. Many cytokines and growth factors activate cell-signaling pathways controlling posttranslational modifications of transcription factors (TF), protein-protein interactions, enzyme activation, binding of proteins to both DNA and RNA, stability of proteins and mRNAs and activation of epigenetic regulators. The ultimate step of this intricate combination of activated and inactivated proteins is the transcription or silencing of genes. To reach this status of differentiated cells, a continuous fine-tuning of genetic programs is essential for the control of the rhythm of cellular divisions, which decreases with cell differentiation, to prevent early cell death of progenitors and precursors as well as to express specific phenotypes. Besides key factors required for self-renewal and commitment of HSCs, complex regulatory networks of TFs mark the cell paths and participate up until their destiny. During hematopoiesis, TF activities depend on interactions between themselves and with cofactors. In the myeloid and lymphoid branches of hematopoiesis, several TFs are differentially involved in divergent pathways. Further TF families are crucial in hematopoietic cell fate decisions. Lichtinger *et al.*, showed that the combination and interdependent regulation of T-cell acute lymphocytic leukemia protein (TAL)1/ stem cell leukemia (SCL), Friend leukemia integration (FLI)1 and CCAAT-enhancer-binding protein (C/EBP) is required for the correct temporal expression of lineage specific genes [2]. Malinge *et al.*, reported that the TF Ikaros, which is expressed in erythro-megakaryocyte progenitors [3], blocks terminal megakaryocytic maturation through the inhibition of GATA-1 expression in correlation with target gene inhibition including LIM domain only (Lmo)2. This possibly also involves the inhibition of GATA-1 cofactor Friend of GATA (FOG)-1 by Ikaros. As an alternative mechanism, Notch signaling pathway is also affected by Ikaros-mediated inhibition of megakaryocytic differentiation. Indeed, Ikaros was shown to inhibit the Notch-induced megakaryocytic pattern from hematopoietic progenitors [4]. Lmo2 serves as a bridge between GATA-1 and the SCL complex. Tripic *et al.*, showed that SCL increases GATA-1 transcriptional activity in the murine erythroid cell line G1E-ER4, which displays an inducible GATA-1 construct. Conversely, in the absence of Lmo2, GATA-1 plays a repressive role on target genes in correlation with the absence of SCL complex association.

Besides, transcription factors play a central role in hematopoietic development, from HSC commitment to terminal differentiation and death. They are differentially and temporally expressed along the differentiation process. Furthermore, an additional level of regulation has joined the network of regulatory factors with the involvement of non-coding RNAs (ncRNAs). This term defines RNA transcripts without protein-coding capacity including constitutively expressed housekeeping small RNAs, ribosomal (rRNAs), transfer (tRNAs), small nuclear (snRNAs), small nucleolar (snoRNAs), transfer-messenger (tmRNAs) and telomerase RNAs. Furthermore, regulatory ncRNAs have been also described including the family of microRNAs (miRNAs or miRs) and the family of larger regulatory ncRNAs, long non-coding (lncRNAs) (Figure 2). This review updates knowledge about regulatory ncRNAs in hematopoiesis by especially focusing on miRNAs as well as the lncRNAs. Particular attention is given to the miRNAs/transcription factors forming regulatory network.

2. Non-Coding RNA

Genomic analyses determined that the proportion of the genome coding for proteins corresponds approximately to 1.5%, while at least two thirds are transcribed, suggesting that many non-coding sequences are transcribed into ncRNA. In addition, while protein-coding sequences yet represent a minority of the genome of multicellular organisms, their proportion further declines with increasing complexity of the organism, with a concomitant increase in the amount of non-coding regions in intergenic or intronic sequences [5–7]. A non-coding RNA (ncRNA) or non-protein-coding RNA (npcRNA) is a functional RNA molecule not translated into a protein.

Figure 2. Schematic representation of different non coding RNAs.

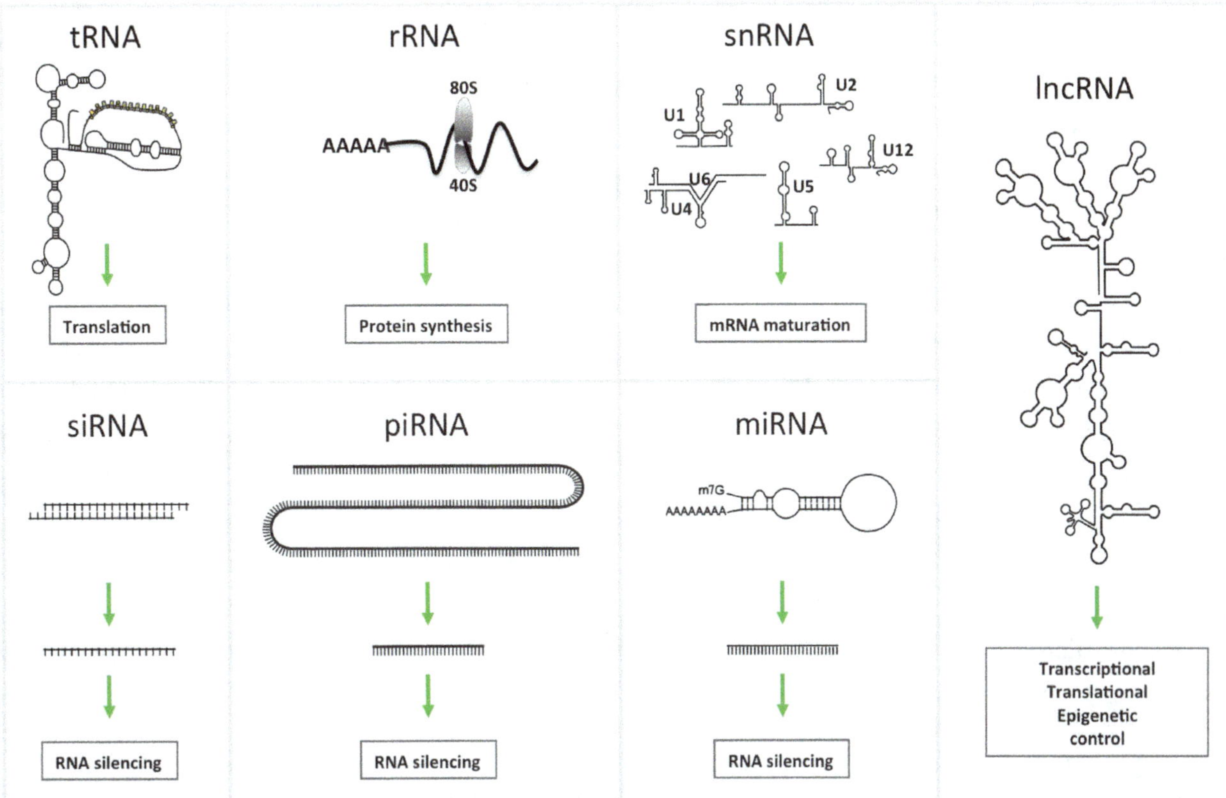

The discovery of RNA interference (RNAi) [8] in *C. elegans* and earlier, the identification of a new class of small RNAs known as miRNAs [9] led to greater attention to ncRNA and their involvement in the regulation of biological processes. The total number of ncRNAs is still unknown, but through transcriptomic and bioinformatic studies, one suspects the existence of several thousands of ncRNAs. A large number of them remain to be fully identified, and the functions of most of them have not yet been validated. NcRNAs are grouped into several RNA families, subclassified according to their function, size, structure and conservation (Figure 2). We thus find the ubiquitous and well-known transfer RNA (tRNA) required as the physical link between the nucleotide sequence of nucleic acids (mRNA) and the amino acid sequence of proteins; The ribosomal RNA (rRNA) as RNA component of the ribosome, and essential for protein synthesis in all living organisms. Small nuclear RNAs (snRNAs), form a class of RNA molecules with an average length of 150 nucleotides localized within the nucleus of eukaryotic cells. Their primary function is pre-mRNA (hnRNA) processing, for which they are always associated with a set of specific proteins and the complexes are referred to as small nuclear ribonucleoproteins (snRNP). A subclass of snRNA is called small nucleolar RNAs (snoRNAs) localized in the nucleolus and implicated in the maturation of RNA molecules by guidance of chemical modifications targeting mainly rRNAs, tRNAs and snRNAs. Piwi-interacting RNAs (piRNAs) form the largest class of small non-coding RNA molecules expressed in animal cells and form the third class of small RNA silencers. They form RNA-protein complexes by interacting with Piwi proteins and are required for both epigenetic and post-transcriptional gene silencing of retrotransposons and other genetic elements in germ line cells, particularly during spermatogenesis. Small Interfering RNAs or silencing RNAs (siRNAs) are a class of short double stranded RNA molecules (20–25 bp) playing a role in the RNA interference (RNAi) pathway. They modulate expression of specific genes by interfering with RNA translation by complementary nucleotide sequences. Post-transcriptional activity of siRNAs was first discovered in plants [10], and their possible use in mammalian cells was rapidly shown [11], evidencing the potential as a biomedical research tool. Interference induced by siRNAs share a large part of the signaling pathways used by miRNA naturally present in mammalian cells. The presence of siRNA encoded by the human genome has never been shown. Extracellular or exosomal RNA (exRNA) designate RNA species present in body fluids (venous blood, saliva, breast milk, urine, semen, menstrual blood or vaginal fluid) composed by mRNA, tRNA, miRNA, siRNA and lncRNA, generally enclosed within vesicular bodies preventing their digestion. Biochemical evidence supports the idea that exRNA uptake is a common process, suggesting new pathways for intercellular communication. The relative abundance of certain exRNAs can be correlated to cellular signaling or specific disease states. Recently a study characterized the population of RNA in human plasma by deep sequencing, and demonstrated that most abundant RNAs are miRNAs (42.32%), followed by ribosomal RNAs (9.16% of all mapable counts), long non-coding RNAs (3.36%), piwi-interacting RNAs (1.31%), transfer RNAs (1.24%), small nuclear RNAs (0.18%), and small nucleolar RNAs (0.01%) [12].

Finally, micro RNAs (miRNAs) and long noncoding RNAs (lncRNAs) form the two families of ncRNA more widely described in our review.

2.1. MiR and Hematopoiesis

2.1.1. MiRNAs, a Family of Regulatory ncRNAs

While the first small endogenous ncRNAs (*lin-4/let-7*) were described in *C. elegans* in 1993 [13], these RNAs became by now a large family of regulatory ncRNAs referred as microRNAs (miRNAs) [14–16]. These single stranded RNAs are characterized by their size of about 17–25 nucleotides and are highly conserved during evolution [14]. Mature and functional miRNAs result from a multistage process and their expression leads to post-transcriptional silencing of target genes through epigenetic regulatory functions by repressing mRNA translation [17]. Therefore miRNAs are also involved in a complex cellular network of gene regulation. On one hand, most miRNAs are able to target several mRNAs while a specific mRNA can be targeted by several miRNAs [18,19]. On the other hand, miRNAs target TF mRNAs and in turn TFs regulate miRNA gene expression. Expression of miRNAs is then tightly regulated at both transcriptional and post-transcriptional levels. Primary miRNAs (pri-miRNA) result from the transcription of genes requiring RNA polymerase II or III, and critical epigenetic regulation steps [20]. As for TFs, miRNAs display tissue and developmental specificities. In correlation with their spatial and temporal expression, miRNAs contribute to the regulation of embryogenesis, cell proliferation, differentiation and death [21]. Therefore, deregulation of miRNA expression, caused by genetic alterations, transcriptional or processing failures is incriminated in the development of many human diseases, including cancer.

Transcriptional regulation of miRNAs is not yet fully understood while processing of the transcript in mature miRNA has been described in detail [20,22]. Briefly, in the canonical miRNA biosynthetic pathway, the transcription of miRNA genes by RNA polymerase II (or III for miRNAs encoded within

Alu repeat sequences) [23] generates a pri-miRNA displaying hairpin structures. In fact, pri-miRNA transcripts can be organized in clusters encoding multiple miRNA sequences [24]. In the nucleus, pri-miRNAs are then cleaved between the hairpin structures to generate pre-miRNAs of 60 to 110 nt. This cleavage is performed by the nuclear endoribonuclease Drosha (RNaseIII) and DiGeorge critical region 8 (DGCR8) cofactor (Pasha in Drosophila)-forming complex. Pre-miRNAs are then exported to the cytoplasm by Exportin-5 in a Ran-GTP dependent manner, where they undergo further processing. Cytoplasmic endonuclease RNase III Dicer associated to the transactivating response RNA-binding protein (TRBP) and the protein activator of PKR (PACT), cleave the stem loop of the hairpin structure to form an asymmetric RNA duplex of about 22 nt. Double strand RNA interacts with Argonaute within the RNA-induced silencing complex (RISC) in order to select the guide strand matching with the 3'-untranslated region (UTR) of the target mRNA, thus forming the miRISC. The complementary "passenger" strand of the duplex is released and degraded. The guide strand allows the RISC-mediated inhibition of translation of the target mRNA as the main mechanism of action. Nevertheless, destabilization of target mRNAs through the accelerated deadenylation was reported as an alternative mechanism for miRNAs action [25,26]. Recently, an additional mode of action was proposed in which endogenous miRNAs act at the transcriptional level to silence genes, involving chromatin remodeling complexes [27].

2.1.2. MiRNAs Regulate the Regulatory Network during Hematopoiesis

The crucial role of miRNAs in the regulation of specific steps of hematopoiesis is now well-documented based on the modulation of their expression during HSC differentiation processes. Furthermore, in the hematopoietic system, numerous miRNAs were described to affect translation or stability of mRNAs including mRNAs encoding TFs. Thus, interplay of TFs and miRNAs are considered essential regulators of gene expression in hematopoiesis (Figure 3).

2.1.2.1. MiR-223 Is a Key Hematopoietic miRNA

One of the most extensively studied miRNA in hematopoiesis is miR-223 also described as the "fine-tuner" of granulocytic differentiation, maturation and function while its expression decreases during monocytic, erythroid [28] and mast-cell differentiation [29–32]. Nevertheless, up-regulation of miR-223 was observed by Lu *et al.* [33] and Choong *et al.*, during EPO-mediated erythroid differentiation of human CD34$^+$ HSPCs [34], whereas its expression was down-regulated in K562 cells. Finally, a functional study revealed a role for miR-223 in erythroid differentiation [35]. According to the various expression levels in different types of hematopoietic cells, miR-223 is linked to several major specific transcription factors.

Expression of miR-223 is associated with granulocyte differentiation and was shown to increase in retinoic acid (RA)-induced granulocytic differentiation of leukemic cells. C/EBPα is a key factor for the commitment of hematopoietic cells towards the granulocytic lineage and its RA-mediated upregulation was anticipating miR-223 activation in NB4 and HL60 promyelocytic cells, suggesting a role for this TF in the transcriptional regulation of miR-223. In correlation, two putative C/EBPα binding sites were identified upstream of the region encoding the pre-miR-223 sequence. Results were confirmed by transfection of expression constructs, which demonstrated the region containing the

C/EBPα binding sites to be required for miR-223 responsiveness to RA. Moreover, chromatin immunoprecipitation (ChiP) assays showed that C/EBPα binds *in vivo* to miR-223 promoter. Besides C/EBPα, nuclear factor I-A (NFI-A) binding site was described in the miR-223 promoter [29], overlapping one of the two C/EBPα sites. This TF is also involved in the regulation of miR-223 expression and maintains its basal level in undifferentiated cells while C/EBPα over-expression substitutes NFI-A in RA-induced granulocytic differentiation eventually leading to miR-223 up-regulation. Furthermore, NFI-A is targeted by miR-223. Altogether, granulocytic differentiation correlates well with miR-223-mediated inhibition of NFI-A mRNA translation, therefore preventing competition with C/EBPα [29].

Figure 3. Partial representation of the network involving microRNAs (mir) and regulatory proteins in hematopoiesis. Network was built using Cytoscape 2.8.2 software [36]. Transcription factors are represented as green rectangles, miRNAs as red diamonds, and other proteins as grey ellipses. Arrows depict relationships: arrows (activation), T (inhibition), circle (undetermined/binding). RARB, Retinoic acid receptor B, ARF, Alternate Reading Frame, CSF1R, colony-stimulating factor 1 receptor; BCR-ABL, breakpoint cluster region-Abelson; CBFB, Core-binding factor subunit beta; IL6, interleukine 6; LMO2, LIM domain only 2; HBA1, hemoglobin A1; KLFD, Krüppel like factor D; RUNX1, Runt-related transcription factor 1; CEBPA, CCAAT/enhancer-binding protein alpha; CEBPB, CCAAT/enhancer-binding protein beta; GM-CSF, granulocyte macrophage-colony stimulating factor; NFIA, nuclear factor I-A.

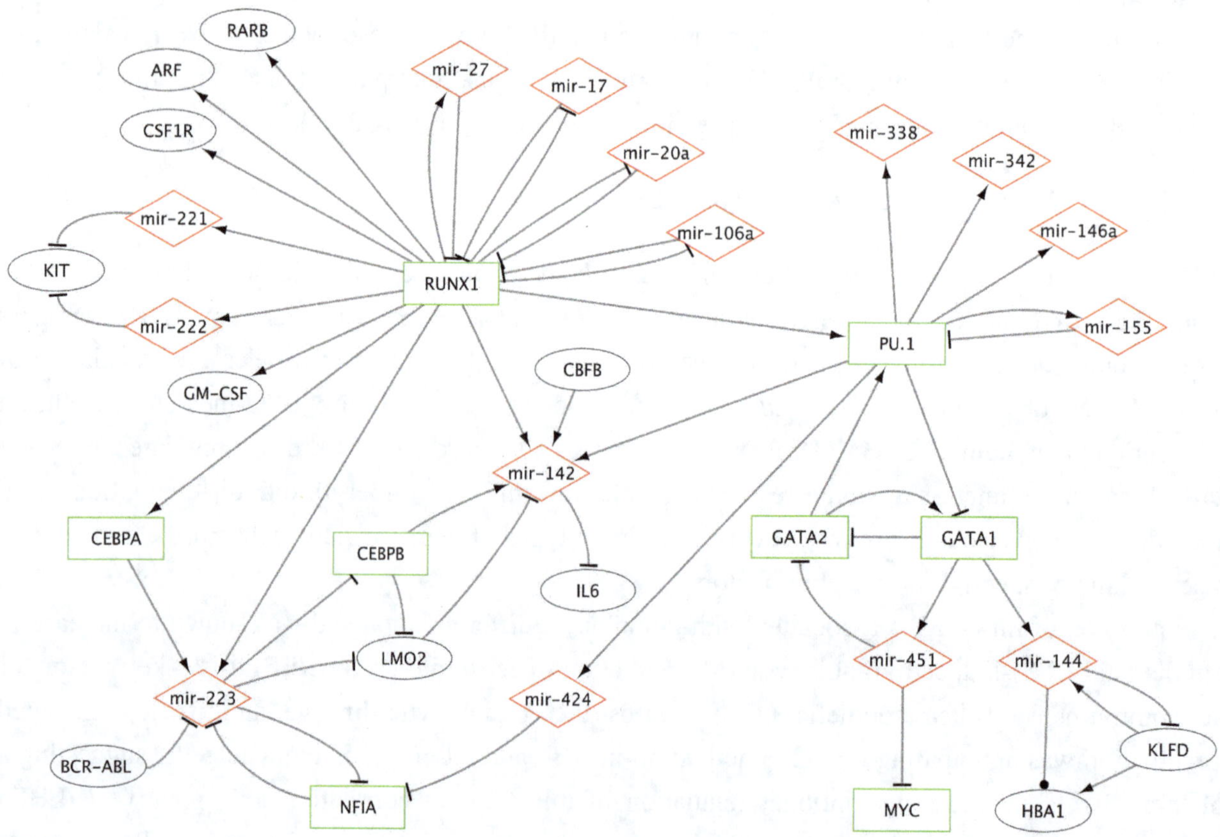

Interestingly, a recent study reported an alternative mechanism for NFI-A regulation by endogenous miR-223. Authors demonstrated the ability of miR-223 to affect transcription of NFI-A gene in addition to a post-transcriptional effect [27]. Authors first observed that nuclear import of miR-223 increased during granulocytic differentiation of myeloid precursors. Using confocal microscopy, results showed that the nuclear compartmentalization of miR-223 was increased in RA treated HL60 cells and primary acute promyelocytic leukemia (APL) blasts cells. ChIP assays performed in HL60 cells revealed *in vivo* presence of miR-223 at complementary sequences of the NFI-A promoter. Moreover, mutation of miR-223 binding sites abolished miR-223-mediated repression of NFI-A promoter reporter constructs [27]. Beyond the increasing variety of miRNA function in hematopoiesis, these findings reveal a novel mechanism of transcriptional control.

According to the pivotal role of miR-223 in hematopoiesis, its deregulation contributes to leukemia development. Notably, it was recently established that the fusion oncokinase breakpoint cluster region (Bcr)-Abelson (Abl) repressed miR-223 expression in chronic myeloid leukemia (CML). miR-223 expression was also significantly decreased in B cell lymphoproliferative disorders [37] and could serve as a prognostic factor for CLL and CML patients.

Transcription factors Satb1 and Runt-related transcription factor 1 (RUNX1) or acute myeloid leukemia 1 protein (AML1) were recently described as central regulators of the early stage of HSCs commitment and self-renewal. Many publications report RUNX1 as an essential TF for hematopoiesis initiation at both adult and embryonic levels [38–40]. In mice embryos, expression of this TF has been localized in all sites giving rise to hematopoietic cells. Hemogenic sites such as intra-aortic hematopoietic clusters are absent in RUNX1$^{-/-}$ embryos [41] and the hematopoietic defect in RUNX1$^{-/-}$ embryonic stem cells was rescued by ectopic expression of RUNX1 in mice [39,42]. Recently, it was also shown that the isoform RUNX1a enhances hematopoietic lineage commitment from human embryonic stem cells [43]. RUNX1 is required for both embryonic and post-natal hematopoiesis. It is also involved in the control of genes essential for myeloid differentiation. A connection between this TF and several miRNAs related to specific myeloid genes has been reported. Notably, RUNX1 also regulates miR-223 gene transcription in myeloid precursors, maintaining chromatin in a transcriptionally active state. Consequently, the fusion protein RUNX1/ETO (AML1/ETO), in which RUNX1 is inactivated, specifically triggers transcriptional silencing of miR-223 [44]. RUNX1/ETO, which causes acute myeloid leukemia (AML), acts as a dominant-negative repressor of RUNX1 target genes [45], including colony-stimulating factor 1 receptor (*CSF1R*/c-fms), GM-CSF [46,47], tumor suppressor p14(ARF) and the retinoic acid receptor β (RARB) [44].

Although up-regulation of miR-223 was correlated to decreased NFI-A expression and granulocytic differentiation, regulation of specific myeloid genes has not been yet established. Nevertheless, miR-223 together with constitutively expressed miR-142 was shown to decrease proliferation required for myeloid differentiation [48]. It is well established that miR-142 is implicated in the regulation of T cell function and development and B cell lymphoma.

2.1.2.2. MiRs-223/142/155 and Specific TFs as a Miniature Regulatory Network

Besides targeting NFI-A, miR-223 also regulates Lmo2-L/-S isoforms and C/EBPβ expression in myeloid cells. A miR-223-mediated decrease of cell proliferation occurred through a miR223-C/

EBPβ-Lmo2-miR142 pathway. This model is based on the findings that miR-223 targets C/EBPβ as well as Lmo2 mRNAs and that C/EBPβ regulates transcription of Lmo2, which is then down-regulated at both transcriptional and posttranscriptional levels. This leads to the expression of miR-142 and cell proliferation attenuation since Lmo2 negatively regulates miR-142 gene transcription [48]. Lmo2 is also involved in erythroid-specific gene expression [49] through the formation of a DNA binding complex involving also TAL1, E2A, Ldb/LNI-1 and GATA1 [50]. In agreement with the regulation of Lmo2 expression by miR-223, it has been reported that forced expression of miR-223 inhibited erythroid development of $CD34^+$ HSCs [35]. Altogether these results hint for a negative role of this miRNA in erythropoiesis whereas is positive in myeloid differentiation. Similarly, NFI-A up-regulation promotes erythropoiesis while its silencing promotes granulopoiesis in correlation with miR-223 modulation [29,51,52].

Besides Lmo2, Sun *et al.*, recently demonstrated that PU.1, C/EBPβ RUNX1 and the co-factor CBFβ also regulated miR-142 gene transcription. Specific binding sites for these three TFs exist in the miR-142 gene promoter with PU.1 acting predominantly. Indeed, C/EBPβ and RUNX1 alone led to miR-142 deficiency. However, miR-142 expression levels within hematopoietic cells depend on different combinations of PU.1 together with C/EBPβ RUNX1 [53]. By regulating PU.1 expression, miR-155, another relevant miRNA in hematopoiesis, is involved in miR-142 gene transcription. miR-155 targets PU.1 mRNA triggering its repression and consequently the reduction of miR-142 production. A good illustration of the interrelationship between miR-155, PU.1 and miR-142 is the pathway leading to IL-6 expression. Lipopolysaccharide (LPS) stimulates Toll-like receptor 4 (TLR4), which induces miR-155 expression, and down-regulation of PU.1, subsequently leading to decreased miR-142 expression and thus inducing expression of its target, IL-6. This regulatory mechanism is relevant for IL-6–mediated immunological processes and also the functions of miR-142 in the lymphoid system [53].

2.1.2.3. RUNX1 Connection with miRs-222/221, 17-5p, 20a, 106a and 27a

RUNX1 is implicated in the control of miR-222/221 gene cluster transcription as demonstrated by ChiP and luciferase reporter gene assays in U937 monocytic leukemia cells. Indeed, RUNX1 binds to two regions containing RUNX1-consensus sequences in the miR-222/221 gene promoter displaying four consensus sequences. Moreover, RUNX1 dose-dependently activates miR-222/221 promoter in the reporter gene constructs [54]. Expression of miR-222/221 was increased during GM-CSF-mediated myeloid differentiation of normal bone marrow $CD133^+$ stem progenitor cells. These results correlate with the down-regulation of the stem cell factor (SCF) receptor KIT expression whose mRNA is targeted by miR-222/221. Conversely, the expression of miR-222/221 and miR-223 was lower in leukemia cells expressing RUNX1 fusion oncoproteins, in correlation with higher levels of KIT oncogene expression and inhibition of myeloid differentiation [54].

On the other side, RUNX1 binds to the promoters of miRNA 17-5p–92 and 106a–92 gene clusters leading to the inhibition of miRNA 17-5p, 20a and 106a transcription. These miRNAs are upregulated in undifferentiated CD34+ hematopoietic stem/progenitor cells (HSPCs) while their expression decreases during monocytic differentiation and maturation. This down-regulation is correlated to the increasingly expressed RUNX1 protein in monocytic cells. In turn, miRNAs 17-5p, 20a and 106a inhibit RUNX1 translation by interacting with the 3' UTR generating a regulatory loop. RUNX1 was also shown to act as a regulator of miR-27 transcription, which is involved in erythro-megakaryocytic and granulocytic differentiation pathways. Generating a negative regulatory loop, miR-27 targets RUNX1 mRNA. This leads to RUNX1 down-regulation allowing granulocytic differentiation of myeloblasts [55]. Similarly, TPA (12-o-tetradecanoylphorbol-13-acetate)-induced megakaryocytic differentiation of K562 cells was concomitant with increased miR-27a expression and RUNX1 down-regulation [56]. RUNX1 appears thus as a crucial transcription factor in hematopoiesis, given its expression in the different cell lineages and its ability to directly regulate a wide panel of miRNAs as well as TFs including C/EBP and PU.1, which in turn target promoters of miRNA genes [57].

2.1.2.4. MiR-146a and miR-155 Genes Are Regulated by PU.1

During differentiation progress, TFs integrate the extensive regulatory network insuring hematopoietic homeostasis. GATA-1 and GATA-2 are the major TFs regulating erythro-megakaryocytic pathways whereas PU.1 and C/EBP control myelo-lymphoid differentiation. Nevertheless, GATA-1 and PU.1 proteins physically interact and inhibit each other. PU.1, a member of the Ets(E-twenty six) TF family, is involved in differentiation from HSCs to multipotent progenitors at different levels [58]. PU.1 is required for myeloid and lymphoid [59] differentiation and plays a role in cell fate decision. Its expression levels determine the fate of early T-cell progenitors since over-expression of PU.1 was shown to reorient differentiation to the myeloid lineage [60–63] depending on Notch. In the absence of Notch signaling, PU.1 promotes the myeloid pathway, whereas activation of Notch signaling is observed in T-cell lineage pathways. Del Real *et al.*, demonstrated that in absence of Notch signaling, PU.1 regulates expression of TF genes essential for T-cells, including Myb, Tcf7 and Gata3.

PU.1 also plays a pivotal role in lympho-myeloid development through its ability to regulate transcription of miRNA genes as reported by Ghani *et al.* These authors highlighted transcriptional regulation of miR-146a, miR-342, miR-338 and miR-155 genes by PU.1. Results revealed that miR-146a was the most robustly PU.1-induced miRNA. miR-146a and miR-155 up-regulation was independent of *de novo* protein synthesis suggesting that PU.1 is able to regulate transcription of these miRNA genes. Conversely, miR-342 and miR-338 regulation could require cooperative factors linked to PU.1 [64]. Authors also showed that miR-146a directed selective differentiation of HSCs into peritoneal macrophages during adult hematopoiesis and concluded that PU.1 temporally controls the expression of miRNAs required for correct HSCs differentiation.

PU.1 up-regulation was also shown to control human monocyte-macrophage differentiation through the activation of miR-424 [65]. The authors validated NFI-A mRNA as a true functional target of miR-424. Similarly to the C/EBPα-miR223-NFI-A regulatory circuit that regulates granulopoiesis, PU.1-miR-424-NFI-A regulates monocytic differentiation.

2.1.2.5. Erythroid Specific Expression of miRs-144 and 451

GATA factors play a central role in the network of hematopoiesis regulation through interactions with co-factors. Predominant expression of GATA-1 results in the erythroid differentiation. Moreover, PU.1 activity is also involved in the lineage-specific fate decisions of erythroid or myeloid differentiation through its physical interaction with GATA-1 leading to reciprocal inhibition of their transcriptional activities. Especially, suppression of GATA1 activity by PU.1 leads to a shift in cell fate towards the myeloid-lymphoid lineages [66,67]. Upon myeloid differentiation, PU.1 is overexpressed in correlation with GATA-1 and -2 down-regulation. Using the G1ME hematopoietic cell line derived from *in vitro* differentiation of murine GATA-1^{-} embryonic stem cells (ESC) [68], Chou *et al.*, demonstrated that GATA factors act sequentially to control lineage determination during hematopoiesis, through modulation of repressive effects at key regulatory elements of the PU.1 gene [69]. So far, no link between PU.1 expression and the production of miRNAs regulated by GATA-1 has been demonstrated. GATA-1 was shown to regulate *miR-144/451* gene transcription. GATA-1 up-regulation induced *miR-144* and *miR-451* in G1E cells as well as in human CD34^{+} cells and murine erythroleukemia (MEL) cells in correlation with erythroid maturation [70]. The crucial role of miRs-451/144 in physiological erythropoiesis was also demonstrated in zebrafish and mouse models [71,72]. Expression of *miR-451* is restricted to the erythroid lineage and its induced silencing affected neither megakaryopoiesis nor granulopoiesis in zebrafish embryos. GATA-2, which is able to bind the *miR-144/451* locus in the absence of GATA-1, does not activate transcription. A couple of predicted miR-144 and miR-451 target genes, that were down-regulated after GATA-1 activation, were reported [70]. Interestingly, proliferation regulator c-myc, whose overexpression inhibits erythropoiesis, emerged as a predicted target gene of miR-451. Furthermore, it was shown in zebrafish that GATA-2 3′UTR was targeted by miR-451 but not by miR-144 [72]. *GATA-2* is known to preserve the immaturity of hematopoietic precursor cells [73] and induces overexpression of GATA-1, which in turn triggers GATA-2 down-regulation as a negative regulatory feedback loop, required for normal erythropoiesis. Considering that GATA-2 positively regulates PU.1 gene expression, it seems plausible that miR-451 could play a role in the GATA-1/GATA-2 balance and PU.1 down-regulation to warrant erythroid maturation according to a GATA-1 > miR451 > GATA-2 > PU.1 axis. Besides, this would be in agreement with decreased miR-155 expression observed during erythropoiesis since PU.1 regulates gene expression of this miRNA [31,74]. Mir-144 was reported to selectively regulate embryonic α-globin gene expression during primitive erythropoiesis in zebrafish through negative feedback regulation involving erythroid-specific Krüppel-like (KLF)-D TF, which selectively binds to promoters of both α-globin and miR-144 genes to activate their transcriptions [38,75].

Except for miRs-144/451 that are erythroid-specific miRNAs, most miRNAs are ubiquitously expressed at different stages of hematopoietic differentiation in correlation with TF expression and activities involved in the regulation of lineage specific genes.

2.2. LncRNA and Hematopoietic Lineage

2.2.1. LncRNA

Among all sub-categories of ncRNAs, long non-coding RNAs (lncRNAs) constitute the most recent and least characterized family. LncRNAs include all ncRNA larger than 200 nucleotides and not yet categorized in one of the other RNA families. In contrast to small ncRNAs, which are highly conserved among species and which are involved in transcriptional and posttranscriptional gene silencing, lncRNAs are poorly conserved [76] and their modes of regulation are diverse and not yet totally elucidated. The process of lncRNA transcription and maturation is similar to that of mRNA even if their genes are not subject to the same histone modifications (H3K4me3 and H3K36me). As for mRNA, lncRNAs are transcribed by RNA polymerase II. Processing of lncRNA involves 3' poly(A) tailing and 5'-end capping as well as splicing. LncRNAs have small open reading frames without any protein-coding potential, but can be sometimes be associated with ribosomes in the cytoplasm, suggesting an additional role in mRNA metabolism.

Analysis of the human genome and transcriptome estimated that there are about 23,000 lncRNAs, comparable to the number of protein-coding RNAs and greatly exceeding the number of miRNAs (close to 2000) [77,78]. LncRNAs sequences are spread over the entire genome and can be found on all chromosomes [79]. During the last decade, the number of studies on this type of RNA has been increasing, but the general knowledge remains low given the large number of existing lncRNAs. However, it has already been shown that these lncRNAs are tissue specific and differently expressed under both normal and pathological conditions, implying that they may play important biological roles. Indeed they can regulate biological processes, including cell division, survival, and differentiation, as well as some processes related to cancer development. The first schemes of control or mechanism of action were also recently revealed. Several long ncRNA regulate gene expression by modifying chromatin structure. Different studies demonstrate a wide diversity of mechanisms by which a lncRNA regulate chromatin of a single promoter, a gene cluster, or an entire chromosome [6], in order to activate or silence genes in *cis* or in *trans* [80]. Until now, only a small number of lncRNAs were identified and fully characterized.

LncRNAs play different roles in transcription [81]. For example MALAT1 can regulate preRNA splicing by influencing distribution and phosphorylation of the serine/arginine (SR) splicing factor [82]. They can also increase mRNA stability by interaction with exonuclease XRN1, as shown in a recent study of the Moon [83]. Perfect hybridization between mRNA and lncRNA is another system to stabilize RNA: BACE1-AS fully hybridized to exon 6 of BACE1, which presents a binding site for miRNA-485-5p, thus leading to a competition between lncRNA and miRNA for this site [84]. On the opposite, it has also been shown that these ncRNAs can decrease mRNA stability by promoting degradation of mRNA [85]. LncRNA involvement in the regulation of translation was initially suggested by the observation of an association between lncRNA and ribosomes, even though lncRNA are not translated. This kind of regulation seems to be negative in some cases, for example when lncRNA-p21 annealed imperfectly but throughout the coding sequence of the mRNA of β-catenin or JunB, blocking their translation [86]. Positive regulation was also reported, for example in the case of lncRNA AS-Uchl1 that promoted polysome formation around Uchl1 mRNA [87]. In a less specific way, lncRNA BC1 was shown to inhibit the whole process of translation by preventing the assembly of translation initiation complex in neurons [88].

2.2.1.1. Expression and Regulation of lncRNA

As mentioned previously, lncRNA are tissue specific and differently expressed under normal and pathological conditions, implicating that their transcripts are regulated. While information about this regulation is constantly expending, the way lncRNA abundance is controlled is still widely unknown. Intronic sequences represent 30% of the human genome and represent the major source for lncRNAs (intronic lncRNAs). In this case, regulation of non-coding RNA expression is usually linked to host gene expression. LncRNA sequence can also be independent of genes and its expression is then under promoter dependence. This second subcategory is commonly called Long Intergenic Non-Coding RNA (lincRNA). In addition to their regulation at a transcriptional level, lncRNAs can also be regulated post-transcriptionally [81]. A study about the overall evaluation of ncRNA half-life (involving about 800 transcripts) in mouse neuronal cells showed that all lncRNA differed in stability with a minority of "unstable" lncRNA. Stability appears to depend on localization (intronic lncRNA have a lower stability compared to lincRNA), splicing (spliced non-coding transcripts are more stable than the non-spliced ones), and finally subcellular localization (nuclear lncRNA may have shorter half-lifes compared to cytoplasmic ones) [89]. Most of the lncRNAs are stabilized with a poly(A) tail similar to mRNAs. Another type of stabilization involves a triple helix at the 3' end of lncRNAs protecting against 3'→5' exonucleolytic cleavage. This structure was recently shown for lncRNA-MALAT1 in multiple endocrine neoplasia β [90,91].

If lncRNAs present differential stability depending on their location or their origin, stability can also be modulated, thereby altering the turnover of these regulators. It has been recently shown that lincRNA-p21 could be destabilized by human antigen R (HuR)/Argonaute (Ago)-2 complex as well as by the miR Let-7b. Silencing of HuR or Ago2 increases lincRNA-p21 stability, whereas overexpression of let-7b induces destabilization. Degradation of RNA requires decapping and deadenylation, and these aspects of lncRNA degradation are still not well known and need to be further elucidated in order to improve the understanding of lncRNA abundance control [86].

2.2.1.2. Mechanisms

Unlike miRNAs, which follow a well-established mechanism of action, causing inhibition of translation or degradation of mRNA, lncRNAs act almost as diversely as proteins (Figure 4). LncRNAs can interact with proteins as well as with RNA or DNA. Interaction of lncRNA with other RNA occurs through total or partial hybridization of complementary sequences. This mechanism is common for anti-sense lncRNAs such as p15AS. It was recently published that interaction of lncRNA with miRNA happens with a preferential hybridization of the miRNA to the 3' end of the lncRNA [92]. RNA-RNA interaction can lead to translational regulation, but also to splicing or to inhibition of RNA function. LncRNA-DNA interaction can occur similarly to a RNA-RNA interaction, by the sequence complementarily forming a RNA-DNA duplex, or in a more complex configuration such as the hybridization of lncRNA and the dihydrofolate reductase (DHFR) promoter that forms a DNA-DNA-RNA triplex [93], leading to sequence specific transcriptional repression [94]. DNA-RNA interactions recruit DNA regulating factors that can affect transcription, histone or DNA modifications including methylation or acetylation. Finally, the ability of lncRNA to form secondary and tertiary structures allows them to establish complex interactions enabling them to catch and sequester proteins as well as to join distant areas.

In an effort to summarize the different modes of regulation involving lncRNAs, Wang and Chang propose four classes or archetypes of lncRNA regulation [95], further summarized in the recent publication of Da Sacco *et al.* [96]. LncRNAs usually follow several modes of action, and so most of them belong to several of these archetypes.

Figure 4. Scheme of regulations involving lncRNAs and miRNAs, influencing transcription, maturation or translation processes and example of interaction between lncRNAs, DNA, proteins and miRNAs. Blue arrows correspond to physiological maturation of RNA. Green lines correspond to inhibition processes and dotted black lines correspond to variable regulations (positive or negative depending on the mechanism). (−) Negative effect; (+/−) negative or positive effect.

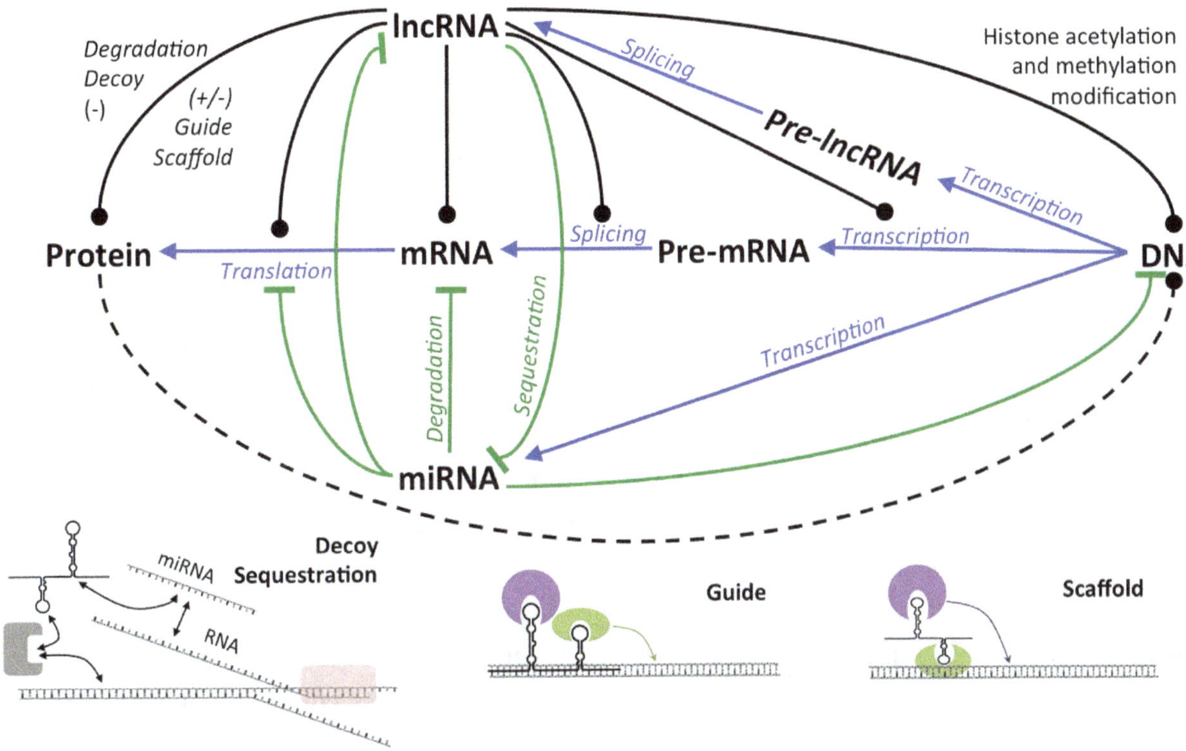

Signal: Expression of these lncRNA can be used as a marker of intracellular signaling or response to stimuli, as it can induce responses *via* transcription of RNA without translation or posttranslational modifications. More a tool than a mechanism, these lncRNAs essentially serve as biomarkers. This archetype therefore concerns all lncRNA with a strong relationship in spatiotemporal response, such as rapid reaction to temperature changes regulated by lncRNAs COLDAIR and COOLAIR [97,98] or imprinting controlled by Xist.

Decoy: In molecular decoy, lncRNA compete with another nucleotide sequence or structure for the binding of a TF, chromatin modifier, or other regulatory factor such as miRNA thereby preventing the miRNA to inhibit translation of their targets [99,100]. These lncRNAs are therefore considered negative regulators of effectors.

Guide: The third archetype is the guide RNA, which can be described as a connection between lncRNAs and proteins, further guided by the lncRNA to a target. This pattern of regulation stems from the observation that lncRNAs are able to induce changes in gene expression in *cis* (neighboring) or *trans* (distant) genes. Target proteins recognized by the lncRNA may be repressor or activator complexes, or TFs, with the final goal of controlling the expression of a target gene, causing changes in the epigenome, whether the control takes place in *cis* or in *trans*.

Scaffold (protein linker): In the fourth archetype proposed by Wang *et al.*, the lncRNA becomes a docking structure for regulatory proteins, in order to assure accurate assembly of ribonucleoprotein complexes. In many biological signaling processes, this control feature is essential for specificity and dynamics of molecular interactions and signaling events. Until now, it was thought that specificity of such complexes was based essentially on intrinsic protein properties, but recently it was hypothesized in addition that lncRNA could also contribute. LncRNA have specific RNA sequences that recognize individual protein effectors leading eventually to ribonucleoprotein complexes that combine the regulative properties of both RNA and protein components. The idea of protein assembly dependent on a nucleotidic scaffold provides novel insights into therapeutic strategies to artificially modify specificity or activity of these bipartite signaling complexes.

2.2.2. LncRNAs in Hematopoiesis

2.2.2.1. LncRNAs in Erythropoiesis

The knowledge about involvement of lncRNA in hematopoietic differentiation or self-renewal of hematopoietic stem cells remains poor, and a part of the observed current results are derived from murine cellular models. However, unlike miRNAs, lncRNAs are very poorly conserved between species so that results cannot readily be extrapolated to human. However, it may be possible that the overall operating mode is preserved and that it would be possible to transpose a process from one species to another, considering that functions of lncRNA are mainly mediated by their secondary and tertiary structure and not only through their sequence [101]. Currently, such a transposition has yet to be demonstrated, but the importance of the secondary structures is shown by the joint action of human Alu, and mouse B2 that both show no direct primary sequence similarity but are both able to sequester RNA polymerase II [93,102,103].

In murine Ter-199 cells, a direct link between differentiation of hematopoietic cells towards the erythroid pathway and the lncRNA EPS (erythroid prosurvival) regulates inhibition of apoptosis during differentiation [104,105] and terminal erythropoiesis. Interference with the expression of this lncRNA (even at 50%) induced a significant decrease in cell proliferation and increased annexin V expression. At least one of the targets of lncRNA-EPS could be Pycard, which is an antiapoptotic gene whose protein product is able to activate caspases during apoptosis. Many examples of Pycard regulation by lncRNA-EPS exist. First of all, expression of Pycard and EPS are inversely correlated during erythropoiesis. Moreover, overexpression of EPS leads essentially to repression of Pycard (among those studied by the team of Hu *et al.*). In addition, Pycard overexpression generates the same phenotype than inhibition of EPS during erythroid terminal differentiation: inhibition of proliferation, induction of cell death and inhibition of enucleation, the final step of differentiation. Finally, overexpression of Pycard abrogates anti-apoptotic effects induced by ectopic expression of EPS.

We mentioned earlier possible interactions between miRNAs and lncRNAs, so lncRNAs could modulate miRNAs *via* their transcription or their sequestration, and conversely, a miRNA can regulate lncRNA indirectly by acting at least on translation of the TF involved in the expression of this lncRNA. A recent study focused on the interaction between lncRNA and miRNA, using Photoactivatable-Ribonucleoside-Enhanced Crosslinking and Immunoprecipitation (PAR-CLIP) and targeting Ago (Argonaute) proteins [92], the catalytic components of the RISC complex. Thus, the miRNA bound to Ago, guides it to its complementary sequence, allowing to identify the targets of miRNA and then to hypothesize the lncRNA involved in the action of the miRNA. This study mentioned four miRNAs targeting lncRNAs related to hematopoietic differentiation: miR-196A, miR-196B, miR-9-1 and miR-210. MiR196A and miR196B target HOXA5 [106] implicated in erythropoiesis and myelopoiesis [107,108]. MiR-9-1 and miR-210 [75] target ALK4, which can be related to erythropoiesis [109]. MiR-196A and B seem to interact with three ncRNAs referenced as ENST00000523790.1, ENST00000519935.1, ENST00000489695.1; miR-210 interacts with ENST00000525865.1 and miR-9-1 with ENST00000511014.1, ENST00000509783.1, ENST00000505030.1, ENST00000504246.1, and ENST00000500197.2. LncRNAs interact with miRNAs which themselves modulate genes involved in hematopoietic differentiation. This does not automatically mean that these lncRNA are involved in differentiation but could be targets for future investigations.

A better understanding of the interactions between miRNA and lncRNA will allow deducing the effects of an ncRNA when the effect of its interacting molecule is known. To simplify these analyses, the miRcode database was implemented [110], using GENCODE database and providing a comprehensive map of putative miRNA target sites across the GENCODE long non-coding transcriptome.

2.2.2.2. LncRNAs in Hematopoietic Differentiation

H19 was one of the first imprinted noncoding RNAs to be identified [111,112]. It is transcribed from chromosome 11, contains the sequence of miR-675 in its first exon, and is one of the few lncRNAs conserved between species. H19 is co-regulated with insulin-like growth factor (IGF)-2, which is expressed from the same locus. In hematopoietic cells, this pair of growth regulatory genes is expressed in precursor cells and downregulated during normal and pathological differentiation [113]. However, expression of this ncRNA is not restricted to hematopoietic tissue, as it seems also expressed in ES cells, fetal and adult tissues [114]. Recent results show that H19 RNA could play a role in *trans* repression of Igf2 expression [115]. Abundant data about the implication of this lncRNA in cancer development show that H19 is essential for growth of some human tumor types, and it appears that its aberrant expression in cancer cells is due to modification of the methylation of imprinting control regions (ICR) of H19/IGF2 during differentiation. Recent data on bladder cancer documented how upregulated H19 increased cancer cell proliferation by increasing inhibitor of DNA binding 2 (ID2) expression [116].

In 2007, a study mentioned a new essential lncRNA implicated in hematopoietic differentiation [117]. Authors showed in several steps that a transcript is highly over-expressed during eosinophilic differentiation, and that the intronic region that encodes two transcript variants is highly conserved between human, mouse and chicken, which could reflect an evolutionary pressure leading to conservation of this potentially essential sequence. Moreover, the lack of a large ORF and paradoxically a poor amino acid conservation of small ORF strongly suggests that this transcript is an ncRNA. Results were confirmed by the lack of association of this transcript with ribosomes and absence of translation. An association with other types of proteins was proved nevertheless. This ncRNA was termed eosinophil Granule Ontogeny (EGO) and exists in two isoforms (EGO-A and EGO-B) of 535 and 1460 bases, respectively. Tissue-specific expression patterns suggest that EGOs act on bone marrow hematopoietic cell development but not on lymphoid development. Its silencing proved its requirement for major basic protein (MBP) and eosinophil derived neurotoxin (EDN) mRNA expression, but not for GATA-1. Its precise mechanism of action thus remains to be elucidated.

It has been shown recently in mice that HOXA6 and HOXA7 genes are indirectly regulated by lncRNA Mistral (MIRA) *via* recruitment of Histone-lysine *N*-methyltransferase (MLL)1 to chromatin, allowing transcription of HOXA6 and 7 genes. Results obtained in the murine system demonstrate involvement of ncRNAs in regulation of genes involved in mouse embryonic stem cell (mESC) differentiation [118]. These results can be extended beyond regulation of mESC, as HOXA6 plays an important role in the regulation of HSC self-renewal in human, and its overexpression is also involved in leukemia [119]; HOXA7 acts as an intermediate in the regulation of granulocytic differentiation repressed by the Polycomb group RING finger protein 2 (PCGF2) [120].

The transcription factor TAL1 (SCF) is a major regulator of hematopoietic differentiation [121]. The locus containing this gene also encodes two sense and antisense lncRNAs. Transcriptional inhibition of a lincRNA can lead to decreased, transcriptionally independent, expression of neighboring protein-coding genes in multiple human loci. Orom *et al.* showed that the lincRNA ENST00000444042.2 (CYP4A22-AS1-001) encoded by a sequence downstream of TAL1, on the same DNA strand, is a positive regulator of this gene's expression [76]. The study shows that these effects are independent of lncRNA orientation towards their target sequences. Depletion of this lincRNA induces a specific, strong and significant decrease of TAL1 expression, but does not affect expression of other genes on the same locus. Depletion of ENST00000429328, a lincRNA located on the same locus, affects the expression of another closely located gene, deoxycytidylate kinase (CMPK1) but not the expression of TAL1. Possible mechanisms of action could be: interaction through sequence or structural homology with the encoding target gene, through the recruitment of transcriptional activator or basal TFs, or by the removal of a repressor. Finally, chromatin remodeling could also be involved.

Zhang *et al.* identified an intergenic transcriptional activity, located between the human HOXA1 and HOXA2 genes, which presents a myeloid-specific expression with specific up-regulation during granulocytic differentiation [122]. This lncRNA is called HOX antisense intergenic RNA myeloid 1 (HOTAIRM1). Its induction during RA-induced granulocytic differentiation acts through the RA receptor and depends on the expression of myeloid cell development factors targeted by RA signaling. Extinction of this gene attenuated the expression of HOXA1 and HOXA2 genes induced by RA, without affecting expression of more distal HOXA genes. This knock out also affects transcription of CD11b and CD18 involved in myeloid differentiation. Finally, authors suggest that HOTAIRM1 plays a role in myelopoiesis through modulation of HOXA cluster gene expression.

The lncRNA HOXA cluster antisense RNA 2 (HOXA-AS2) is transcribed from a gene located between genes HOXA3 and HOXA4, and is expressed in human peripheral blood neutrophils. This ncRNA plays an anti-apoptotic role in All trans RA (ATRA)-induced myeloid differentiation, protecting cells against ATRA-induced apoptosis probably through the inhibition of the TNFα-related apoptosis-inducing ligand (TRAIL) pathway [123]: in the promyelocytic leukemia cell model NB4, knockdown of HOXA-AS2 increases number of both apoptotic cells and TRAIL. Conversely, ATRA-induced NB4 cells treated with TRAIL show an increase in HOXA-AS2 expression.

The lncRNA Xist triggers X chromosome inactivation in female mammals and is expressed at an early embryonic stage. Xist is important for the development of mice embryos particularly between 3.5 and 12.5 days post-coitum. Moreover, the adults that already committed hematopoietic progenitors, rather than HSC, were also able to express this lncRNA [124]. Overexpression of Xist led to the loss of a majority of blood cell types besides pre-B and pre-T lymphocytes, which are the cell types able to reactivate Xist expression. It is interesting to note that this reactivation occurs in pre-B cells and pre-T lymphoid cells that undergo allelic exclusion of immunoglobulin heavy chain (*Igh*) or TCRβ loci. This observation led to the hypothesis of a link between allelic exclusion of antigen receptor genes in lymphocytes and X chromosome inactivation in embryonic cells.

A few years ago, a meta-analysis was conducted by Gibb *et al.* on 272 human serial analyses of gene expression (SAGE) involving lncRNA sequences. These analyses included 26 different normal and 19 cancer tissue types in order to establish a first global profiling of lncRNA [79]. Results show tissue-specific lncRNA expression in normal tissues and a systematic abnormal lncRNA expression in human cancer. The paper essentially discusses the difference between normal and cancerous tissues (breast, brain, lung and blood) and discusses lncRNA expression that could become interesting for future studies. Four lncRNA have an expression pattern specific to white blood cells: ENSG00000232192 (Transcribed ENST00000446321, 810 bases) also named RP11-62I21.1 is an lncRNA transcribed by the antisense strand of the protein-coding gene KIF26B-005, coding for the kinesin family member 26B which is not implicated in any hematopoietic differentiation. The second lncRNA gene, ENSG00000246100 (CTC-774J1.2), is located in an intergenic region and is transcribed to five lncRNA variants of 3322, 982, 504, 693 and 995 bases. ENSG00000227712 (RP11-418J17.3), transcribes into an lncRNA of 898 bases. The most interesting one, ENSG00000256910 (AL034397.1), corresponds to a sequence close to the gene encoding the protein Z39Ig associated with monocyte and macrophage cells and linked to inflammatory reaction [125,126]. The AL034397.1 gene produces two transcripts by alternative splicing containing respectively, two and three exons: ENST00000540516 and ENST00000538676. For both, the last exon contains the sequence of miRNA-223, cited previously for its essential implication in hematopoietic differentiation. No links have yet been established between these two non-coding RNAs, but it is tempting to speculate about an interaction between these two non-coding RNA.

Altogether there are less than a dozen lncRNAs with a confirmed involvement in hematopoietic differentiation. For some others, involvement may be assumed. It is not yet possible to define a network of global interactions between lncRNAs, but we strongly believe that such a regulatory network involving TFs, miRNA and lncRNA exists and will be defined in the near future.

3. Conclusions

Long term studies of TFs involved in the regulation of genes driving hematopoietic cell fate decision, proliferation, survival, differentiation and death, designed a complex regulation network from HSC to differentiated blood cells. During the last decade, an additional level of regulation was added to this TF network with the discovery and understanding of a new class of regulatory non-coding RNAs, the miRNAs. Understanding of functional and reciprocal interactions with TFs significantly improved our knowledge about the molecular biology of hematopoiesis and hematological diseases. Nowadays, the complexity of the regulatory network is even increasing through the emergence of lncRNAs as novel relevant regulatory elements in all biological processes. Even though the roles of lncRNAs remain partially undetermined, their ability to interact with miRNAs, regulatory proteins and DNA has been evidenced, including in the hematopoietic system. Together, TFs, miRNAs and lncRNAs most likely constitute a wide and complex regulatory network contributing to physiological hematopoietic development as well as to pathological alterations.

Acknowledgments

Sébastien Chateauvieux and Anthoula Gaigneaux were supported by fellowships from Télévie-Luxembourg. Research at LBMCC is financially supported by the Fondation de Recherche Cancer et Sang, the Recherches Scientifiques Luxembourg association, the Een Haerz fir kriibskrank Kanner association, the Action Lions Vaincre le Cancer association, the European Union (ITN "RedCat" 215009, Interreg Iva project "Corena") and the Télévie Luxembourg. MD is supported by the National Research Foundation (NRF) by the MEST of Korea for Tumor Microenvironment Global Core Research Center (GCRC) grant, [grant number 2012-0001184]; by the Seoul National University Research grant and by the Research Settlement Fund for the new faculty of SNU.

References

1. Inaba, M.; Yamashita, Y.M. Asymmetric stem cell division: Precizsion for robustness. *Cell Stem Cell* **2012**, *11*, 461–469.
2. Lichtinger, M.; Ingram, R.; Hannah, R.; Muller, D.; Clarke, D.; Assi, S.A.; Lie, A.L.M.; Noailles, L.; Vijayabaskar, M.S.; Wu, M.; *et al.* RUNX1 reshapes the epigenetic landscape at the onset of haematopoiesis. *EMBO J.* **2012**, *31*, 4318–4333.
3. Papathanasiou, P.; Attema, J.L.; Karsunky, H.; Hosen, N.; Sontani, Y.; Hoyne, G.F.; Tunningley, R.; Smale, S.T.; Weissman, I.L. Self-renewal of the long-term reconstituting subset of hematopoietic stem cells is regulated by Ikaros. *Stem Cells* **2009**, *27*, 3082–3092.
4. Malinge, S.; Thiollier, C.; Chlon, T.M.; Dore, L.C.; Diebold, L.; Bluteau, O.; Mabialah, V.; Vainchenker, W.; Dessen, P.; Winandy, S.; *et al.* Ikaros inhibits megakaryopoiesis through functional interaction with GATA-1 and NOTCH signaling. *Blood* **2013**, *121*, 2440–2451.
5. Mattick, J.S. Non-coding RNAs: The architects of eukaryotic complexity. *EMBO Reports* **2001**, *2*, 986–991.
6. Mattick, J.S.; Makunin, I.V. Non-coding RNA. *Hum. Mol. Gen.* **2006**, *15*, R17–R29.
7. Pheasant, M.; Mattick, J.S. Raising the estimate of functional human sequences. *Genome Res.* **2007**, *17*, 1245–1253.
8. Fire, A.; Xu, S.; Montgomery, M.K.; Kostas, S.A.; Driver, S.E.; Mello, C.C. Potent and specific genetic interference by double-stranded RNA in Caenorhabditis elegans. *Nature* **1998**, *391*, 806–811.
9. Wightman, B.; Ha, I.; Ruvkun, G. Posttranscriptional regulation of the heterochronic gene *lin-14* by *lin-4* mediates temporal pattern formation in C. elegans. *Cell* **1993**, *75*, 855–862.
10. Hamilton, A.J.; Baulcombe, D.C. A species of small antisense RNA in posttranscriptional gene silencing in plants. *Science* **1999**, *286*, 950–952.

11. Elbashir, S.M.; Harborth, J.; Lendeckel, W.; Yalcin, A.; Weber, K.; Tuschl, T. Duplexes of 21-nucleotide RNAs mediate RNA interference in cultured mammalian cells. *Nature* **2001**, *411*, 494–498.
12. Huang, X.; Yuan, T.; Tschannen, M.; Sun, Z.; Jacob, H.; Du, M.; Liang, M.; Dittmar, R.L.; Liu, Y.; Liang, M.; *et al.* Characterization of human plasma-derived exosomal RNAs by deep sequencing. *BMC Genomics* **2013**, *14*, 319.
13. Lee, R.C.; Feinbaum, R.L.; Ambros, V. The C. elegans heterochronic gene *lin-4* encodes small RNAs with antisense complementarity to *lin-14*. *Cell* **1993**, *75*, 843–854.
14. Lau, N.C.; Lim, L.P.; Weinstein, E.G.; Bartel, D.P. An abundant class of tiny RNAs with probable regulatory roles in Caenorhabditis elegans. *Science* **2001**, *294*, 858–862.
15. Lagos-Quintana, M.; Rauhut, R.; Lendeckel, W.; Tuschl, T. Identification of novel genes coding for small expressed RNAs. *Science* **2001**, *294*, 853–858.
16. Lee, R.C.; Ambros, V. An extensive class of small RNAs in Caenorhabditis elegans. *Science* **2001**, *294*, 862–864.
17. Pillai, R.S.; Bhattacharyya, S.N.; Artus, C.G.; Zoller, T.; Cougot, N.; Basyuk, E.; Bertrand, E.; Filipowicz, W. Inhibition of translational initiation by Let-7 MicroRNA in human cells. *Science* **2005**, *309*, 1573–1576.
18. Friedman, R.C.; Farh, K.K.; Burge, C.B.; Bartel, D.P. Most mammalian mRNAs are conserved targets of microRNAs. *Genome Res.* **2009**, *19*, 92–105.
19. Lewis, B.P.; Shih, I.H.; Jones-Rhoades, M.W.; Bartel, D.P.; Burge, C.B. Prediction of mammalian microRNA targets. *Cell* **2003**, *115*, 787–798.
20. Wang, Z.; Yao, H.; Lin, S.; Zhu, X.; Shen, Z.; Lu, G.; Poon, W.S.; Xie, D.; Lin, M.C.; Kung, H.F. Transcriptional and epigenetic regulation of human microRNAs. *Cancer Lett.* **2013**, *331*, 1–10.
21. Kloosterman, W.P.; Plasterk, R.H. The diverse functions of microRNAs in animal development and disease. *Dev. Cell* **2006**, *11*, 441–450.
22. Yates, L.A.; Norbury, C.J.; Gilbert, R.J. The long and short of microRNA. *Cell* **2013**, *153*, 516–519.
23. Borchert, G.M.; Lanier, W.; Davidson, B.L. RNA polymerase III transcribes human microRNAs. *Nat. Struct. Mol. Biol.* **2006**, *13*, 1097–1101.
24. Altuvia, Y.; Landgraf, P.; Lithwick, G.; Elefant, N.; Pfeffer, S.; Aravin, A.; Brownstein, M.J.; Tuschl, T.; Margalit, H. Clustering and conservation patterns of human microRNAs. *Nucleic Acids Res.* **2005**, *33*, 2697–2706.
25. Guo, H.; Ingolia, N.T.; Weissman, J.S.; Bartel, D.P. Mammalian microRNAs predominantly act to decrease target mRNA levels. *Nature* **2010**, *466*, 835–840.
26. Wu, L.; Fan, J.; Belasco, J.G. MicroRNAs direct rapid deadenylation of mRNA. *Proc. Natl. Acad. Sci. USA* **2006**, *103*, 4034–4039.
27. Zardo, G.; Ciolfi, A.; Vian, L.; Starnes, L.M.; Billi, M.; Racanicchi, S.; Maresca, C.; Fazi, F.; Travaglini, L.; Noguera, N.; *et al.* Polycombs and microRNA-223 regulate human granulopoiesis by transcriptional control of target gene expression. *Blood* **2012**, *119*, 4034–4046.

28. Johnnidis, J.B.; Harris, M.H.; Wheeler, R.T.; Stehling-Sun, S.; Lam, M.H.; Kirak, O.; Brummelkamp, T.R.; Fleming, M.D.; Camargo, F.D. Regulation of progenitor cell proliferation and granulocyte function by microRNA-223. *Nature* **2008**, *451*, 1125–1129.
29. Fazi, F.; Rosa, A.; Fatica, A.; Gelmetti, V.; De Marchis, M.L.; Nervi, C.; Bozzoni, I. A minicircuitry comprised of microRNA-223 and transcription factors NFI-A and C/EBPalpha regulates human granulopoiesis. *Cell* **2005**, *123*, 819–831.
30. Fukao, T.; Fukuda, Y.; Kiga, K.; Sharif, J.; Hino, K.; Enomoto, Y.; Kawamura, A.; Nakamura, K.; Takeuchi, T.; Tanabe, M. An evolutionarily conserved mechanism for microRNA-223 expression revealed by microRNA gene profiling. *Cell* **2007**, *129*, 617–631.
31. Masaki, S.; Ohtsuka, R.; Abe, Y.; Muta, K.; Umemura, T. Expression patterns of microRNAs 155 and 451 during normal human erythropoiesis. *Biochem. Biophys. Res. Commun.* **2007**, *364*, 509–514.
32. Monticelli, S.; Ansel, K.M.; Xiao, C.; Socci, N.D.; Krichevsky, A.M.; Thai, T.H.; Rajewsky, N.; Marks, D.S.; Sander, C.; Rajewsky, K.; *et al.* MicroRNA profiling of the murine hematopoietic system. *Genome Biol.* **2005**, *6*, R71.
33. Lu, J.; Getz, G.; Miska, E.A.; Alvarez-Saavedra, E.; Lamb, J.; Peck, D.; Sweet-Cordero, A.; Ebert, B.L.; Mak, R.H.; Ferrando, A.A.; *et al.* MicroRNA expression profiles classify human cancers. *Nature* **2005**, *435*, 834–838.
34. Choong, M.L.; Yang, H.H.; McNiece, I. MicroRNA expression profiling during human cord blood-derived CD34 cell erythropoiesis. *Exp. Hematol.* **2007**, *35*, 551–564.
35. Felli, N.; Pedini, F.; Romania, P.; Biffoni, M.; Morsilli, O.; Castelli, G.; Santoro, S.; Chicarella, S.; Sorrentino, A.; Peschle, C.; *et al.* MicroRNA 223-dependent expression of LMO2 regulates normal erythropoiesis. *Haematologica* **2009**, *94*, 479–486.
36. Cline, M.S.; Smoot, M.; Cerami, E.; Kuchinsky, A.; Landys, N.; Workman, C.; Christmas, R.; Avila-Campilo, I.; Creech, M.; Gross, B.; *et al.* Integration of biological networks and gene expression data using Cytoscape. *Nat. Protoc.* **2007**, *2*, 2366–2382.
37. Zhou, K.; Yi, S.; Yu, Z.; Li, Z.; Wang, Y.; Zou, D.; Qi, J.; Zhao, Y.; Qiu, L. MicroRNA-223 expression is uniformly down-regulated in B cell lymphoproliferative disorders and is associated with poor survival in patients with chronic lymphocytic leukemia. *Leukemia Lymphoma* **2012**, *53*, 1155–1161.
38. Chen, M.J.; Yokomizo, T.; Zeigler, B.M.; Dzierzak, E.; Speck, N.A. Runx1 is required for the endothelial to haematopoietic cell transition but not thereafter. *Nature* **2009**, *457*, 887–891.
39. Okuda, T.; Takeda, K.; Fujita, Y.; Nishimura, M.; Yagyu, S.; Yoshida, M.; Akira, S.; Downing, J.R.; Abe, T. Biological characteristics of the leukemia-associated transcriptional factor AML1 disclosed by hematopoietic rescue of AML1-deficient embryonic stem cells by using a knock-in strategy. *Mol. Cell. Biol.* **2000**, *20*, 319–328.
40. Wang, Q.; Stacy, T.; Binder, M.; Marin-Padilla, M.; Sharpe, A.H.; Speck, N.A. Disruption of the Cbfa2 gene causes necrosis and hemorrhaging in the central nervous system and blocks definitive hematopoiesis. *Proc. Natl. Acad. Sci. USA* **1996**, *93*, 3444–3449.
41. Yokomizo, T.; Ogawa, M.; Osato, M.; Kanno, T.; Yoshida, H.; Fujimoto, T.; Fraser, S.; Nishikawa, S.; Okada, H.; Satake, M.; *et al.* Requirement of Runx1/AML1/PEBP2alphaB for the generation of haematopoietic cells from endothelial cells. *Genes Cells* **2001**, *6*, 13–23.

42. Nishimura, M.; Fukushima-Nakase, Y.; Fujita, Y.; Nakao, M.; Toda, S.; Kitamura, N.; Abe, T.; Okuda, T. VWRPY motif-dependent and -independent roles of AML1/Runx1 transcription factor in murine hematopoietic development. *Blood* **2004**, *103*, 562–570.
43. Ran, D.; Shia, W.J.; Lo, M.C.; Fan, J.B.; Knorr, D.A.; Ferrell, P.I.; Ye, Z.; Yan, M.; Cheng, L.; Kaufman, D.S.; *et al.* RUNX1a enhances hematopoietic lineage commitment from human embryonic stem cells and inducible pluripotent stem cells. *Blood* **2013**, *121*, 2882–2890.
44. Fazi, F.; Racanicchi, S.; Zardo, G.; Starnes, L.M.; Mancini, M.; Travaglini, L.; Diverio, D.; Ammatuna, E.; Cimino, G.; Lo-Coco, F.; *et al.* Epigenetic silencing of the myelopoiesis regulator microRNA-223 by the AML1/ETO oncoprotein. *Cancer Cell* **2007**, *12*, 457–466.
45. Nimer, S.D.; Moore, M.A. Effects of the leukemia-associated AML1-ETO protein on hematopoietic stem and progenitor cells. *Oncogene* **2004**, *23*, 4249–4254.
46. Frank, R.; Zhang, J.; Uchida, H.; Meyers, S.; Hiebert, S.W.; Nimer, S.D. The AML1/ETO fusion protein blocks transactivation of the GM-CSF promoter by AML1B. *Oncogene* **1995**, *11*, 2667–2674.
47. Zhang, D.E.; Fujioka, K.; Hetherington, C.J.; Shapiro, L.H.; Chen, H.M.; Look, A.T.; Tenen, D.G. Identification of a region which directs the monocytic activity of the colony-stimulating factor 1 (macrophage colony-stimulating factor) receptor promoter and binds PEBP2/CBF (AML1). *Mol. Cell. Biol.* **1994**, *14*, 8085–8095.
48. Sun, W.; Shen, W.; Yang, S.; Hu, F.; Li, H.; Zhu, T.H. miR-223 and miR-142 attenuate hematopoietic cell proliferation, and miR-223 positively regulates miR-142 through LMO2 isoforms and CEBP-beta. *Cell Res.* **2010**, *20*, 1158–1169.
49. Warren, A.J.; Colledge, W.H.; Carlton, M.B.; Evans, M.J.; Smith, A.J.; Rabbitts, T.H. The oncogenic cysteine-rich LIM domain protein rbtn2 is essential for erythroid development. *Cell* **1994**, *78*, 45–57.
50. Wadman, I.A.; Osada, H.; Grutz, G.G.; Agulnick, A.D.; Westphal, H.; Forster, A.; Rabbitts, T.H. The LIM-only protein Lmo2 is a bridging molecule assembling an erythroid, DNA-binding complex which includes the TAL1, E47, GATA-1 and Ldb1/NLI proteins. *EMBO J.* **1997**, *16*, 3145–3157.
51. Starnes, L.M.; Sorrentino, A.; Ferracin, M.; Negrini, M.; Pelosi, E.; Nervi, C.; Peschle, C. A transcriptome-wide approach reveals the key contribution of NFI-A in promoting erythroid differentiation of human CD34(+) progenitors and CML cells. *Leukemia* **2010**, *24*, 1220–1223.
52. Starnes, L.M.; Sorrentino, A.; Pelosi, E.; Ballarino, M.; Morsilli, O.; Biffoni, M.; Santoro, S.; Felli, N.; Castelli, G.; de Marchis, M.L.; *et al.* NFI-A directs the fate of hematopoietic progenitors to the erythroid or granulocytic lineage and controls beta-globin and G-CSF receptor expression. *Blood* **2009**, *114*, 1753–1763.
53. Sun, Y.; Sun, J.; Tomomi, T.; Nieves, E.; Mathewson, N.; Tamaki, H.; Evers, R.; Reddy, P. PU.1-dependent transcriptional regulation of miR-142 contributes to its hematopoietic cell-specific expression and modulation of IL-6. *J. Immunol.* **2013**, *190*, 4005–4013.
54. Brioschi, M.; Fischer, J.; Cairoli, R.; Rossetti, S.; Pezzetti, L.; Nichelatti, M.; Turrini, M.; Corlazzoli, F.; Scarpati, B.; Morra, E.; *et al.* Down-regulation of microRNAs 222/221 in acute myelogenous leukemia with deranged core-binding factor subunits. *Neoplasia* **2010**, *12*, 866–876.

55. Feng, J.; Iwama, A.; Satake, M.; Kohu, K. MicroRNA-27 enhances differentiation of myeloblasts into granulocytes by post-transcriptionally downregulating Runx1. *Br. J. Haematol.* **2009**, *145*, 412–423.
56. Ben-Ami, O.; Pencovich, N.; Lotem, J.; Levanon, D.; Groner, Y. A regulatory interplay between miR-27a and Runx1 during megakaryopoiesis. *Proc. Natl. Acad. Sci. USA* **2009**, *106*, 238–243.
57. Rossetti, S.; Sacchi, N. RUNX1: A microRNA hub in normal and malignant hematopoiesis. *Int. J. Mol. Sci.* **2013**, *14*, 1566–1588.
58. Wontakal, S.N.; Guo, X.; Will, B.; Shi, M.; Raha, D.; Mahajan, M.C.; Weissman, S.; Snyder, M.; Steidl, U.; Zheng, D.; *et al.* A large gene network in immature erythroid cells is controlled by the myeloid and B cell transcriptional regulator PU.1. *PLoS Genet.* **2011**, *7*, e1001392.
59. Houston, I.B.; Kamath, M.B.; Schweitzer, B.L.; Chlon, T.M.; DeKoter, R.P. Reduction in PU.1 activity results in a block to B-cell development, abnormal myeloid proliferation, and neonatal lethality. *Exp. Hematol.* **2007**, *35*, 1056–1068.
60. Anderson, M.K.; Weiss, A.H.; Hernandez-Hoyos, G.; Dionne, C.J.; Rothenberg, E.V. Constitutive expression of PU.1 in fetal hematopoietic progenitors blocks T cell development at the pro-T cell stage. *Immunity* **2002**, *16*, 285–296.
61. Dionne, C.J.; Tse, K.Y.; Weiss, A.H.; Franco, C.B.; Wiest, D.L.; Anderson, M.K.; Rothenberg, E.V. Subversion of T lineage commitment by PU.1 in a clonal cell line system. *Dev. Biol.* **2005**, *280*, 448–466.
62. Laiosa, C.V.; Stadtfeld, M.; Xie, H.; de Andres-Aguayo, L.; Graf, T. Reprogramming of committed T cell progenitors to macrophages and dendritic cells by C/EBP alpha and PU.1 transcription factors. *Immunity* **2006**, *25*, 731–744.
63. Lefebvre, J.M.; Haks, M.C.; Carleton, M.O.; Rhodes, M.; Sinnathamby, G.; Simon, M.C.; Eisenlohr, L.C.; Garrett-Sinha, L.A.; Wiest, D.L. Enforced expression of Spi-B reverses T lineage commitment and blocks beta-selection. *J. Immunol.* **2005**, *174*, 6184–6194.
64. Ghani, S.; Riemke, P.; Schonheit, J.; Lenze, D.; Stumm, J.; Hoogenkamp, M.; Lagendijk, A.; Heinz, S.; Bonifer, C.; Bakkers, J.; *et al.* Macrophage development from HSCs requires PU.1-coordinated microRNA expression. *Blood* **2011**, *118*, 2275–2284.
65. Rosa, A.; Ballarino, M.; Sorrentino, A.; Sthandier, O.; de Angelis, F.G.; Marchioni, M.; Masella, B.; Guarini, A.; Fatica, A.; Peschle, C.; *et al.* The interplay between the master transcription factor PU.1 and miR-424 regulates human monocyte/macrophage differentiation. *Proc. Natl. Acad. Sci. USA* **2007**, *104*, 19849–19854.
66. Rhodes, J.; Hagen, A.; Hsu, K.; Deng, M.; Liu, T.X.; Look, A.T.; Kanki, J.P. Interplay of pu.1 and gata1 determines myelo-erythroid progenitor cell fate in zebrafish. *Dev. Cell* **2005**, *8*, 97–108.
67. Zhang, P.; Zhang, X.; Iwama, A.; Yu, C.; Smith, K.A.; Mueller, B.U.; Narravula, S.; Torbett, B.E.; Orkin, S.H.; Tenen, D.G. PU.1 inhibits GATA-1 function and erythroid differentiation by blocking GATA-1 DNA binding. *Blood* **2000**, *96*, 2641–2648.
68. Stachura, D.L.; Chou, S.T.; Weiss, M.J. Early block to erythromegakaryocytic development conferred by loss of transcription factor GATA-1. *Blood* **2006**, *107*, 87–97.
69. Chou, S.T.; Khandros, E.; Bailey, L.C.; Nichols, K.E.; Vakoc, C.R.; Yao, Y.; Huang, Z.; Crispino, J.D.; Hardison, R.C.; Blobel, G.A.; *et al.* Graded repression of PU.1/Sfpi1 gene transcription by GATA factors regulates hematopoietic cell fate. *Blood* **2009**, *114*, 983–994.

70. Dore, L.C.; Amigo, J.D.; Dos Santos, C.O.; Zhang, Z.; Gai, X.; Tobias, J.W.; Yu, D.; Klein, A.M.; Dorman, C.; Wu, W.; *et al.* A GATA-1-regulated microRNA locus essential for erythropoiesis. *Proc. Natl. Acad. Sci. USA* **2008**, *105*, 3333–3338.
71. Papapetrou, E.P.; Korkola, J.E.; Sadelain, M. A genetic strategy for single and combinatorial analysis of miRNA function in mammalian hematopoietic stem cells. *Stem Cells* **2010**, *28*, 287–296.
72. Pase, L.; Layton, J.E.; Kloosterman, W.P.; Carradice, D.; Waterhouse, P.M.; Lieschke, G.J. miR-451 regulates zebrafish erythroid maturation *in vivo* via its target gata2. *Blood* **2009**, *113*, 1794–1804.
73. Minegishi, N.; Suzuki, N.; Yokomizo, T.; Pan, X.; Fujimoto, T.; Takahashi, S.; Hara, T.; Miyajima, A.; Nishikawa, S.; Yamamoto, M. Expression and domain-specific function of GATA-2 during differentiation of the hematopoietic precursor cells in midgestation mouse embryos. *Blood* **2003**, *102*, 896–905.
74. Bruchova, H.; Yoon, D.; Agarwal, A.M.; Mendell, J.; Prchal, J.T. Regulated expression of microRNAs in normal and polycythemia vera erythropoiesis. *Exp. Hematol.* **2007**, *35*, 1657–1667.
75. Mizuno, Y.; Tokuzawa, Y.; Ninomiya, Y.; Yagi, K.; Yatsuka-Kanesaki, Y.; Suda, T.; Fukuda, T.; Katagiri, T.; Kondoh, Y.; Amemiya, T.; *et al.* miR-210 promotes osteoblastic differentiation through inhibition of AcvR1b. *FEBS Lett.* **2009**, *583*, 2263–2268.
76. Orom, U.A.; Derrien, T.; Beringer, M.; Gumireddy, K.; Gardini, A.; Bussotti, G.; Lai, F.; Zytnicki, M.; Notredame, C.; Huang, Q.; *et al.* Long noncoding RNAs with enhancer-like function in human cells. *Cell* **2010**, *143*, 46–58.
77. Carninci, P.; Kasukawa, T.; Katayama, S.; Gough, J.; Frith, M.C.; Maeda, N.; Oyama, R.; Ravasi, T.; Lenhard, B.; Wells, C.; *et al.* The transcriptional landscape of the mammalian genome. *Science* **2005**, *309*, 1559–1563.
78. Carninci, P.; Hayashizaki, Y. Noncoding RNA transcription beyond annotated genes. *Curr. Opin. Genetics Dev.* **2007**, *17*, 139–144.
79. Gibb, E.A.; Vucic, E.A.; Enfield, K.S.; Stewart, G.L.; Lonergan, K.M.; Kennett, J.Y.; Becker-Santos, D.D.; MacAulay, C.E.; Lam, S.; Brown, C.J.; *et al.* Human cancer long non-coding RNA transcriptomes. *PLoS One* **2011**, *6*, e25915.
80. Whitehead, J.; Pandey, G.K.; Kanduri, C. Regulation of the mammalian epigenome by long noncoding RNAs. *Biochim. Biophys. Acta* **2009**, *1790*, 936–947.
81. Yoon, J.H.; Abdelmohsen, K.; Gorospe, M. Posttranscriptional gene regulation by long noncoding RNA. *J. Mol. Biol.* in press, 2012.
82. Tripathi, V.; Ellis, J.D.; Shen, Z.; Song, D.Y.; Pan, Q.; Watt, A.T.; Freier, S.M.; Bennett, C.F.; Sharma, A.; Bubulya, P.A.; *et al.* The nuclear-retained noncoding RNA MALAT1 regulates alternative splicing by modulating SR splicing factor phosphorylation. *Mol. Cell* **2010**, *39*, 925–938.
83. Moon, S.L.; Anderson, J.R.; Kumagai, Y.; Wilusz, C.J.; Akira, S.; Khromykh, A.A.; Wilusz, J. A noncoding RNA produced by arthropod-borne flaviviruses inhibits the cellular exoribonuclease XRN1 and alters host mRNA stability. *RNA* **2012**, *18*, 2029–2040.
84. Faghihi, M.A.; Modarresi, F.; Khalil, A.M.; Wood, D.E.; Sahagan, B.G.; Morgan, T.E.; Finch, C.E.; St Laurent, G., 3rd; Kenny, P.J.; Wahlestedt, C. Expression of a noncoding RNA is elevated in Alzheimer's disease and drives rapid feed-forward regulation of beta-secretase. *Nat. Med.* **2008**, *14*, 723–730.

85. Kim, Y.K.; Furic, L.; Parisien, M.; Major, F.; DesGroseillers, L.; Maquat, L.E. Staufen1 regulates diverse classes of mammalian transcripts. *EMBO J.* **2007**, *26*, 2670–2681.
86. Yoon, J.H.; Abdelmohsen, K.; Srikantan, S.; Yang, X.; Martindale, J.L.; De, S.; Huarte, M.; Zhan, M.; Becker, K.G.; Gorospe, M. LincRNA-p21 suppresses target mRNA translation. *Mol. Cell* **2012**, *47*, 648–655.
87. Carrieri, C.; Cimatti, L.; Biagioli, M.; Beugnet, A.; Zucchelli, S.; Fedele, S.; Pesce, E.; Ferrer, I.; Collavin, L.; Santoro, C.; *et al.* Long non-coding antisense RNA controls Uchl1 translation through an embedded SINEB2 repeat. *Nature* **2012**, *491*, 454–457.
88. Kindler, S.; Wang, H.; Richter, D.; Tiedge, H. RNA transport and local control of translation. *Annu. Rev. Cell Dev. Biol.* **2005**, *21*, 223–245.
89. Clark, M.B.; Johnston, R.L.; Inostroza-Ponta, M.; Fox, A.H.; Fortini, E.; Moscato, P.; Dinger, M.E.; Mattick, J.S. Genome-wide analysis of long noncoding RNA stability. *Genome Res.* **2012**, *22*, 885–898.
90. Brown, J.A.; Valenstein, M.L.; Yario, T.A.; Tycowski, K.T.; Steitz, J.A. Formation of triple-helical structures by the 3'-end sequences of MALAT1 and MENbeta noncoding RNAs. *Proc. Natl. Acad. Sci. USA* **2012**, *109*, 19202–19207.
91. Wilusz, J.E.; Jnbaptiste, C.K.; Lu, L.Y.; Kuhn, C.D.; Joshua-Tor, L.; Sharp, P.A. A triple helix stabilizes the 3' ends of long noncoding RNAs that lack poly(A) tails. *Genes Dev.* **2012**, *26*, 2392–2407.
92. Jalali, S.; Bhartiya, D.; Lalwani, M.K.; Sivasubbu, S.; Scaria, V. Systematic transcriptome wide analysis of lncRNA-miRNA interactions. *PLoS One* **2013**, *8*, e53823.
93. Hung, T.; Chang, H.Y. Long noncoding RNA in genome regulation: Prospects and mechanisms. *RNA Biol.* **2010**, *7*, 582–585.
94. Martianov, I.; Ramadass, A.; Serra Barros, A.; Chow, N.; Akoulitchev, A. Repression of the human dihydrofolate reductase gene by a non-coding interfering transcript. *Nature* **2007**, *445*, 666–670.
95. Wang, K.C.; Chang, H.Y. Molecular mechanisms of long noncoding RNAs. *Mol. Cell* **2011**, *43*, 904–914.
96. Da Sacco, L.; Baldassarre, A.; Masotti, A. Bioinformatics tools and novel challenges in long non-coding RNAs (lncRNAs) functional analysis. *Int. J. Mol. Sci.* **2012**, *13*, 97–114.
97. Heo, J.B.; Sung, S. Vernalization-mediated epigenetic silencing by a long intronic noncoding RNA. *Science* **2011**, *331*, 76–79.
98. Swiezewski, S.; Liu, F.; Magusin, A.; Dean, C. Cold-induced silencing by long antisense transcripts of an Arabidopsis Polycomb target. *Nature* **2009**, *462*, 799–802.
99. Cesana, M.; Cacchiarelli, D.; Legnini, I.; Santini, T.; Sthandier, O.; Chinappi, M.; Tramontano, A.; Bozzoni, I. A long noncoding RNA controls muscle differentiation by functioning as a competing endogenous RNA. *Cell* **2011**, *147*, 358–369.
100. Poliseno, L.; Salmena, L.; Riccardi, L.; Fornari, A.; Song, M.S.; Hobbs, R.M.; Sportoletti, P.; Varmeh, S.; Egia, A.; Fedele, G.; *et al.* Identification of the miR-106b~25 microRNA cluster as a proto-oncogenic PTEN-targeting intron that cooperates with its host gene MCM7 in transformation. *Sci. Signal.* **2010**, *3*, ra29.

101. Torarinsson, E.; Sawera, M.; Havgaard, J.H.; Fredholm, M.; Gorodkin, J. Thousands of corresponding human and mouse genomic regions unalignable in primary sequence contain common RNA structure. *Genome Res.* **2006**, *16*, 885–889.
102. Espinoza, C.A.; Allen, T.A.; Hieb, A.R.; Kugel, J.F.; Goodrich, J.A. B2 RNA binds directly to RNA polymerase II to repress transcript synthesis. *Nat. Struct. Mol. Biol.* **2004**, *11*, 822–829.
103. Mariner, P.D.; Walters, R.D.; Espinoza, C.A.; Drullinger, L.F.; Wagner, S.D.; Kugel, J.F.; Goodrich, J.A. Human Alu RNA is a modular transacting repressor of mRNA transcription during heat shock. *Mol. Cell* **2008**, *29*, 499–509.
104. Hu, J.; Xu, T.; Zhu, T.; Lou, Q.; Wang, X.; Wu, Y.; Huang, R.; Liu, J.; Liu, H.; Yu, F.; *et al.* Monoclonal antibodies against accumulation-associated protein affect EPS biosynthesis and enhance bacterial accumulation of Staphylococcus epidermidis. *PLoS One* **2011**, *6*, e20918.
105. Paralkar, V.R.; Weiss, M.J. A new 'Linc' between noncoding RNAs and blood development. *Genes Dev.* **2011**, *25*, 2555–2558.
106. McGlinn, E.; Yekta, S.; Mansfield, J.H.; Soutschek, J.; Bartel, D.P.; Tabin, C.J. *In ovo* application of antagomiRs indicates a role for miR-196 in patterning the chick axial skeleton through Hox gene regulation. *Proc. Natl. Acad. Sci. USA* **2009**, *106*, 18610–18615.
107. Crooks, G.M.; Fuller, J.; Petersen, D.; Izadi, P.; Malik, P.; Pattengale, P.K.; Kohn, D.B.; Gasson, J.C. Constitutive HOXA5 expression inhibits erythropoiesis and increases myelopoiesis from human hematopoietic progenitors. *Blood* **1999**, *94*, 519–528.
108. Moore, M.A.; Dorn, D.C.; Schuringa, J.J.; Chung, K.Y.; Morrone, G. Constitutive activation of Flt3 and STAT5A enhances self-renewal and alters differentiation of hematopoietic stem cells. *Exp. Hematol.* **2007**, *35*, 105–116.
109. Wang, Q.; Huang, Z.; Xue, H.; Jin, C.; Ju, X.L.; Han, J.D.; Chen, Y.G. MicroRNA miR-24 inhibits erythropoiesis by targeting activin type I receptor ALK4. *Blood* **2008**, *111*, 588–595.
110. Jeggari, A.; Marks, D.S.; Larsson, E. miRcode: A map of putative microRNA target sites in the long non-coding transcriptome. *Bioinformatics* **2012**, *28*, 2062–2063.
111. Goshen, R.; Rachmilewitz, J.; Schneider, T.; de-Groot, N.; Ariel, I.; Palti, Z.; Hochberg, A.A. The expression of the H-19 and IGF-2 genes during human embryogenesis and placental development. *Mol. Reproduct. Dev.* **1993**, *34*, 374–379.
112. Nunez, C.; Bashein, A.M.; Brunet, C.L.; Hoyland, J.A.; Freemont, A.J.; Buckle, A.M.; Murphy, C.; Cross, M.A.; Lucas, G.; Bostock, V.J.; *et al.* Expression of the imprinted tumour-suppressor gene H19 is tightly regulated during normal haematopoiesis and is reduced in haematopoietic precursors of patients with the myeloproliferative disease polycythaemia vera. *J. Pathol.* **2000**, *190*, 61–68.
113. Tessema, M.; Langer, F.; Bock, O.; Seltsam, A.; Metzig, K.; Hasemeier, B.; Kreipe, H.; Lehmann, U. Down-regulation of the IGF-2/H19 locus during normal and malignant hematopoiesis is independent of the imprinting pattern. *Int. J. Oncol.* **2005**, *26*, 499–507.
114. Lustig, O.; Ariel, I.; Ilan, J.; Lev-Lehman, E.; De-Groot, N.; Hochberg, A. Expression of the imprinted gene H19 in the human fetus. *Mol. Reproduct. Dev.* **1994**, *38*, 239–246.
115. Borensztein, M.; Monnier, P.; Court, F.; Louault, Y.; Ripoche, M.A.; Tiret, L.; Yao, Z.; Tapscott, S.J.; Forne, T.; Montarras, D.; *et al.* Myod and H19-Igf2 locus interactions are required for diaphragm formation in the mouse. *Development* **2013**, *140*, 1231–1239.

116. Luo, M.; Li, Z.; Wang, W.; Zeng, Y.; Liu, Z.; Qiu, J. Upregulated H19 contributes to bladder cancer cell proliferation by regulating ID2 expression. *FEBS J.* **2013**, *280*, 1709–1716.
117. Wagner, L.A.; Christensen, C.J.; Dunn, D.M.; Spangrude, G.J.; Georgelas, A.; Kelley, L.; Esplin, M.S.; Weiss, R.B.; Gleich, G.J. EGO, a novel, noncoding RNA gene, regulates eosinophil granule protein transcript expression. *Blood* **2007**, *109*, 5191–5198.
118. Bertani, S.; Sauer, S.; Bolotin, E.; Sauer, F. The noncoding RNA Mistral activates Hoxa6 and Hoxa7 expression and stem cell differentiation by recruiting MLL1 to chromatin. *Mol. Cell* **2011**, *43*, 1040–1046.
119. Jacob, E.; Hod-Dvorai, R.; Ben-Mordechai, O.L.; Boyko, Y.; Avni, O. Dual function of polycomb group proteins in differentiated murine T helper (CD4+) cells. *J. Mol. Signal.* **2011**, *6*, 5.
120. Jo, S.; Lee, H.; Kim, S.; Hwang, E.M.; Park, J.Y.; Kang, S.S.; Chung, H. Inhibition of PCGF2 enhances granulocytic differentiation of acute promyelocytic leukemia cell line HL-60 via induction of HOXA7. *Biochem. Biophys. Res. Commun.* **2011**, *416*, 86–91.
121. Porcher, C.; Swat, W.; Rockwell, K.; Fujiwara, Y.; Alt, F.W.; Orkin, S.H. The T cell leukemia oncoprotein SCL/tal-1 is essential for development of all hematopoietic lineages. *Cell* **1996**, *86*, 47–57.
122. Zhang, X.; Lian, Z.; Padden, C.; Gerstein, M.B.; Rozowsky, J.; Snyder, M.; Gingeras, T.R.; Kapranov, P.; Weissman, S.M.; Newburger, P.E. A myelopoiesis-associated regulatory intergenic noncoding RNA transcript within the human HOXA cluster. *Blood* **2009**, *113*, 2526–2534.
123. Zhao, H.; Zhang, X.; Frazao, J.B.; Condino-Neto, A.; Newburger, P.E. HOX antisense lincRNA HOXA-AS2 is an apoptosis repressor in all trans retinoic acid treated NB4 promyelocytic leukemia cells. *J. Cell. Biochem.* **2013**, doi:10.1002/jcb.24586.
124. Savarese, F.; Flahndorfer, K.; Jaenisch, R.; Busslinger, M.; Wutz, A. Hematopoietic precursor cells transiently reestablish permissiveness for X inactivation. *Mol. Cell. Biol.* **2006**, *26*, 7167–7177.
125. Ahn, J.H.; Lee, Y.; Jeon, C.; Lee, S.J.; Lee, B.H.; Choi, K.D.; Bae, Y.S. Identification of the genes differentially expressed in human dendritic cell subsets by cDNA subtraction and microarray analysis. *Blood* **2002**, *100*, 1742–1754.
126. Lee, M.Y.; Kim, W.J.; Kang, Y.J.; Jung, Y.M.; Kang, Y.M.; Suk, K.; Park, J.E.; Choi, E.M.; Choi, B.K.; Kwon, B.S.; *et al.* Z39Ig is expressed on macrophages and may mediate inflammatory reactions in arthritis and atherosclerosis. *J. Leukocyte Biol.* **2006**, *80*, 922–928.

3

Analysis of Transcriptional Regulation of the Human miR-17-92 Cluster: Evidence for Involvement of Pim-1

Maren Thomas [1,†], Kerstin Lange-Grünweller [1,†], Dorothee Hartmann [1], Lara Golde [1], Julia Schlereth [1], Dennis Streng [1], Achim Aigner [2], Arnold Grünweller [1,*] and Roland K. Hartmann [1,*]

1. Institut für Pharmazeutische Chemie, Philipps-Universität Marburg, 35032 Marburg, Germany; E-Mails: thomasm@staff.uni-marburg.de (M.T.); langegru@staff.uni-marburg.de (K.L.-G.); Dorothee_Hartmann@web.de (D.H.); laragolde@gmx.de (L.G.); julia.schlereth@staff.uni-marburg.de (J.S.); dennis.streng@pharmazie.uni-marburg.de (D.S.)
2. Medizinische Fakultät, Rudolf-Boehm-Institut für Pharmakologie und Toxikologie, Klinische Pharmakologie, Universität Leipzig, 04107 Leipzig, Germany; E-Mail: Achim.Aigner@medizin.uni-leipzig.de

† These authors contributed equally to this work.

* Authors to whom correspondence should be addressed; E-Mails: arnold.gruenweller@staff.uni-marburg.de (A.G.); roland.hartmann@staff.uni-marburg.de (R.K.H.);

Abstract: The human polycistronic miRNA cluster miR-17-92 is frequently overexpressed in hematopoietic malignancies and cancers. Its transcription is in part controlled by an E2F-regulated host gene promoter. An intronic A/T-rich region directly upstream of the miRNA coding region also contributes to cluster expression. Our deletion analysis of the A/T-rich region revealed a strong dependence on c-Myc binding to the functional E3 site. Yet, constructs lacking the 5'-proximal ~1.3 kb or 3'-distal ~0.1 kb of the 1.5 kb A/T-rich region still retained residual specific promoter activity, suggesting multiple transcription start sites (TSS) in this region. Furthermore, the protooncogenic kinase, Pim-1, its phosphorylation target HP1γ and c-Myc colocalize to the E3 region, as inferred from chromatin immunoprecipitation. Analysis of pri-miR-17-92 expression levels in K562 and

HeLa cells revealed that silencing of E2F3, c-Myc or Pim-1 negatively affects cluster expression, with a synergistic effect caused by c-Myc/Pim-1 double knockdown in HeLa cells. Thus, we show, for the first time, that the protooncogene Pim-1 is part of the network that regulates transcription of the human miR-17-92 cluster.

Keywords: miRNA; miR-17-92 cluster; Pim-1; miRNA promoter; c-Myc; HP1γ; RNAi

---

1. Introduction

MicroRNAs (miRNAs) are important post-transcriptional riboregulators of gene expression with high relevance to cancer formation and metastasis [1]. In general, miRNAs are derived from RNA polymerase II (RNAPII) primary transcripts (pri-miRNA) that are further processed to ~70 nt precursors (pre-miRNA) and after nuclear export to mature miRNAs by the activity of the two endonucleases, DROSHA/DGCR8 and DICER [2–5]. MiRNAs are incorporated into the miRNA-induced silencing complex (miRISC) and act as repressors of translation by imperfect base-pairing to their target sites in mRNAs [3]. The majority of miRNAs is encoded in intronic regions, either individually or as "polycistronic" miRNA clusters that are cotranscribed [3,6].

Several deregulated miRNAs or miRNA clusters are involved in tumorigenesis, accounting for their designation as tumor-suppressing or as oncogenic miRNAs [7]. Such miRNAs can downregulate targets involved in the regulation of apoptosis or cell cycle progression [1]. One well-characterized polycistronic cluster is the miR-17-92 cluster, also known as OncomiR-1 or C13*orf*25. This cluster encodes six miRNAs belonging to four different seed families: (i) the miR-17 family with miR-17 and miR-20a, (ii) the miR-18 family with miR-18a, (iii) the miR-19 family with miR-19a and miR-19b-1 and (iv) the miR-92 family [8–11]. The human miR-17-92 cluster is encoded in the chromosomal region, 13q31.3, and is amplified in several solid tumors, as well as in some hematopoietic malignancies [8,12]. Because of numerous known targets of its individual miRNAs, the miR-17-92 locus exerts pleiotropic functions during development, proliferation, apoptosis and angiogenesis in different cell systems [13–15]. In mice, deletion of the cluster prevents normal B-cell development as a consequence of premature cell death [14]. In a mouse B-cell lymphoma model, simultaneous overexpression of c-Myc and the miR-17-92 cluster accelerated lymphomagenesis [9]. This oncogenic effect could later be assigned primarily to miR-19a/b, which dampens expression of the tumor suppressor PTEN, thereby repressing apoptosis [13,15].

Analyses of transcriptional regulation of oncogenic miRNAs and miRNA clusters are of great importance for strategies aiming at cancer prevention. Unfortunately, most miRNA promoters have not been characterized or identified yet [16]. In the case of the miR-17-92 cluster, expression is thought to be promoted from a host gene promoter region upstream of exon 1, with transcription starting at a consensus initiator sequence downstream of a non-consensus TATA box [17,18]. Additionally, this core promoter region contains functional E2F transcription factor binding sites. E2F1-3 were shown to activate C13*orf*25 expression from this promoter and chromatin immunoprecipitation assays (ChIP) identified E2F3 to be the main E2F variant associated with the host gene promoter [17,18]. No E2F binding was detected in the region between the host gene promoter and the miR-17-92 cluster [18].

Furthermore, nucleosome mapping combined with chromatin signatures for transcriptionally active promoters [19–21] indicated that transcriptional activity of the miR-17-92 cluster also originates from the intronic A/T-rich region directly upstream of the miRNA coding sequences [16]. This is in line with the finding that cluster expression is activated by c-Myc binding to a conserved E-box element (E3) ~1.5 kb upstream of the miRNA coding sequence [9,10,20]. Indeed, luciferase reporter assays confirmed that both the host gene promoter and the intronic region confer transcriptional activity [16,21].

Here, we subjected the intronic A/T-rich region to deletion analysis using luciferase reporter constructs. Transcription was found to strongly depend on c-Myc binding to the E3 site, but even shorter fragments (<0.3 kb) of sequences directly preceding the miR-17-92 coding sequence still promoted residual, but substantial and specific transcriptional activity. Interestingly, we identified the protooncogene Pim-1 and one of its phosphorylation targets, HP1γ [22], to be associated with the chromatin region containing the E3 site, suggesting that the human C13*orf*25 locus belongs to the set of genes that are regulated by c-Myc and Pim-1 [23,24]. SiRNA-mediated Pim-1 knockdown indeed resulted in reduced pri-miR-17-92 levels, as did a knockdown of c-Myc or E2F3. In Hela cells, a double knockdown of c-Myc/Pim-1 decreased the pri-miR-17-92 levels more than single knockdowns, consistent with a synergism of c-Myc and Pim-1 at the intronic C13*orf*25 promoter.

2. Results and Discussion

2.1. Results

2.1.1. c-Myc-Dependent Intronic Transcriptional Activity within the Human miR-17-92 Locus

The 3.4 kb upstream genomic region of the miR-17-92 coding sequence can be subdivided into a G/C-rich and an A/T-rich part. The former is a CpG island (~1.9 kb, 78% GC content; see http://genome.ucsc.edu [25], GRCh37/hg19 assembly) that has its 5'-boundary ~0.4 kb upstream of the TSS of the host gene promoter [20] and its 3'-boundary ~1.4 kb upstream of the miR-17-5p coding sequence, representing the 5'-terminal miRNA of the cluster. The A/T-rich region (~64% A/T content) following the CpG island begins immediately downstream of a functional and highly conserved c-Myc binding site (5'-CATGTG, E-box E3), which is localized ~1.5 kb upstream of the miR-17-5p coding sequence [10] (Figures 1A and S1).

We have analyzed the intronic region of C13*orf*25, including the preceding functional c-Myc box E3 [20] and truncated segments of the A/T-rich region (Figure 1B) for transcriptional activation. For this purpose, luciferase reporter constructs were transfected into K562 (a human myelogenous erythroleukemia cell line from a CML patient) and HeLa cells (an epithelial human cell line from a cervical carcinoma). We selected these two cell lines as a starting point to study transcription of the miR-17-92 cluster in the context of different cellular expression levels (Figure S2).

Figure 1. (**A**) Genomic organization of C13*orf*25. The locus consists of four exons and three introns; the six miRNAs of the miR-17-92 cluster are encoded in intron 3. Sequences upstream of the cluster can be subdivided into a G/C-rich CpG island and an A/T-rich downstream part. The host gene promoter thought to be activated by E2F3 is located in the CpG island about 3.4 kb upstream of the miR-17-5p coding sequence. The functional c-Myc site (E3) is located ~1.5 kb upstream of miR-17-5p. Sequence numbering is based on the NCBI reference sequence NG_032702.1 and the GRCh37/hg19 assembly [25]. Note that previous related studies referred to the numbering system of the previous hg18 assembly [16,17,20,21]. The numbering of the hg18 and hg19 assemblies is correlated as follows: nt 92,002,872 (0 kb in Figure 1A) of hg19 is nt 90,800,873 of hg18; (**B**) Schematic representation of the different C13*orf*25 portions fused to the luciferase structural gene. The functional E3 box for c-Myc binding is indicated in the 1.5 kb construct (white vertical line); (**C**) Promoter activities of the different luciferase reporter constructs in K562 and HeLa cells. Obtained luciferase activities were measured as relative light units (RLU) and normalized to the pGL3 control plasmid carrying the SV40 promoter (Promega). A reporter construct lacking the SV40 promoter, as well as a construct harboring the 339 bp fragment of the A/T-rich intronic region in inverted orientation were used as controls. RLU values of the individual constructs were derived from 5 to 16 experiments (+/− S.E.M.).

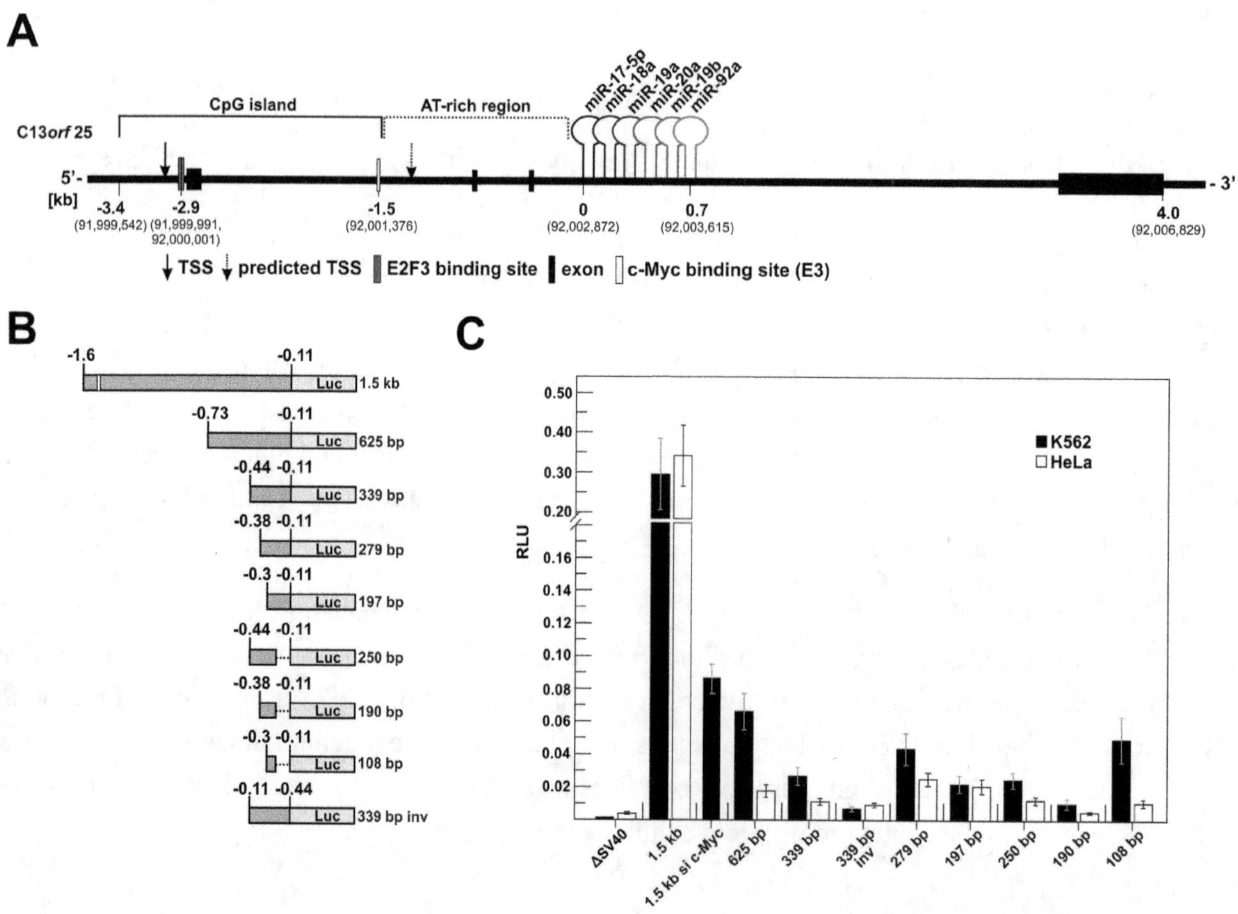

The ~1.5 kb reporter construct, comprising c-Myc box E3 (Figures 1A and S1) and the A/T-rich region (lacking the 113 bp preceding the mature miR-17-5p coding sequence for reasons of PCR feasibility), showed substantial transcriptional activity, amounting to 30%–35% in both cell lines relative to the pGL3 control plasmid harboring an SV40 promoter (Figure 1C). This is in line with results of a similar study of the mouse miR-17-92 locus [21]. Furthermore, Ozsolak *et al.* [16] predicted an intronic TSS to be localized ~0.2 kb downstream of the E3 site. Indeed, truncating the 1.5 kb fragment to 625 bp, which deletes the E3 site, strongly reduced reporter activity by ~4.5-fold in K562 and by almost 20-fold in HeLa cells compared to the activity of the ~1.5 kb construct (Figure 1C). To substantiate this finding, we tested the ~1.5 kb construct in K562 cells under conditions of a siRNA-mediated knockdown of c-Myc. This reduced reporter expression to a similar extent as the truncation to 625 bp, supporting the notion that c-Myc binding to the E3 site plays a key role in activating transcription from this intronic region (Figure 1C). SiRNA-mediated c-Myc knockdown in HeLa cells also suggests a ~four-fold decrease in transcription originating from the ~1.5 kb reporter construct (data not shown), again consistent with the crucial role of c-Myc binding to the E3 site. As the 625 bp fragment still conferred basal promoter activity, we further shortened this region to ~340 bp, ~280 bp and ~200 bp. Additionally, we included short fragments with their 3'-boundary ~290 bp upstream of the mature miR-17-5p coding sequence (250, 190 and 108 bp in Figure 1B). We also inversed the orientation of the ~340 bp fragment in front of the luciferase gene (Figure 1C, 339 bp inverse (inv)) to include a control fragment with comparable A/T content. This inversed fragment conferred reporter activity 5.3-fold (K562) or 2.4-fold (HeLa) higher than that of the pGL3 control vector lacking the SV40 promoter (ΔSV40, Figure 1C).

All the fragments≤ 340 bp conferred residual promoter activities, some clearly to a higher extent than the inverted 339 bp fragment in both cell lines (see the 279 and 197 bp fragments, Figure 1C). This indicates that parts of the intronic A/T-rich region promote specific transcriptional activity, the extent partly differing between the two cell lines (Figure 1C). Notably, despite using a variety of web-based promoter prediction tools (see Suppl. Material), no correlation between fragment activity and promoter elements predicted in this region was identified. In K562 cells, the smaller fragments, including the 625 bp fragment, showed an overall trend towards stronger expression relative to HeLa cells.

2.1.2. Pim-1 and HP1γ Are Associated with the Intronic c-Myc Binding Site

We next asked if other factors beyond c-Myc may be involved in human miR-17-92 cluster expression from the A/T-rich region. Transcriptional regulation by c-Myc is associated with Pim-1-dependent H3S10 phosphorylation in about 20% of all genes regulated by c-Myc [24]. Moreover, Pim-1 and c-Myc act synergistically in severe forms of B-cell lymphomas and Pim-1, as well as the miR-17-92 cluster are overexpressed in K562 cells [26]. We performed ChIP assays to test whether Pim-1 localizes to the internal promoter region of the miR-17-92 cluster. For this analysis, we amplified a ~90 bp DNA fragment (segment A1 in Figure 2A) 0.1 kb downstream of the functional c-Myc E3 site. The same DNA segment has been analyzed in a previous study on c-Myc [10]. Our ChIP analysis revealed that not only c-Myc, as expected, but also Pim-1 localizes to this genomic region (Figure 2B, left lanes in upper and middle panels). Indeed, this is consistent with the finding that Pim-1-catalyzed H3S10 phosphorylation is required for c-Myc-dependent transcriptional activation [24]. We further analyzed another known phosphorylation target of Pim-1, the heterochromatin protein-1 gamma (HP1γ) [22], for its association with the E3 region. HP1γ localized to this genomic area, as well (Figure 2B, lower panel). Moreover, we were able to identify an association of HP1γ along the miRNA coding region, which is indicative of active transcription (see Figure S3 and Discussion section).

Figure 2. (**A**) Schematic representation of the intronic A/T-rich region preceding the miR-17-92 coding sequence. The region A1 (blue box) defines the genomic sequence 0.1 kb downstream of the functional c-Myc binding site (E3; yellow box) that was amplified in ChIP analyses; (**B**) ChIP analysis of the intronic region A1 in K562 cells, using antibodies specific for c-Myc, Pim-1 and HP1γ. +AB: with antibody; −AB: without antibody; Mock: buffer only without cell lysate; Input: supernatant of the −AB -sample after immunoprecipitation and centrifugation (for details, see Supplementary Materials).

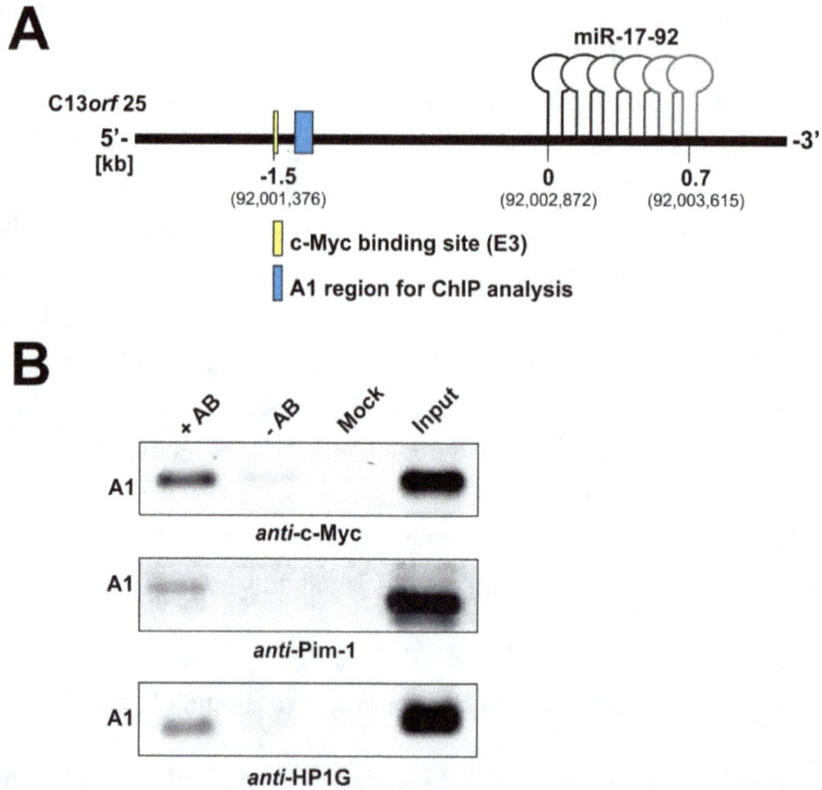

2.1.3. Transcriptional Activity of the Human miR-17-92 Cluster Depends on c-Myc and Pim-1

To further substantiate the role of Pim-1 in miR-17-92 cluster expression, we quantified the cellular pri-miR-17-92 levels by qRT-PCR (see Figure 3A for primer positions) after siRNA-mediated Pim-1 knockdown relative to a c-Myc knockdown in K562 and HeLa cells. Since combined ChIP and reporter gene assays suggested that the transcription factor E2F3 is a major activator of transcription initiated at the host gene promoter [17,18], we included E2F3 in our knockdown experiments as a possible measure for the contribution of the host gene promoter to miR-17-92 expression. We also quantified the levels of c-Myc, E2F3 and Pim-1 mRNAs after knockdown by qRT-PCR to evaluate knockdown efficiencies (Supplementary Table S1). For Pim-1, we have shown good correlation between mRNA and protein levels [26], suggesting that reduced mRNA levels will also entail decreased protein levels. A corresponding parallel analysis of protein levels was inconclusive, owing to a non-interpretable pattern obtained with the used E2F3 antibody [18]. In the study presented here, only experiments with a knockdown efficiency >50% were included (Supplementary Table S1). Single knockdowns of either c-Myc or E2F3 decreased the pri-miR-17-92 levels in HeLa cells to ~35 and 60%, respectively, relative to the control siRNA (Figure 3B). Notably, a 40% reduction of pri-miR-17-92 levels was also observed upon Pim-1 knockdown. Similar results were obtained in K562 cells, with decreases in pri-miR-17-92 levels to ~30%, 30% and 45%, respectively (Figure 3C). However, double knockdowns had additive suppression effects on pri-miR-17-92 levels in the case of c-Myc/E2F3 (HeLa and K562), c-Myc/Pim-1 (Hela) and E2F3/Pim-1 (HeLa). To shed more light on the role of Pim-1, we further analyzed luciferase activity of the 1.5 kb construct harboring the functional c-Myc E3 site in K562 and HeLa cells upon Pim-1 knockdown. We did not observe a substantial decrease in reporter expression after Pim-1 knockdown in K562 cells (data not shown), but a three-fold reduction (Supplementary Figure S4) in HeLa cells (see Discussion).

Figure 3. (**A**) Illustration of the primers (red arrows) used for the qRT-PCR quantification of pri-miR-17-92 transcript levels; (**B**,**C**) qRT-PCR-based quantitation of pri-miR-17-92 transcript levels in HeLa (**B**) or K562 cells (**C**) after siRNA-mediated knockdown of c-Myc, E2F3 or Pim-1 or after combined knockdown of c-Myc/E2F3, c-Myc/Pim-1 or E2F3/Pim-1. 2^-ΔΔpri-17-92 values were normalized against 5S rRNA and an internal control siRNA (siVR1), representing mean values from at least three independent experiments (+/− S.E.M.). Statistical analyses were done using the software, R.

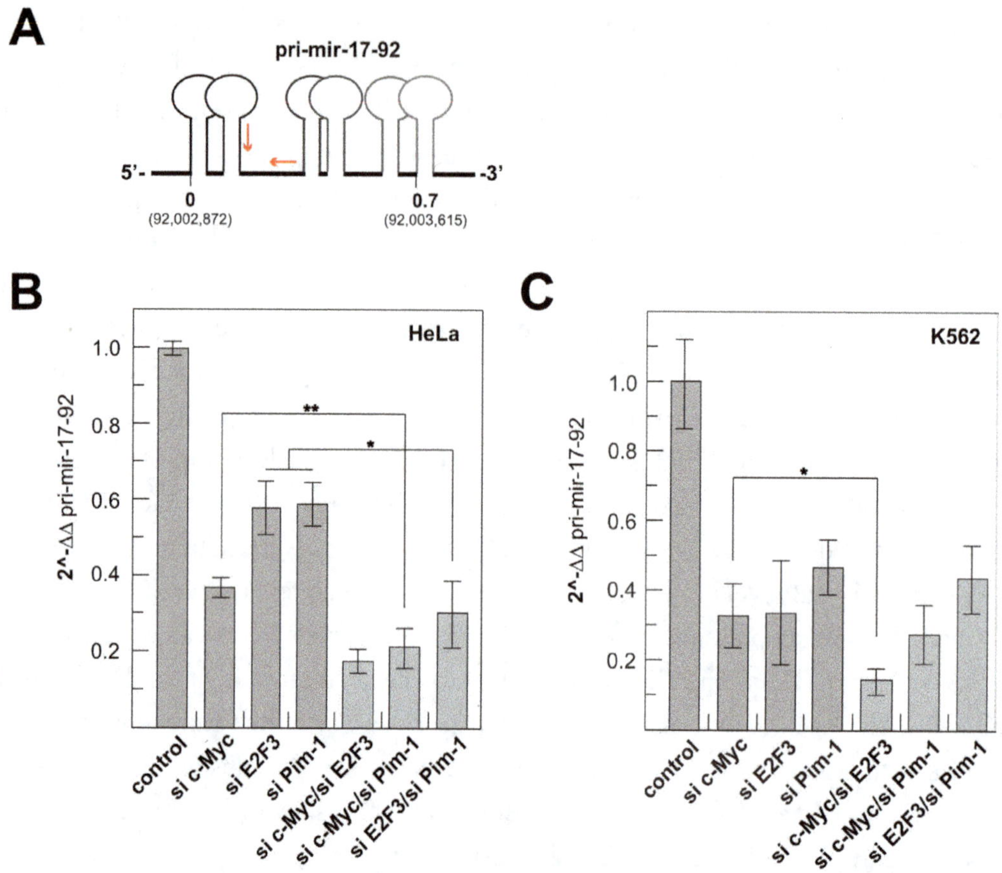

2.2. Discussion

The transcription of the oncogenic miR-17-92 cluster is thought to originate from two different TSSs: one is localized in close proximity to the host gene promoter element [17] (Supplementary Figure S1), and the other TSS was predicted to map to the region ~200 bp downstream of the functional c-Myc site E3 (Figures 1A and S1). The latter prediction was based on nucleosome mapping and chromatin signatures for active promoters. The derived algorithm identified 175 human promoters proximal to miRNA coding sequences and was reported to correctly predict transcription initiation regions to a resolution of 150 bp with high sensitivity and specificity. The majority of predictions were also consistent with known "expressed sequence tag" (EST) TSSs or cDNA 5'-ends [16].

Beyond previous studies [16,20,21], we investigated the intronic A/T-rich region preceding the human miR-17-92 cluster in more detail and compared it to siRNA-mediated knockdown of c-Myc. Similar effects were obtained, substantiating the notion that c-Myc and the c-Myc E3 site play a crucial role in activating transcription from the intronic promoter region. However, the 625 bp and even some of the further truncated fragments (~280 and ~200 bp) of the A/T-rich region conferred residual specific promoter activity in both cell types (Figure 1), indicating that parts of the A/T-rich region, downstream of the c-Myc E3 site, contribute to cluster expression. This E3 box-independent transcriptional activity was more pronounced for K562 relative to HeLa cells, which correlates with the particularly high cluster expression in K562 cells (Supplementary Figure S2). As a possible explanation, transcriptional activity of the ~1.5 kb fragment may be dominated by the recruitment of c-Myc to the E3 site region, while differential activity mediated by the smaller fragments in K562 *vs.* HeLa cells may report that their residual transcriptional activation is mechanistically different from that of the ~1.5 kb fragment. This could mean that regulatory factors of the transcription machinery are differentially expressed in the two cell lines.

ChIP assays revealed that not only c-Myc, but also the protooncogene Pim-1 and its phosphorylation target, HP1γ, associate with the chromatin region harboring the c-Myc E3 site (Figure 2B). Importantly, Pim-1-catalyzed phosphorylation of H3S10 at c-Myc target genes is necessary to regulate key genes required for c-Myc-dependent oncogenic transformation [27].

In mammals, three paralogs of HP1 (α, β and γ) regulate heterochromatin formation, gene silencing or gene activation [28,29]. HP1α and β proteins are mainly recruited to heterochromatin regions harboring H3K9me2,3 modifications, whereas HP1γ is found in association with euchromatin [30] and active genes [29]. Furthermore, HP1c, the *Drosophila* homolog of HP1γ, associates with transcriptionally active chromatin containing H3K4me3 and H3K36me3 histone marks [28]. HP1γ can further be recruited to inducible promoters, where it replaces HP1β, thereby inducing a switch from the repressive to the active transcriptional state. This replacement with HP1γ requires H3 phospho-acetylation [31]. In this context, a transient phosphorylation of H3S10 (via Aurora B kinase) was shown to be necessary for the dissociation of HP1 proteins from chromatin during the M phase of the cell cycle [32]. In the induced state, HP1γ can also be localized within coding regions of protein genes, together with elongating RNA polymerase II [31].

Our data, showing that HP1γ colocalizes with Pim-1 and c-Myc (Figure 2), is in line with the aforementioned activating role of HP1γ during transcription. We extended our ChIP assays to the miRNA coding region of C13*orf*25 to analyze HP1γ association with this part of the cluster. Indeed, ChIP analysis along the miRNA coding sequence identified HP1γ at all four analyzed subregions (A2–A5, Supplementary Figure S3). To our knowledge, this is the first indication that HP1γ is involved in activating the transcription of miRNAs.

The association of Pim-1 with the intronic chromosomal region near the c-Myc E3 site led us to the assumption that Pim-1 plays an important role in the transcriptional activation of the miR-17-92 cluster. This was tested by RNAi also, including E2F3 as an assumed indicator of host gene promoter activity. The strong negative effects of individual knockdowns of c-Myc, Pim-1 and E2F3 on pri-miR-17-92 levels indicate that all three proteins are important for cluster expression by affecting transcription from the host gene promoter (E2F3) or the intronic promoter region (c-Myc, Pim-1). This raises the question about the mechanistic role of Pim-1 in cluster expression from the intronic promoter. In contrast to HeLa cells Supplementary (Figure S4), a Pim-1 knockdown in K562 cells failed to significantly decrease reporter activity from the ~1.5 kb fragment. Among other possibilities, Pim-1 may be recruited to the functional c-Myc E3 site in the context of the cellular chromatin structure in K562 cells, but not in the context of the plasmid-encoded reporter gene. Alternatively, Pim-1 recruitment to the E3 site occurs, as shown by the ChIP assays, but is not a crucial prerequisite for transcriptional activation in K562 cells. On the other hand, the three-fold decrease in ~1.5 kb reporter activity observed in HeLa cells upon Pim-1 depletion adds evidence in support of a crucial role for Pim-1 in miR-17-92 cluster expression, but simultaneously points to cell type-specific differences. For future investigations, other cell lines will be tested, particularly ones that express c-Myc, but not Pim-1. Clearly, decreases in pri-miR-17-92 levels upon c-Myc, E2F3 and/or Pim-1 knockdown (Figure 3) may include indirect effects, e.g., originating from inhibition of cell proliferation (Pim-1), changes in the kinetics of pri-miR-17-92 processing, global changes in transcriptional networks (E2F3, c-Myc) or mutual transactivation (E2F3 and c-Myc) [33–35]. Further complication may arise from the fact that miR-17-5p and miR-20a of the cluster are negative regulators of E2F1-3 mRNAs [10,18].

As c-Myc, HP1γ and histone H3 are known phosphorylation targets of Pim-1, future studies may address the influence of Pim-1 on the phosphorylation status of these proteins at the E3 site, utilizing antibodies that are highly specific for the phosphorylated *versus* unphosphorylated state.

The siRNA-mediated c-Myc knockdown, decreasing c-Myc mRNA levels on average by 65% (HeLa) or 81% (K562; see Supplementary Table S1), resulted in a 60%–70% reduction in pri-miR-17-92 levels in HeLa and K562 cells (Figure 3). This effect may report a rough estimate of the proportion of cluster transcripts normally initiated in the intronic promoter region in these two cell lines, for the following reasons: the C13*orf*25 region contains four c-Myc binding sites (boxes E1-4) and two additional ones with lower c-Myc occupancy (relative to E1) upstream of the host gene promoter [20]. Box E1, immediately downstream of host gene promoter's TSS, was shown by deletion analysis to be inhibitory, which correlates with c-Myc forming heterodimers with MXI or MNT at this site to repress transcription [20]. Thus, host gene promoter activity may even somewhat increase under conditions of a c-Myc knockdown, although such an effect could, in turn, be neutralized by reduced c-Myc-mediated transactivation of E2F [35]. ChIP-Seq data for K562 and HeLa-S3 cells revealed the by far highest c-Myc occupancy at site E3 (little at E2 and E4), where c-Myc forms heterodimers with MAX to activate transcription [20]. A straightforward interpretation of the additive effect of a c-Myc/E2F3 double knockdown in Hela and K562 cells is that this combination negatively affected transcription from the host gene and intronic promoter regions.

A major finding of our study is the recruitment of Pim-1 to the intronic c-Myc E3 site (Figure 2) and the strong negative effect of a Pim-1 knockdown on cluster expression (Figure 3B,C). Interestingly, Pim-1 knockdown efficiencies are comparable in K562 (73%) and HeLa (71%) cells, whereas the effect of the knockdown on cluster expression is stronger in K562 cells (55% reduction compared to 40% in HeLa cells) with the higher Pim-1 expression level. This might be due to cell type-dependent indirect effects of Pim-1 on the regulation of the miR-17-92 cluster. Moreover, double knockdown experiments in HeLa cells revealed a synergistic effect relative to individual c-Myc and Pim-1 knockdowns (Figure 3B), which was not seen for K562 cells. The siRNA-mediated reduction of c-Myc and Pim-1 mRNAs were on average 86% and 77% in HeLa and 86% and 52%, respectively, in K562 cells (Supplementary Table S1). The somewhat weaker suppression of Pim-1 in the c-Myc/Pim-1 double knockdown context (cf. with single knockdowns, Table S1) in K562 *versus* HeLa cells may have contributed to the absence of a clear additive effect upon c-Myc/Pim-1 double knockdown in K562 cells.

3. Experimental Section

3.1. Oligonucleotides

Small interfering RNAs (siRNAs) were purchased from Dharmacon (Boulder, CO, USA):

VR1 siRNA [36] was used as an unrelated negative control, with the following sequences of sense and antisense strand.

VR1 siRNA sense 5'-GCG CAU CUU CUA CUU CAA CdTdT and antisense 5'-GUU GAA GUA GAA GAU GCG CdTdT.

The sequences of all other siRNAs used in this study are:

Pim-1 siRNA sense 5'-GAU AUG GUG UGU GGA GAU AdTdT and antisense 5'-UAU CUC CAC ACA CCA UAU CdTdT; Pim-1 siRNA 2 sense→5'-GGA ACA ACA UUU ACA ACU CdTdT and antisense 5'-GAG UUG UAA AUG UUG UUC CdTdT; c-Myc siRNA sense 5'-CAG GAA CUA UGA CCU CGA CUA dTdT and antisense 5'-UAG UCG AGG UCA UAG UUC CUG dTdT; E2F3 siRNA sense 5'-ACA GCA AUC UUC CUU AAU AdTdT and antisense 5'-UAU UAA GGA AGA UUG CUG UdTdT.

3.2. Antibodies

Antibodies against c-Myc (sc-40) and Pim-1 (sc-13513), as well as the secondary antibody (sc-2005: goat anti-mouse IgG HRP conjugated) were purchased from Santa Cruz Biotechnology (Heidelberg, Germany) except for the Phospho HP1γ (Ser83) antibody (2600S), which was obtained from Cell Signaling Technology (Danvers, MA, USA).

3.3. Plasmid Construction and Seed Mutagenesis

For the construction of promoter-luciferase fusions, the SV40 promoter of plasmid "pGL3 control" (Promega, Mannheim, Germany) was removed via digestion with *Bgl*II and *Hind*III (Thermo Fisher Scientific, Schwerte, Germany) and replaced with fragments derived from the intronic A/T-rich region of C13*orf*25 (reference nucleotide sequence NG_032702.1). Promoter fragments were amplified from genomic DNA of the human cell line K562 using primers specified in the Supplementary Material. PCR products were purified using the Wizard® SV Gel and PCR Clean-Up System (Promega, Mannheim, Germany) and digested with *Bgl*II and *Hind*III for insertion into pGL3. All constructs were cloned in *E. coli* DH5α cells and verified by DNA sequencing. The pGL3 vector lacking the SV40 promoter, as well as the pGL3 construct carrying the C13*orf*25-derived 339 bp fragment in inverse orientation (pGL3 339 bp inv), were used as negative controls.

3.4. Transfection Procedures and Luciferase Reporter Assays

Assays are described in detail in the Supplementary Materials.

3.5. Chromatin Immunoprecipitation

Chromatin immunoprecipitation (ChIP) was performed according to a protocol from the Or Gozani lab at Stanford University (http://www.stanford.edu/group/gozani) [37], with some modifications. For details, see the Supplementary Material.

4. Conclusions

We report here that miR-17-92 cluster expression from the intronic A/T-rich promoter region, although critically depending on c-Myc binding, includes some specific contribution of sequences within ~0.7 kb upstream of the mature miR-17-5p coding sequence. Our reporter expression data suggest multiple TSSs within this A/T-rich region, although the transcription initiation region predicted by Ozsolak *et al.* [16], ~0.2 kb downstream of the c-Myc E3 box, may well be the major one. E3 site-independent transcription initiation within ~0.7 kb upstream of the mature miR-17-5p coding sequence was more pronounced in K562 *versus* HeLa cells (Figure 1), indicating cell type-specific differences in cluster expression from the intronic promoter region. By RNAi and ChIP, we establish that Pim-1 acts in concert with c-Myc at the E3 site to activate transcription from the intronic promoter region.

Acknowledgments

We are grateful to Lisa Schemberger, Nicole Bürger and Moana Klein for technical assistance, Marcus Lechner for statistical analyses and Markus Gößringer for fruitful discussions. The work was supported by grants of the Fritz Thyssen Stiftung (reference no. 10.06.1.186 to A.G. and R.K.H.) and the German Cancer Aid (Deutsche Krebshilfe, grants 106992 and 109260 to A.G., R.K.H. and A.A.).

Appendix

Cell Culture

Cells (K562 and HeLa) were cultivated under standard conditions in a humidified atmosphere (37 °C, 5% CO_2) supplemented with RPMI 1640 (K562) or IMDM (HeLa) containing 10% FCS (PAA, Cölbe, Germany).

Chromatin Immunoprecipitation (ChIP)

2×10^7 K562 cells in 13 mL RPMI medium were crosslinked with 1% formaldehyde for 10 min at 37 °C. Reactions were stopped by adjusting to 0.125 M glycine, and cells were collected by centrifugation at 400*g* for 5 min at room temperature. For cell lysis, cells were resuspended in 750 µL RIPA-buffer (10 mM Tris-HCl pH 7.4, 150 mM NaCl, 1% deoxycholate, 1% NP40, 0.1% SDS, 0.2 mM PMSF) supplemented with the complete Mini Protease Inhibitor Mix from Roche (Mannheim, Germany) according to the manufacturer's instructions. Lysed cells were sonicated in a Branson Sonifier 250 (duty cycle 50%, output control 2, for 3.5 min, on ice water) (Heinemann, Schwäbisch Gmünd, Germany) and centrifuged at 16,000*g* for 10 min at 4 °C. The supernatant was pre-cleared with 10 µL of blocked *Staphylococcus aureus* cells (Pansorbin® Cells, Calbiochem/Merck, Darmstadt, Germany) for 15 min at 4 °C on a rotor wheel. After a second centrifugation step (16,000g, 5 min, room temperature), the supernatant was split into two samples (each ~350 µL, representing + and- specific antibody (AB)), which were adjusted to buffer D (16.7 mM Tris-HCl pH 8.1, 167 mM NaCl, 1.2 mM

EDTA, 1.1% Triton-X 100, 0.01% SDS) and a total volume of 500 µL. 1 µg (1 to 5 µL) of the respective antibody was added to "+AB" samples, whereas the same volume buffer D was added to "–AB" samples. At this point, also the "Mock" control was prepared, consisting of 500 µL buffer D. "+AB", "–AB" and "Mock" samples were then incubated for at least 3 h at 4 °C. The following antibodies were used: monoclonal anti-c-Myc (sc-40) and anti-Pim-1 (sc-13513) (Santa Cruz Biotechnology (Heidelberg, Germany), anti-Phospho HP1γ (Ser83) polyclonal antibody (2600S) (Cell Signaling Technology, Danvers, MA, USA). In the case of mouse monoclonal antibodies (c-Myc and Pim-1), samples were additionally incubated for 1 h with a second monoclonal goat anti-mouse IgG antibody (sc-2005, Santa Cruz Biotechnology, Heidelberg, Germany). Immunoprecipitation was initiated by adding 10 µL of Pansorbin® cells (see above) to the "+AB", "–AB" and "Mock" samples, followed by incubation for 15 min at room temperature. Samples were centrifuged (16,000× *g*, 3 min, room temperature); the supernatant of the "–AB" sample was saved, later serving as the input control. Pellets were washed twice with dialysis buffer (50 mM Tris-HCl pH 8.0, 2 mM EDTA) and four times with IP-wash buffer (100 mM Tris-HCl pH 9.0, 500 mM LiCl, 1% NP40, 1% deoxycholate). Antibody-bound material was eluted from Pansorbin® cells by adding 150 µL elution buffer (50 mM $NaHCO_3$, 1% SDS), vortexing for 3 s, and centrifugation (16,000 *g*, 3 min, room temperature). The supernatant was collected and the procedure was repeated. Reverse crosslinking and RNA digestion was performed in 280 µL buffer [0.3 M NaCl, and 1 µL RNase A (10 mg/mL)] for 5 h at 67 °C. Chromatin was precipitated with ethanol, followed by a Proteinase K digest (Thermo Fisher Scientific, Bremen, Germany). DNA was purified by phenol/chloroform extraction and ethanol precipitation in the presence of 0.3 M NaOAc, pH 5.2. PCR amplification using Taq DNA polymerase and the co-immunoprecipitated DNA as template was done under the following conditions: 2 min at 95 °C in the absence of enzyme, followed by 35 amplification cycles of 45 s at 95 °C/45 s at 60 °C/45 s at 72 °C using the following primers:

A1 forward 5'-AAA GGC AGG CTC GTC GTT G
A1 reverse 5'-CGG GAT AAA GAG TTG TTT CTC CAA
A2 forward 5'-ACA TGG ACT AAA TTG CCT TTA AAT G
A2 reverse 5'-AAT CTT CAG TTT TAC AAG GTG ATG
A3 forward 5'-ACT GCA GTG AAG GCA CTT GT
A3 reverse 5'-TGC CAG AAG GAG CAC TTA GG
A4 forward 5'-CCA ATA ATT CAA GCC AAG CAA
A4 reverse 5'-AAA TAG CAG GCC ACC ATC AG
A5 forward 5'-GCC CAA TCA AAC TGT CCT GT
A5 reverse 5'-CGG GAC AAG TGC AAT ACC AT

Transfection of Reporter Constructs

For transfection of the suspension cell line K562, cells were washed in medium without serum, followed by electroporation in a BioRad Gene Pulser XCell (München, Germany) with a single pulse in a 4 mm cuvette, using 5 μg of the respective pGL3 derivative plasmid per million cells. 48 h after transfection, cells were washed in PBS, lysed and prepared for luciferase reporter assay measurements.

Transfection of siRNAs

Transfection of the suspension cell line K562 was performed by electroporation. After a washing step in medium without serum, 10^6 cells were mixed with 1 μg of siRNA (VR1 siRNA, Pim-1 siRNA, c-Myc siRNA or E2F3 siRNA). For the double knockdown experiments, 1 μg of each siRNA, respectively, was used. Cells were electroporated at 330 V with a single pulse in a BioRad Gene Pulser XCell (München, Germany) using a 4 mm cuvette. After electroporation, K562 cells were resuspended in medium containing 10% FCS and cultivated in a humidified atmosphere at 37 °C for 24 h. Cells were washed with PBS and prepared for total RNA extraction.

Transfection of HeLa cells was done using the transfection agent Lipofectamine™ 2000 (Life Technologies Invitrogen, Darmstadt, Germany). One day before transfection, 8×10^4 cells were seeded into 24-well plates and cultivated under standard conditions. SiRNA complexes (VR1 siRNA, Pim-1 siRNA, c-Myc siRNA or E2F3 siRNA) were prepared according to the manufacturer's protocol. To perform single knockdown experiments, 40 pmol of the respective siRNA were used. In case of the double knockdown experiments, 20 pmol of each siRNA were mixed in Opti-MEM® 1 (Life Technologies Invitrogen, Darmstadt, Germany). 4 to 6 h after transfection, the medium was replaced with IMDM containing 10% FCS. Cells were cultivated 48 h under standard conditions until preparation for total RNA extraction.

RNA Preparation and Quantitative Real-Time PCR

For total RNA isolation, transfected cells (K562 or HeLa) were lysed (vortexing or mixing by pipetting up and down) in 750 μL lysis solution (0.8 M guanidinium-thiocyanate, 0.4 M ammonium-thiocyanate, 0.1 M sodium acetate pH 5.0, 5% glycerin, 38% phenol pH 4.5–5.0 (Roth®-Aqua-Phenol, Roth, Karlsruhe, Germany), 1 pellet 8-hydroxychinolin). Then, 200 μL of chloroform was added and phases were separated by centrifugation. The aqueous phase was mixed with 2 volumes of isopropanol, followed by incubation for 15 min at room temperature and centrifugation. The air-dried RNA pellet was dissolved in RNase-free water and incubated for 30 min at 37 °C with 1 U DNase I per µg RNA in 100 µL 1× DNase I buffer (DNase I, Thermo Fisher Scientific, Schwerte, Germany) according to the manufacturer's instructions. Then, another identical aliquot of DNase I was added, followed by incubation at 37 °C for another 30 min. Samples were extracted with an equal volume of Roth®-Aqua-Phenol (see above), followed by extraction of the aqueous phase with chloroform and isopropanol precipitation as above. RNA pellets were finally washed with 75% ethanol, air-dried and redissolved in 10 µL RNase-free water. 0.5 to 1 µg of total RNA were reverse-transcribed with RevertAid H Minus RT Polymerase (Thermo Fisher Scientific) according to the manufacturer's protocol. For determination of KD efficiencies a random hexamer primer was used to generate cDNA samples. In case of calculating pri-mir-17-92 levels the gene-specific reverse primer specified below was used for cDNA synthesis. Quantitative RT-PCR was performed in duplicate in a BioRad iQ™5 (BioRad, München, Germany) with the Absolute qPCR SYBR Green Capillary Mix (Thermo Scientific AbGene, Hamburg, Germany); cDNAs were diluted 1:5 or 1:10 and 4 µL of the reaction mixture used for determining RNA transcription levels. Quantitative PCR assays for miRNA detection were conducted as follows: Thermo-Start™ DNA polymerase was activated for 15 min at 95 °C followed by 55 amplification cycles of 10 s at 95 °C/20 s at 60 °C/12 s at 72 °C. Subsequently, melting curves of the PCR products were generated: samples were cooled from 95 to 65 °C (20 °C per s), kept at 65 °C for 20 s, followed by heating steps of 1 °C per cycle up to 95 °C and kept for 10 s at 95 °C.

Quantitative RT-PCR assays for mRNA detection were changed as follows: Thermo-Start™ DNA polymerase was activated for 15 min at 95 °C followed by 55 amplification cycles with a denaturation step for 10 s, primer annealing for 10 s at 55 °C and amplification at 72 °C for 10 s. Subsequently, a melting curve was generated for the PCR products; samples cooled from 95 to 65 °C (20 °C per s), kept at 65 °C for 20 s, followed by heating steps of 1 °C per cycle up to 95 °C and kept for 10 s at 95 °C.

The mRNA and pri-miR-17-92 levels were calculated from the crossing points by the $2^{-\Delta\Delta C_T}$ method [38] using β-Actin mRNA or 5S rRNA as internal controls. Knockdown efficiency was quantitated by qRT-PCR and only those experiments showing more than 65% reduction in protein levels were used for quantification of pri-miR-17-92 levels. All primers for qPCR measurements were purchased from Metabion (Martinsried, Germany) and designed using the software tool Universal ProbeLibrary (Roche Applied Biosystems, Mannheim, Germany). A list of all primer sequences according to the human sequences is shown underneath:

Pri-miR-17-92: forward primer 5'-CAT CTA CTG CCC TAA GTG CTC CTT and reverse primer 5'-GCT TGG CTT GAA TTA TTG GAT GA;

5S rRNA: 5'-TCT CGT CTG ATC TCG GAA GC and 5'-AGC CTA CAG CAC CCG GTA TT;

c-Myc mRNA: forward primer 5'-CCT TGC AGC TGC TTA GAC and reverse primer 5'-GAG TCG TAG TCG AGG TCA T;
E2F3 mRNA: forward primer 5'-GAG ACT GAA ACA CAC AGT CC and reverse primer 5'-CCT GAG TTG GTT GAA GCC;
Pim-1 mRNA: forward primer 5'-ATC AGG GGC CAG GTT TTC T and reverse primer 5'-GGG CCA AGC ACC ATC TAA T;
Actin mRNA: forward primer 5'-CCA ACC GCG AGA AGA TGA and reverse primer 5'-CCA GAG GCG TAC AGG GAT AG.

Luciferase Reporter Assays

Luciferase reporter assays were performed using the Promega Luciferase Assay System (Promega, Mannheim, Germany). After aspirating the medium, cells were washed with PBS and lysed in 100 μL 1.2× reporter lysis buffer. In a 96-well plate, 10 μL of the respective cell lysate were mixed with 10 μL of luciferase substrate. Chemiluminescence was measured immediately in a Safire$^{2\text{TM}}$ micro-plate reader (Tecan, Crailshaim, Germany).

Plasmid Construction and Seed Mutagenesis

All primers, which were used for plasmid construction and seed mutagenesis, are listed below (restriction sites in italics):
pGL3 1.5 kb: 5'-ATA T*AG ATC T*TG CCG CCG GGA AAC GGG TT and reverse primer R1 5'-ATA T*AA GCT T*CC ATA CAA ATT CAG CAT AAT CCC TAA TGG;
pGL3 625 bp: 5'-ATA TAG ATC TCT TTA GAC AAT GTA CCT TTT CTG and reverse primer R1 (see above);
pGL3 339 bp: 5'-ATA T*AG ATC T*GT GGA AGC CAG AAG AGG AGG A and reverse primer R1 (see above);
pGL3 279 bp: 5'-ATA TAG ATC TGG TAC ACA TGG ACT AAA TTG CC and reverse primer R1 (see above);
pGL3 197 bp: 5'-ATA T*AG ATC T*CT CTA TGT GTC AAT CCA TTT GGG AG and reverse primer R1 (see above);
pGL3 250 bp: 5'-ATA T*AG ATC T*GT GGA AGC CAG AAG AGG AGG A and reverse primer R3 5'- ATA T*AA GCT T*GC CTT AAG AAT TCT TTA CAG AAG GC;
pGL3 190 bp: 5'-ATA T*AG ATC T*GG TAC ACA TGG ACT AAA TTG CC and reverse primer R3 (see above);
pGL3 108 bp: 5'-ATA T*AG ATC T*CT CTA TGT GTC AAT CCA TTT GGG AG and reverse primer R3 (see above);
pGL3 339 inv: 5'-ATA T*AA GCT T*GT GGA AGC CAG AAG AGG AGG A and 5'-ATA T*AG ATC T*CC ATA CAA ATT CAG CAT AAT CCC TAA TGG.

Bioinformatical Promoter Analysis

For promoter analyses of the intronic A/T-rich region, a sequence of 1490 bp (human), beginning at the functional c-Myc site (E3) and ending at the 5'-end of the first mature miRNA sequence, miR-17-5p (for details, see Supplementary Figure S1), was analyzed with several web-based promoter prediction tools. The following tools were used to predict promoter elements or putative TSSs:

Neural Network promoter prediction: (http://www.fruitfly.org/seq_tools/promoter.html) [39];
McPromoter006: (http://tools.genome.duke.edu/generegulation/McPromoter/) [40];
Promoter 2.0 Prediction Server: (http://www.cbs.dtu.dk/services/Promoter/) [41];
PromPredict: (http://nucleix.mbu.iisc.ernet.in/prompredict/prompredict.html) [42].

Putative TSSs were predicted and calculated using the software available at http://rulai.cshl.org/tools/genefinder/CPROMOTER/human.htm [43].

All promoter prediction tools predicted several different promoter elements in the 1.5 kb region upstream of miR-17-5p and only the web-based tool Promoter 2.0 failed. The calculated promoter predictions of the indicated tools did not match in any sequence region, giving the assumption that the A/T-rich 1.5 kb intronic region in front of the miR-17-92 coding sequence has only weak promoter activity itself. Due to the fact that this intronic region has an overall high A/T-content, nearly all software programs were able to detect putative promoter regions.

Statistical Analysis

Statistical analyses were done using the software R [44]. p-Values were calculated with the Welch Two Sample Test.

Supplementary Figures

Figure S1. Relevant sequence region of C13*orf*25, including the CpG island harboring the host gene promoter, the A/T-rich region, and the miR-17-92 cluster; sequence and position of the last exon 4 of C13*orf*25 is indicated at the end. Shown sequences are based on the NCBI reference sequence NG_032702.1 and the GRCh37/hg19 assembly [25]. The boundaries of the CpG island, important previously identified regulatory elements, mature miRNA coding sequences and relevant primer sequences are highlighted in the sequence and annotated at the margins.

```
NCBI Reference Sequence: NG_032702.1

Homo sapiens mir-17-92 microRNA cluster containing MIR17HG, MIR17, MIR18A,
MIR19A, MIR19B1, MIR20A, and MIR92A1 (MIR17-MIR92); and microRNA 17 (MIR17);
and miR-17-92 cluster host gene (non-protein coding) (MIR17HG); and microRNA
18a (MIR18A); and microRNA 19a ...

91,995,073 -

91,999,514  AAGGGCCGGT AAAGTAATAA ACTGTGATCG CACAGCGTAC ACGTCGAGTC CCAGGGAGAG
91,999,574  CAGGGCTCCC AGGAAGCCCC GGGAAGCGTG CGCTTCCAGG AGGGTGTGCG GCGCGCGGGG
91,999,634  GTCTGCACGC GCGCAGAGCT TGTTAACGGA GGGCGTGCCG GAGGCGTGGC GCGAGGTGTT
91,999,694  CGGGCGAGGC GCTCCGCGAG GCTGGCGGCG GTGGGCGGGC CACGGGGGAG CGCGGCCCGG
91,999,754  GGAAGGCTGC GCGGCGCCGC CAGCGGCTCC CGGCTCCCGC TCTCCCGGCC GCCAAGAACG
91,999,814  AGCCGCCGTG GGCGGGGCCT GCGGTGATTG GCGGGCGGGC GGGGAGGTCG GAAGTACTTT
91,999,874  GTTTTTTATG CTAATGAGGG AGTGGGGCTT GTCCGTATTT ACGTTGAGGC GGGAGCCGCC
91,999,934  GCCCTTCATT CACCCACATG GTCCTTCGAG GTGCCGCCGC CGCCGCCCGA CCTGCGCCTT
91,999,994  CGCGCCACTT CGCGCCCTCG GGCGAGGCCG AGGGGGTGGG GACGACTACC GCAGCGCCGC
92,000,054  CGGGGCTCGC GCTCCTCCGC GAAGCTCTCC TCGCGGGGCG GGCCGGCCGG CCGCACCCCC
92,000,114  GGCCTGGGGC CTCCGGTCGT AGTAAAGCGC AGGCGGGCGG GGAGGCGGGA GCAGGAGCCC
92,000,174  GCGGCCGGCC AGCCGAAGAT GGTGGCGGCT ACTCCTCCTG GTGAGTCTGC CCGCCCCTCC
92,000,234  GGCGACGGAG GGAAACCTGT TGTGTGCGGC CCGGGTCTGG CGGGCGGGGC GGAGCGGCCC
92,000,294  GGGGCGGACT GGCCCGGGGC AGCGTGGCGG CGGCGGCGTG GCCGGGGCGG GTCTCGGCCG
92,000,354  TTGGCCGCCC CGGCGTGTGG CAGCCGCATC TGGCTGCCCC CTCGCTCGCC CGCGGGCCGG
92,000,414  CGGAGGGGGG CAGGGCCGGG GCGGGAGGGT GGGAGGGGGC GGCGTGCGCG TGGCGGCCGC
92,000,474  GCCCGGGACC CGCGCAGACC CTGCCTGGGC CGACCCGAAG GCGGGTGGGC GGACGGCGAA
92,000,534  CACAATGGCC CCTCGGGGAG AGGACGTGCG AGGCCCGTGC CTTCTCCGGG GCCCGGGGCG
92,000,594  CGCGCGGGGC GTGGGGTCTC TGGGTAGGAA AGTTTCTCCC GAGGGCGAGA GTTAAAGCGC
92,000,654  CTCCAGAACA AAGCGGCGGC GGCGGCGGCA CATGGGGCAG GCCGCGGGCC GGGAGGGGGC
92,000,714  GCGCCCACGA GGTACCTGCG CGCCAGCGGG CGGCGTGGCG TGGGCGGGAG CCCGCGGTTC
92,000,774  CCCAAACTTT GTACGCGCGA GGGTGGGCGG AGGGGCGCCG AGATCGGCGC GGCCTGGGCG
92,000,834  CCACCCCCGC TCCGCGTGGG CTTTGTTAGC CCGCGTGGGC AGCCTCGGGG CGGGGCCCGC
92,000,894  AACTTCCCCG CCGTGGCCCT CGGAGGAGGC CGCAGTCGGC CTCAGCCGCG GCGTGGAGCC
92,000,954  GCCTGCGCCC GGCCGCTTGC TGGGAGTGTG GCGCGGGAGG GCCAGCCCGG CTCGGCGGGA
92,001,014  GCGGCGTCCC CGCCGCCATG TTCCTGCGGG GCGGGCTGCA CGGGGGTGAG GGCGGGGGAC
92,001,074  ATGGCGGCGA CTGCGCGCCG CCGCCGATTG TTCCCGGCTT AGGCCTCGGG CCGCGTGCGA
92,001,134  CGGGCACCGC GGCCGCGGAG CCCCCGCCCC TCTGGGCCGG GCTCGGGGGG GCTGGGGGAC
92,001,194  ACAAAGGAGG GGCGGCGCGC CCGCGTCCCC GCCGCACTCG GGCCTCGGCG CCGCCGGTCG
92,001,254  CCGCGCGGCT GCCGCCGGGA AACGGGTTGG GGGGGTTGCC GCGTCCGGCG GGGCCTGACT
92,001,314  CTGACCCGCC GCCCCCTGGC GGCTACGCGG AGAATCGCAG GGCCGCGCTC CCCCTTGTGC
92,001,374  GACATGTGCT GCCGGCCCGG GCTCCATGAG CGTGGCGGGC ACTTTGCAGT CTCGGGTGTT
92,001,434  CCTGCCCGGT CTTCTGTTCC TAAACTGCAG CAAAGGGAAA AGGAACTGAA AAAGGCAGGC
92,001,494  TCGTCGTTGC AATATCACCA AAAGAGAAAA TTAACGGCAT GCCATCAGGA CCACAGCAGT
92,001,554  TGGAGAAACA ACTCTTTATC CCGGCTTGCA GCCACGAGGT CTTGATTGGG GGAGGGGTGG
92,001,614  TGAAGAATAG TCTGTGGGCT GCTTTTTTTT TTTCCTTTTA CTGGAGCTGT ACAGTGGAGT
92,001,674  CGGTGATTGC TGCTGATCAT AATCAAGTAT TTTAGGAGCT TATTTAGACA TGTATCTGAT
92,001,734  AGCTAAGGAT TTTTCAACTT TATTCTCTTA CGTATTTTTC AACTGTAAAT TATTGGGCTT
92,001,794  TTAAATCCTG CTAGTATTGC TCGACTCTTA CTCTCACAAA TGGATGGAAT TAATTGCTGT
92,001,854  TAGGAGGTTG GAAAATAGCA AATATAGATT TGGACGGTGG TAGTAATTTT GAGCAAATAA
92,001,914  TGTTTTATCT TTTTTTTCCT TATTTTTCCC TATTCCAGTC ATACACGTGG ACCTAACTGC
92,001,974  ACCAGTAGCT TTTCTGAGAA TACTTGCTGA AAAGGAAGTT TTCTGGAATG GGTAAGTGTA
92,002,034  TTCTGATTTT CTTGAACTTT TCTTAAAAAC AAATTTTTCT TGCTATTAAA GTTGAATAAA
92,002,094  TAGGATTGGT TTCTTAGAGA GTAAAAGTAG GTGTTTCTTT CTTTAGACAA TGTACCTTTT
92,002,154  CTGAAAAACT AACTCATTAA GTACGGATTT GCTAATTTTA AGGTAGTAAA ATTACAGTGT
92,002,214  AAATATTCCT GTACATTTTT GGAAACTGGC TTATGCAGTT TACGAAATAT AATTTTAGAC
92,002,274  CCTCTTTTAA GTTGGGTGAT AAAGTAGATA TAACCTGAGA TGATAGATTT AAACAGGATA
92,002,334  TTTACGTTCT GCTACAATTG ACTGATAACA CTTGAAGTGT AGTCTGAACA GTAATTTTGT
92,002,394  TAATCATTTC AACAAGTATT TGCTAAGTGG AAGCCAGAAG AGGAGGAAAA TGTTTTGCCA
92,002,454  CGTGGATGTG AAGATTTCCT CTAAAAGGTA CACATGGACT AAATTGCCTT TAAATGTTCC
92,002,514  AAAATTAGTT CTCATTTATT TGCAGTCTCA TTTTGTTTTG TTTTTTTTCT CTATGTGTCA
92,002,574  ATCCATTTGG GAGAGGCCAG CCATTGGAAG AGCCACCACT TCCAGTGCTA GTTGGATGGT
92,002,634  TGGTTATGAT TGCCTTCTGT AAAGAATTCT TAAGGCATAA ATACGTGTCT AAATGGACCT
92,002,694  CATATCTTTG AGATAATTAA ACTAATTTTT TCTTCCCCAT TAGGGATTAT GCTGAATTTG
92,002,754  TATGGTTTAT AGTTGTTAGA GTTTGAGGTG TTAATTCTAA TTATCTATTT CAAATTTAGC
92,002,814  AGGAAAAAAG AGAACATCAC CTTGTAAAAC TGAAGATTGT GACCAGTCAG AATAATGTCA
92,002,874  AAGTGCTTAC AGTGCAGGTA GTGATATGTG CATCTACTGC AGTGAAGGCA CTTGTAGCAT
92,002,934  TATGGTGACA GCTGCCTCGG GAAGCCAAGT TGGGCTTTAA AGTGCAGGGC CTGCTGATGT
92,002,994  TGAGTGCTTT TTGTTCTAAG GTGCATCTAG TGCAGATAGT GAAGTAGATT AGCATCTACT
```

Margin annotations:

- CpG island 91,999,542
- SP1 site 91,999,823
- atypical TATA box 91,999,909
- initiator TSS 91,999,941
- c-Myc site 91,999,948
- E2F3 site 1 91,999,991
- E2F3 site 2 92,000,001
- Exon 1 92,000,074-92,000,213
- miR-17-92 f1 92,001,263
- functional c-Myc site 92,001,376
- CpG island 92,001,441
- ChIP sequence A1 92,001484-92,001,576
- Exon 2 92,001,952-92,002,024
- c-Myc site 92,001,957
- miR-17-92 f3 92,002,134
- Exon 3 92,002,410-92,002,479
- miR-17-92 f4 92,002,420
- c-Myc site 92,001,452
- miR-17-92 f5 92,002,480
- miR-17-92 f6 92,002,562
- miR-17-92 r3 end 92,002,670
- internal promoter 92,002,663-92,002,719
- miR-17-92 r1 end 92,002,758
- miR-17 92,002,872
- miR-18a 92,003,010

Figure S1. *Cont.*

```
                                                                                        qPCR sequence
                                                                                        pri-17-92
                                                                                        92,003,046-
92,003,054  GCCCTAAGTG CTCCTTCTGG CATAAGAAGT TATGTATTCA TCCAATAATT CAAGCCAAGC  92,003,113
92,003,114  AAGTATATAG GTGTTTTAAT AGTTTTTGTT TGCAGTCCTC TGTTAGTTTT GCATAGTTGC  miR-19a
92,003,174  ACTACAAGAA GAATGTAGTT GTGCAAATCT ATGCAAAACT GATGGTGGCC TGCTATTTCC  92,003,193
92,003,234  TTCAAATGAA TGATTTTTAC TAATTTTGTG TACTTTTATT GTGTCGATGT AGAATCTGCC
92,003,294  TGGTCTATCT GATGTGACAG CTTCTGTAGC ACTAAAGTGC TTATAGTGCA GGTAGTGTTT  miR-20a
92,003,354  AGTTATCTAC TGCATTATGA GCACTTAAAG TACTGCTAGC TGTAGAACTC CAGCTTCGGC  92,003,326
92,003,414  CTGTCGCCCA ATCAAACTGT CCTGTTACTG AACACTGTTC TATGGTTAGT TTTGCAGGTT
92,003,474  TGCATCCAGC TGTGTGATAT TCTGCTGTGC AAATCCATGC AAAACTGACT GTGGTAGTGA  miR-19b
92,003,534  AAAGTCTGTA GAAAAGTAAG GGAAACTCAA ACCCCTTTCT ACACAGGTTG GGATCGGTTG  92,003,498
92,003,594  CAATGCTGTG TTTCTGTATG GTATTGCACT TGTCCCGGCC TGTTGAGTTT GGTGGGGATT  miR-92a
92,003,654  GTGACCAGAA GATTTTGAAA ATTAAATATT ACTGAAGATT TCGACTTCCA CTGTTAAATG  92,003,615
-
92,006,074  CTTCATTTTA CAGGCAGACC TGTCTAACTA CAAGCCAGAC TTGGGTTTTC TCCTGTAGTT  Exon 4
            TGAAGACACA CTGACTCCTG ACAAAATGCA GCCTGCAACT TCCTGGAGAA CAACTCAGTG  92,006,187-
            TCACATTAAA GTTTATTATG TATTTAATGA TACACTGTTT AATTGACAGT TTTGCATAGT  92,006,829
            TTGTCTAACT TTAGAGAATT AAGAGCCTCT CAACTGAGCA GTAAAGGTAA GGAGAGCTCA
            ATCTGCACAG AGCCAGTTTT TAGTGTTTGA TGGAAATAAG ATCATCATGC CCACTTGAGA
            CTTCAGATTA TTCTTTAGCT TAGTGGTTGT ATGAGTTACA TCTTATTAAA GTCGAAATTA
            ATGTAGTTTT CTGCCTTGAT AACATTTCAT ATGTGGTATT AGTTTTAAAG GGTCATTAGG
            AAAATGCACA TATTCCATGA ATTTTAAGAC CCATAGAAAA GTTGAAGAAT GCTTAATTTT
            CTTATCCAGT AATGTAAACA CAGAGACAGA ACATTGAGAT GTGCCTAGTT CTGTATTTAC
            AGTTTGGTCT GGCTGTTTGA GTTCTAGCGC ATTTAATGTT AATAAATAAA ATACTGCATT
92,006,774  TTAAAGCTGT TAAGAAATTG TCCAGAACGA GAATATTGAA ATAAAAACTT CAAGGTTATT
-
92,008,829
```

Figure S2. Quantitative RT-PCR of the pri-mir-17-92 transcription levels in the human cell lines K562, HeLa and HUH7 (hepato cellular carcinoma cells). 2^-ΔΔpri-mir-17-92 values are normalized against 5S rRNA and obtained from at least 3 independent experiments (+/−S.E.M.). The amount of pri-mir-17-92 transcript in K562 cells was set to 1.

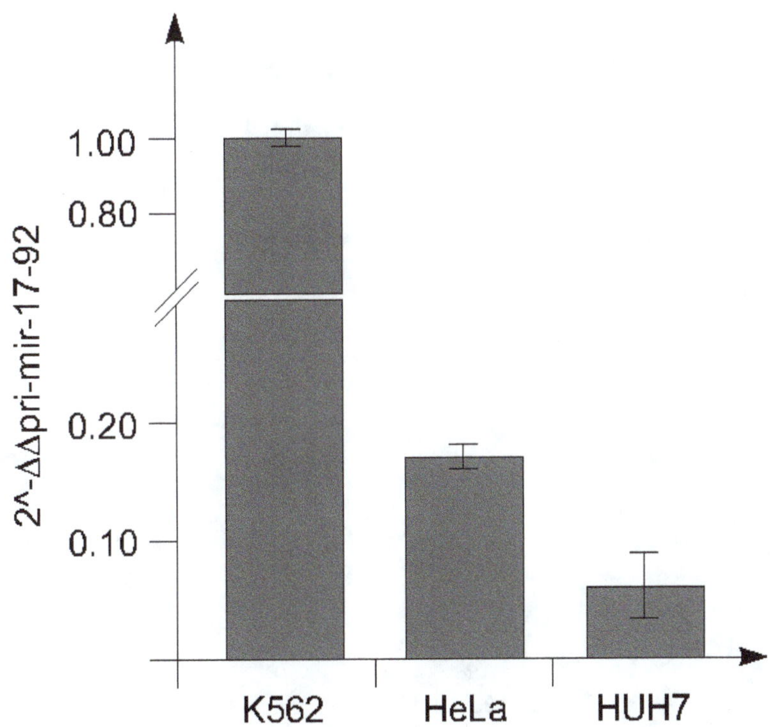

Figure S3. (**A**) Schematic representation of the intronic A/T-rich region preceding the miR-17-92 coding sequence. The region A1 defines the genomic sequence 0.1 kb downstream of the functional c-Myc binding site (E3 box) that was amplified in ChIP analyses. A2 covers a segment immediately upstream of the miRNA-coding region; A3–A5 are located along the coding sequence of the human miR-17-92 cluster. The length (bp) of each amplicon is indicated at the top; (**B**) ChIP analysis of the regions A1 to A5 in K562 cells, using an antibody specific for HP1γ or RNA polymerase II (only A2 analyzed). +AB, with antibody; −AB, without antibody; Mock, buffer only without cell lysate; Input, supernatant of the "−AB" sample after immunoprecipitation and centrifugation (for details, see Supplementary Material); (**C**) ChIP analysis of the A2 region in HeLa cells using the antibody specific for HP1γ.

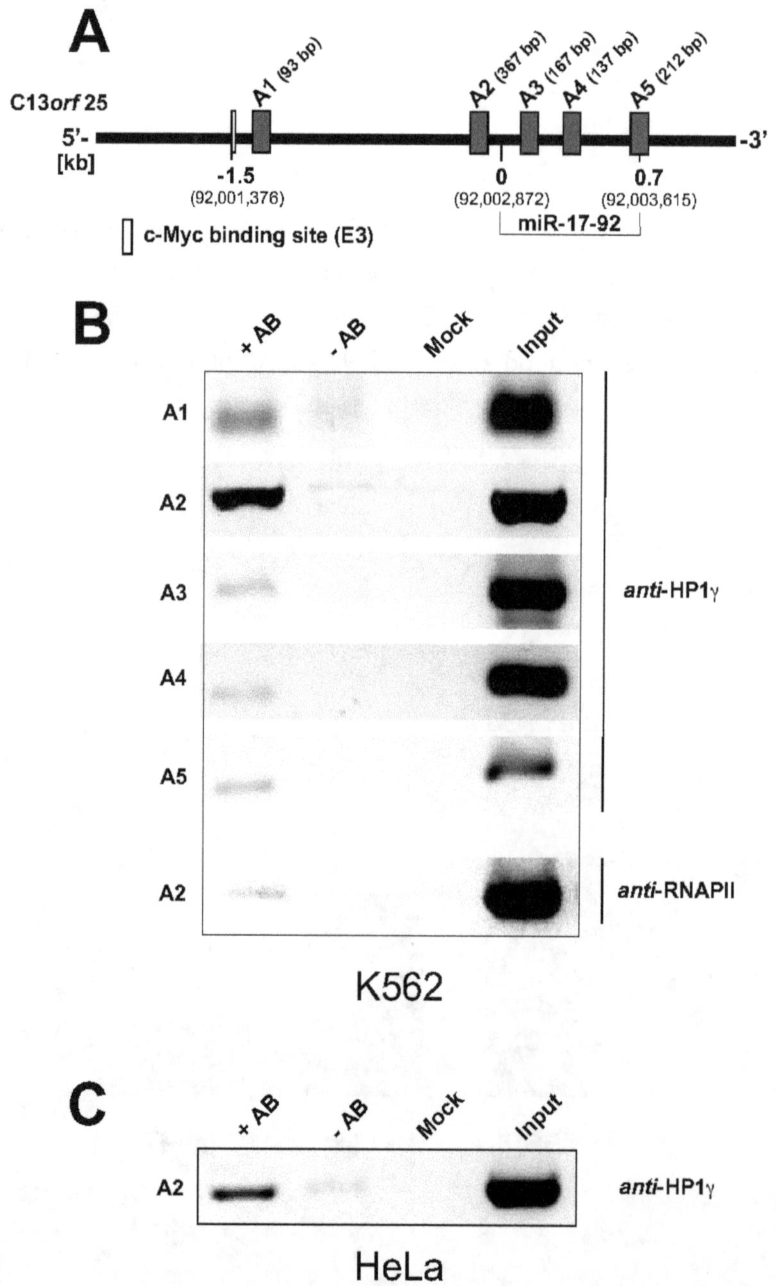

Figure S4. Effect of an siRNA-mediated Pim-1 knockdown on promoter activity of the ~1.5 kb reporter construct in HeLa cells. RLU values were derived from 5 independent triplicate experiments (+/−S.D.). RLU values for the control (left bar, transfected with the reporter plasmid but in the absence of a siRNA) were set to 100%. Lipofectamine transfection of HeLa cells was done as described under Supplementary methods, with the following alterations: 2×10^5 cells were used, and 40 pmol (0.6 µg) siRNA plus 0.5 µg of the reporter plasmid were combined in 50 µL Opti-MEM® I medium and mixed with 1.5 µL Lipofectamine™ 2000 in 50 µL Opti-MEM® I medium. The resulting mixture (~100 µL) was incubated for 20 min at room temperature to allow complex formation before addition to the cells. For the control (left bar), the siRNA was omitted.

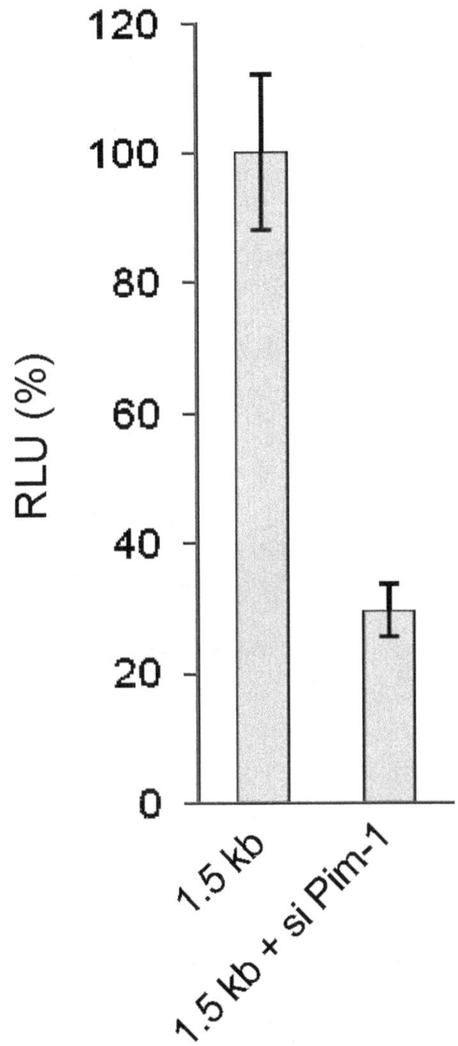

Supplementary Table

Table S1. Quantification of the pri-miR-17-92 levels and the c-Myc, E2F3 and Pim-1 mRNA levels after siRNA-dependent knockdown by qRT-PCR. Expression levels were calculated from the crossing points by the 2^-ΔΔC_T method [38] using β-Actin mRNA or 5S rRNA as internal controls. To determine the c-Myc, E2F3 and Pim-1 knockdown efficiencies, expression levels were normalized to the levels obtained by transfection of K562- or HeLa cells with an unrelated siRNA directed against the vanilloid receptor (siVR1).

K562	c-Myc KD		E2F3 KD		Pim-1 KD	
	2^-ΔΔC_T pri-mir-17-92	KD efficiency	2^-ΔΔC_T pri-mir-17-92	KD efficiency	2^-ΔΔC_T pri-mir-17-92	KD efficiency
	0.13	77	0.21	95	1.23	72
	0.02	87	0.16	98	0.16	69
	0.56	59	0.60	19	1.39	55
	0.75	74			0.76	59
	0.33	75			0.22	49
	0.14	99			2.38	78
	0.20	97			1.84	87
	0.55	80			0.80	88
					0.27	85
					0.40	89
					0.52	67
					0.69	65
					0.37	87
Mean	**0.33**	**81**	**0.34**	**89**	**0.47**	**73**
SEM	0.09	5.0	0.15	8.0	0.08	5.0

K562	c-Myc / E2F3 KD		c-Myc / Pim-1 KD		Pim-1 / E2F3 KD	
	2^-ΔΔC_T pri-mir-17-92	KD efficiency	2^-ΔΔC_T pri-mir-17-92	KD efficiency	2^-ΔΔC_T pri-mir-17-92	KD efficiency
	0.22	82 / 75	0.45	72 / 55	0.69	88 / 86
	0.23	86 / 72	0.19	90 / 54	0.56	68 / 73
	0.15	81 / 48	0.19	95 / 48	0.19	95 / 99
	0.08	99 / 99			0.52	87 / 81
	0.04	99 / 99			0.23	91 / 93
Mean	**0.15**	**89 / 79**	**0.28**	**86 / 52**	**0.44**	**86 / 86**
SEM	0.04	4.0 / 9.6	0.08	7.0 / 2.2	0.09	4.7 / 4.5

HeLa	c-Myc KD		E2F3 KD		Pim-1 KD	
	2^-ΔΔC_T pri-mir-17-92	KD efficiency	2^-ΔΔC_T pri-mir-17-92	KD efficiency	2^-ΔΔC_T pri-mir-17-92	KD efficiency
	0.37	70	0.72	51	0.50	47
	0.29	82	0.66	78	0.34	72
	0.32	61	0.40	76	0.71	63
	0.41	99	0.53	73	0.58	80
	0.48	55	0.64	73	0.54	87
	0.37	64			0.78	72
					0.70	76
Mean	**0.37**	**65**	**0.59**	**70**	**0.59**	**71**
SEM	0.03	6.7	0.07	6.3	0.06	4.9

HeLa	c-Myc / E2F3 KD		c-Myc / Pim-1 KD		Pim-1 / E2F3 KD	
	2^-ΔΔC_T pri-mir-17-92	KD efficiency	2^-ΔΔC_T pri-mir-17-92	KD efficiency	2^-ΔΔC_T pri-mir-17-92	KD efficiency
	0.24	79 / 64	0.24	79 / 84	0.49	81 / 96
	0.13	91 / 81	0.25	77 / 61	0.33	91 / 85
	0.11	93 / 86	0.15	86 / 58	0.10	97 / 94
	0.12	90 / 91	0.12	95 / 91	0.26	92 / 94
	0.23	87 / 74	0.41	84 / 73		
	0.26	83 / 66	0.10	97 / 95		
Mean	**0.18**	**87 / 77**	**0.21**	**86 /77**	**0.30**	**90 / 92**
SEM	0.03	2.2 / 4.4	0.053	3.3 / 6.0	0.08	3.3 / 2.5

References

1. Iorio, M.V.; Croce, C.M. MicroRNAs in cancer: Small molecules with a huge impact. *J. Clin. Oncol.* **2009**, *27*, 5848–5856.
2. Krol, J.; Loedige, I.; Filipowicz, W. The widespread regulation of microRNA biogenesis, function and decay. *Nat. Rev. Genet.* **2010**, *11*, 597–610.
3. Bartel, D.P. MicroRNAs: Genomics, biogenesis, mechanism, and function. *Cell* **2004**, *116*, 281–297.
4. Denli, A.M.; Tops, B.B.J.; Plasterk, R.H.; Ketting, R.F.; Hannon, G.J. Processing of primary microRNAs by the Microprocessor complex. *Nature* **2004**, *432*, 231–235.
5. Yi, R.; Qin, Y.; Macara, I.G.; Cullen, B.R. Exportin-5 mediates the nuclear export of pre-microRNAs and short hairpin RNAs. *Genes Dev.* **2003**, *17*, 3011–3016.
6. Lee, Y.; Jeon, K.; Lee, J.-T.; Kim, S.; Kim, V.N. MicroRNA maturation: Stepwise processing and subcellular localization. *EMBO J.* **2002**, *21*, 4663–4670.
7. Esquela-Kerscher, A.; Slack, F.J. Oncomirs—microRNAs with a role in cancer. *Nat. Rev. Cancer* **2006**, *6*, 259–269.
8. Ota, A.; Tagawa, H.; Karnan, S.; Tsuzuki, S.; Karpas, A.; Kira, S.; Yoshida, Y.; Seto, M. Identification and characterization of a novel gene, C13orf25, as a target for 13q31–q32 amplification in malignant lymphoma. *Cancer Res.* **2004**, *64*, 3087–3095.
9. He, L.; Thomson, J.M.; Hemann, M.T.; Hernando-Monge, E.; Mu, D.; Goodson, S.; Powers, S.; Cordon-Cardo, C.; Lowe, S.W.; Hannon, G.J.; *et al.* A microRNA polycistron as a potential human oncogene. *Nature* **2005**, *435*, 828–833.
10. O'Donnell, K.A.; Wentzel, E.A.; Zeller, K.I.; Dang, C.V.; Mendell, J.T. c-Myc-regulated microRNAs modulate E2F1 expression. *Nature* **2005**, *435*, 839–843.
11. Tanzer, A.; Stadler, P.F. Molecular evolution of a microRNA cluster. *J. Mol. Biol.* **2004**, *339*, 327–335.
12. Lu, J.; Getz, G.; Miska, E.A.; Alvarez-Saavedra, E.; Lamb, J.; Peck, D.; Sweet-Cordero, A.; Ebert, B.L.; Mak, R.H.; Ferrando, A.A.; *et al.* MicroRNA expression profiles classify human cancers. *Nature* **2005**, *435*, 834–838.
13. Olive, V.; Bennett, M.J.; Walker, J.C.; Ma, C.; Jiang, I.; Cordon-Cardo, C.; Li, Q.-J.; Lowe, S.W.; Hannon, G.J.; He, L. miR-19 is a key oncogenic component of mir-17–92. *Genes Dev.* **2009**, *23*, 2839–2849.
14. Ventura, A.; Young, A.G.; Winslow, M.M.; Lintault, L.; Meissner, A.; Erkeland, S.J.; Newman, J.; Bronson, R.T.; Crowley, D.; Stone, J.R.; *et al.* Targeted deletion reveals essential and overlapping functions of the miR-17 through 92 family of miRNA clusters. *Cell* **2008**, *132*, 875–886.
15. Mu, P.; Han, Y.-C.; Betel, D.; Yao, E.; Squatrito, M.; Ogrodowski, P.; de Stanchina, E.; D'Andrea, A.; Sander, C.; Ventura, A. Genetic dissection of the miR-17–92 cluster of microRNAs in Myc-induced B-cell lymphomas. *Genes Dev.* **2009**, *23*, 2806–2811.
16. Ozsolak, F.; Poling, L.L.; Wang, Z.; Liu, H.; Liu, X.S.; Roeder, R.G.; Zhang, X.; Song, J.S.; Fisher, D.E. Chromatin structure analyses identify miRNA promoters. *Genes Dev.* **2008**, *22*, 3172–3183.
17. Woods, K.; Thomson, J.M.; Hammond, S.M. Direct regulation of an oncogenic micro-RNA cluster by E2F transcription factors. *J. Biol. Chem.* **2007**, *282*, 2130–2134.

18. Sylvestre, Y.; de Guire, V.; Querido, E.; Mukhopadhyay, U.K.; Bourdeau, V.; Major, F.; Ferbeyre, G.; Chartrand, P. An E2F/miR-20a autoregulatory feedback loop. *J. Biol. Chem.* **2007**, *282*, 2135–2143.
19. Guenther, M.G.; Levine, S.S.; Boyer, L.A.; Jaenisch, R.; Young, R.A. A chromatin landmark and transcription initiation at most promoters in human cells. *Cell* **2007**, *130*, 77–88.
20. Ji, M.; Rao, E.; Ramachandrareddy, H.; Shen, Y.; Jiang, C.; Chen, J.; Hu, Y.; Rizzino, A.; Chan, W.C.; Fu, K.; *et al.* The miR-17–92 microRNA cluster is regulated by multiple mechanisms in B-cell malignancies. *Am. J. Pathol.* **2011**, *179*, 1645–1656.
21. Pospisil, V.; Vargova, K.; Kokavec, J.; Rybarova, J.; Savvulidi, F.; Jonasova, A.; Necas, E.; Zavadil, J.; Laslo, P.; Stopka, T. Epigenetic silencing of the oncogenic miR-17–92 cluster during PU.1-directed macrophage differentiation. *EMBO J.* **2011**, *30*, 4450–4464.
22. Koike, N.; Maita, H.; Taira, T.; Ariga, H.; Iguchi-Ariga, S.M.M. Identification of heterochromatin protein 1 (HP1) as a phosphorylation target by Pim-1 kinase and the effect of phosphorylation on the transcriptional repression function of HP1. *FEBS Lett.* **2000**, *467*, 17–21.
23. Nawijn, M.C.; Alendar, A.; Berns, A. For better or for worse: The role of Pim oncogenes in tumorigenesis. *Nat. Rev. Cancer* **2011**, *11*, 23–34.
24. Zippo, A.; de Robertis, A.; Serafini, R.; Oliviero, S. PIM1-dependent phosphorylation of histone H3 at serine 10 is required for MYC-dependent transcriptional activation and oncogenic transformation. *Nat. Cell Biol.* **2007**, *9*, 932–944.
25. UCSC Genome Bioinformatics. Available online: http://genome.ucsc.edu (assessed on 1 February 2009).
26. Thomas, M.; Lange-Grünweller, K.; Weirauch, U.; Gutsch, D.; Aigner, A.; Grünweller, A.; Hartmann, R.K. The proto-oncogene Pim-1 is a target of miR-33a. *Oncogene* **2012**, *31*, 918–928.
27. Venturini, L.; Battmer, K.; Castoldi, M.; Schultheis, B.; Hochhaus, A.; Muckenthaler, M.U.; Ganser, A.; Eder, M.; Scherr, M. Expression of the miR-17–92 polycistron in chronic myeloid leukemia (CML) CD34+ cells. *Blood* **2007**, *109*, 4399–4405.
28. Kwon, S.H.; Workman, J.L. HP1c casts light on dark matter. *Cell Cycle* **2011**, *15*, 625–630.
29. Vakoc, C.R.; Mandat, S.A.; Olenchock, B.A.; Blobel, G.A. Histone H3 lysine 9 methylation and HP1gamma are associated with transcription elongation through mammalian chromatin. *Mol. Cell* **2005**, *19*, 381–391.
30. Minc, E.; Allory, Y.; Worman, H.J.; Courvalin, J.; Buendia, B. Localization and phosphorylation of HP1 proteins during the cell cycle in mammalian cells. *Chromosoma* **1999**, *108*, 220–234.
31. Mateescu, B.; Bourachot, B.; Rachez, C.; Ogryzko, V.; Muchardt, C. Regulation of an inducible promoter by an HP1beta-HP1gamma switch. *EMBO Rep.* **2008**, *9*, 267–272.
32. Fischle, W.; Tseng, B.S.; Dormann, H.L.; Ueberheide, B.M.; Garcia, B.A.; Shabanowitz, J.; Hunt, D.F.; Funabiki, H.; Allis, C.D. Regulation of HP1-chromatin binding by histone H3 methylation and phosphorylation. *Nature* **2005**, *438*, 1116–1122.
33. Adams, M.R.; Sears, R.; Nuckolls, F.; Leone, G.; Nevins, J.R. Complex transcriptional regulatory mechanisms control expression of the E2F3 locus. *Mol. Cell Biol.* **2000**, *20*, 3633–3639.
34. Thalmeier, K.; Synovzik, H.; Mertz, R.; Winnacker, E.L.; Lipp, M. Nuclear factor E2F mediates basic transcription and trans-activation by E1a of the human MYC promoter. *Genes Dev.* **1989**, *3*, 527–536.

35. Wong, J.V.; Dong, P.; Nevins, J.R.; Mathey-Prevot, B.; You, L. Network calisthenics: Control of E2F dynamics in cell cycle entry. *Cell Cycle* **2011**, *15*, 3086–3094.
36. Grünweller, A.; Wyszko, E.; Bieber, B.; Jahnel, R.; Erdmann, V.A.; Kurreck, J. Comparison of different antisense strategies in mammalian cells using locked nucleic acids, 2'-*O*-methyl RNA, phosphorothioates and small interfering RNA. *Nucleic Acids Res.* **2003**, *31*, 3185–3193.
37. Gozani Lab. Available online: http://www.stanford.edu/group/gozani (assessed on 1 June 2005).
38. Schmittgen, T.D.; Livak, K.J. Analyzing real-time PCR data by the comparative CT method. *Nat. Prot.* **2008**, *3*, 1101–1108.
39. Reese, M.G. Application of a time-delay neural network to promoter annotation in the *Drosophila melanogaster* genome. *Comput. Chem.* **2001**, *26*, 51–56.
40. Ohler, U. Identification of core promoter modules in *Drosophila* and their application in accurate transcription start site prediction. *Nucleic Acids Res.* **2006**, *34*, 5943–5950.
41. Knudsen, S. Promoter2.0: For the recognition of PolII promoter sequences. *Bioinformatics* **1999**, *15*, 356–361.
42. Rangannan, V.; Bansal, M. Relative stability of DNA as a generic criterion for promoter prediction: whole genome annotation of microbial genomes with varying nucleotide base composition. *Mol. BioSys.* **2009**, *5*, 1758–1769.
43. Zhang, M.Q. Identification of human gene core promoters *in silico*. *Genome Res.* **1998**, *8*, 319–326.
44. The R Development Core Team. *R: A Language and Environment for Statistical Computing*; R Foundation for Statistical Computing: Vienna, Austria, 2012.

4

Emerging Roles of Small Epstein-Barr Virus Derived Non-Coding RNAs in Epithelial Malignancy

Raymond Wai-Ming Lung [1], Joanna Hung-Man Tong [1] and Ka-Fai To [1,2,*]

[1] Department of Anatomical and Cellular Pathology, State Key Laboratory in Oncology in South China, The Chinese University of Hong Kong, Hong Kong, China;
E-Mails: raymond_lung@cuhk.edu.hk (R.W.-M.L.); jtong@cuhk.edu.hk (J.H.-M.T.)

[2] Institute of Digestive Diseases, Li Ka-Shing Institute of Health Sciences, The Chinese University of Hong Kong, Hong Kong, China

* Author to whom correspondence should be addressed; E-Mail: kfto@cuhk.edu.hk;

Abstract: Latent Epstein-Barr virus (EBV) infection is an etiological factor in the progression of several human epithelial malignancies such as nasopharyngeal carcinoma (NPC) and a subset of gastric carcinoma. Reports have shown that EBV produces several viral oncoproteins, yet their pathological roles in carcinogenesis are not fully elucidated. Studies on the recently discovered of EBV-encoded microRNAs (ebv-miRNAs) showed that these small molecules function as post-transcriptional gene regulators and may play a role in the carcinogenesis process. In NPC and EBV positive gastric carcinoma (EBVaGC), 22 viral miRNAs which are located in the long alternative splicing EBV transcripts, named BamH1 A rightward transcripts (BARTs), are abundantly expressed. The importance of several miR-BARTs in carcinogenesis has recently been demonstrated. These novel findings enhance our understanding of the oncogenic properties of EBV and may lead to a more effective design of therapeutic regimens to combat EBV-associated malignancies. This article will review the pathological roles of miR-BARTs in modulating the expression of cancer-related genes in both host and viral genomes. The expression of other small non-coding RNAs in NPC and the expression pattern of miR-BARTs in rare EBV-associated epithelial cancers will also be discussed.

Keywords: Epstein-Barr virus (EBV); nasopharyngeal carcinoma (NPC); EBV-associated gastric carcinoma (EBVaGC); lymphoepithelioma-like carcinoma (LELC) of the lung; lymphoepithelioma-like cholangiocarcinoma; BART; miRNAs; v-snoRNA1

1. Introduction

The Epstein-Barr virus (EBV) is a ubiquitous gamma herpesvirus that was originally identified in Burkitt's lymphoma (BL) by Tony Epstein and his colleagues in 1964 [1]. It is also the first virus known to play an essential role in the induction of human malignancies [2]. Interestingly, EBV infection is common among most population groups worldwide. However, most infected individuals are asymptomatic, and for those who show signs and symptoms of acute EBV infection tend to recover without sequelae [3]. Interestingly, the virus will not be entirely eliminated and will ultimately establish life-long latent infection within memory B cell pool, in which the viral genome circulates as episome and undergoes replication using host cellular expression machinery [4,5]. Latent EBV infection is associated with a wide range of human lymphoid and epithelial malignancies, including Hodgkin disease, Burkitt's lymphoma (BL), nasopharyngeal carcinoma (NPC), and a subset of gastric carcinoma (GC).

While the EBV genome contains around 80 open reading frames (*ORF*s), only a few of viral proteins are selectively expressed during latent infection. Expression of these viral proteins forms four latency patterns [6]. Latency III infection is commonly observed in EBV-associated B-cell lymphoma in the setting of immunosuppression, in which all viral latent genes, including six EBV encoded nuclear antigen proteins (EBNA-1, EBNA-2, EBNA-3A, 3B, 3C and EBNA-LP) and three latent membrane proteins (LMP1, LMP2A and LMP2B), are expressed [7]. The spectrum of lymphomas in latency III includes post-transplant lymphoproliferative diseases (PTLD) and immunodeficiency associated lymphoproliferative diseases. It is likely that the immunosuppressive environment allows the expression of all latent proteins in the absence of host immune responses [8]. EBV protein expression in latency II is limited to EBNA-1 and the three LMPs. This latency type is commonly observed in NPC [9]. EBNA-1, an essential protein for virus persistence, is the only viral protein expressed in latency I. In a healthy carrier, EBV remains dormant in memory B-cells which circulate in the peripheral blood (latency 0). At this stage, expression of the EBNA-1 viral protein is found only during cell division, albeit the LMP transcript is detected at very low level in the peripheral blood [5,10]. Interestingly, two types of non protein-coding RNA, *EBV-encoded RNAs* (*EBER-1* and *EBER-2*) and long alternative splicing non-coding RNA at BamH1 A rightward transcripts (BARTs), are expressed in all forms of latency programs [11,12].

2. EBV Infection in Epithelial Carcinoma

2.1. Nasopharyngeal Carcinoma (NPC)

Nasopharyngeal carcinoma (NPC) is a distinctive type of head and neck cancer arising from the nasopharynx. This is a rare tumor in most part of the world (below 1 per 100,000 persons per year) [13]. However, the incidence and mortality rates of NPC are remarkably high in Southern China including

Hong Kong (about 19.5 and 7.7 per 100,000 persons, respectively) [14]. The World Health Organization (WHO) has classified NPC into three histological categories: type I is keratinizing squamous cell carcinoma (SCC), type II is non-keratinizing carcinoma, and type III is undifferentiated carcinoma. The prevalence of individual type of NPC is dependent on geographic regions. Types I and II NPC are generally found in western population. And type III, which is consistently associated with EBV latent infection, accounts the majority of NPC in Southern China [15]. Because of the unusual prominent lymphocytic infiltration, type III NPC is also commonly described as lymphoepithelioma of nasopharynx. It has been suggested that EBV particles from the infected lymphoid cells are transferred to the nasopharynx cells via cell-cell contact [16,17]. Clonal EBV genome is consistently detected in both high-grade dysplastic lesions of the nasophargnx and invasive NPC, suggesting that EBV may contribute to the neoplastic transformation of nasopharyngeal epithelial cells and facilitate the clonal expansion of malignant cells [13,18].

Although the crucial roles of viral LMPs in NPC pathogenesis have previously been reported [19,20], their expressions in NPC are variable and generally low when compared to those in lymphoid malignancies. In NPC, viral proteins are processed and presented on the cell surface, resulting in the stimulation of host immune surveillance [21,22]. Therefore, low expression of viral proteins is critical for maintenance viral latency. In this regard, EBV may contribute to cancer development by two abundantly expressed non protein-coding viral products, *EBERs* and *BARTs*. Currently, the abundant expressions of 22 *BARTs*-derived microRNAs (miR-BARTs) in NPC have been identified; it is possible that EBV augment cancer development through these viral-encoded miRNAs [23,24].

2.2. EBV-Associated Gastric Carcinoma (EBVaGC)

Gastric carcinoma (GC) is another epithelial cancer reported to be associated with EBV infection. In 1990, Burke *et al.* first reported the presence of EBV DNA in a rare type of gastric carcinoma, namely the lymphoepithelioma-like carcinomas (LELCs) of stomach [25]. Subsequently, Shibata's research team and others demonstrated that EBV is present in more than 80% of gastric LELCs [26,27]. In 1992, Shibta's team further identified the presence of EBV in up to 16% of conventional gastric adenocarcinoma [28]. They collectively described this subtype of gastric cancer as EBV-associated gastric carcinoma (EBVaGC). Unlike NPC, EBVaGC is not an endemic disease and is present in about 10% of all gastric cancer worldwide. The incidence rates are highly similar among North America (9.9%), Europe (9.2%) and Asia (8.3%) [27,29–32]. EBVaGC represents a distinct type of GC as it has special clinicopathological characteristics such as male predominance, predisposition to the proximal stomach, lace pattern in the early GC and more frequent occurrence in gastric remnant carcinoma (GRC). EBVaGC has been usually described as latency I infection although LMP2A was weakly detected in some of the cases [29,33,34]. Previous works performed by our team also demonstrated the importance of LMP2A in gastric cancer development through genome-wide methylation analyses [35–37]. In contrast to the low expression level of viral protein, abundant miR-BARTs expression in the EBVaGC may be implicated in carcinogenesis [38,39].

2.3. Lymphoepithelioma-Like Carcinomas (LELCs)

LELC is histologically defined, a poorly differentiated carcinoma with dense lymphocytic infiltration. This type of cancer can be found not only in type III NPC, but also in other anatomical sites. Interestingly, EBV has been consistently detected in the LELC derived from foregut organs such as salivary gland, lung, thymus, gastric and intra-hepatic biliary epithelium (cholangiocarcinoma). They are rare cancers mainly reported in endemic regions [40–43].

LELC of the lung was first reported in 1987, and fewer than 200 cases have been reported in the literature [44]. This rare tumor occupies less than 1% of the overall lung cancer incidence in southeast China and Japan [45]. Nevertheless, EBER is consistently detected in almost all LELCs of lung in Asia. Latency program of EBV in LELC of the lung is unclear as only ~50% of the cases showed the expression of LMP1 viral oncoprotein. EBV positive carcinomas have been rarely developed in liver, some cases of these carcinomas reported as LELCs but the majority of them are EBV positive cholangiocarcinoma (CCA) [41]. In this review, we will address the expression pattern of EBV miRNAs in LELC of the lung and CCA and will discuss the possible pathogenic roles of EBV in LELC.

3. EBV Non Protein-Coding RNAs

3.1. Epstein-Barr Virus-Encoded RNAs (EBERs)

The EBERs contain two species, EBER1 and EBER2 with sizes of 166 and 172 nucleotides, respectively [46]. These two genes are separated by 161 nucleotides in the EBV genome and transcribed individually in the same direction using RNA polymerase III [46–48]. They are known to be abundantly expressed in all latent EBV infected cells with a copy number of up to 5×10^6 per infected cell [46,49]. Faint EBER signals in the cytoplasm can be detected using high-resolution fluorescent *in situ* hybridization (ISH) technique with confocal laser scanning microscopy [50]. However, these transcripts are dominantly and stably confined to the nucleus, where they are associated with several ribonucleoproteins [46,51], including lupus antigen (LA) [46], ribosomal protein L22 (formerly EBER-associated protein EAR) [52–54], interferon-inducible, double-stranded RNA activated protein kinase R (PKR) [55] and retinoic acid-inducible gene I (RIG-I) [56]. As a result, EBER transcripts can be detected in all EBV positive cells by ISH and are routinely used as a probe to determine EBV infection in clinical samples. The functional role of EBERs in epithelial cancer development is not clear. Despite the demonstration of oncogenic properties of EBERs on B-cells in *in vitro* and *in vivo* experiments, EBERs can neither induce cellular transformation nor increase tumorigenicity in epithelial cell models. However, several reports indicated that EBERs can confer an apoptotic-resistant phenotype in immortalized epithelial cells and can support cell growth by stimulating insulin-like growth factor (IGF-1) secretion in EBV-positive gastric carcinoma and NPC cell lines [57–59] (for more details about EBER1 function, see [47,51]).

Early work from Steitz's research team has shown that the predicted folding structure of EBERs is comprised of several stem-loops which are highly similar to two non-coding adenovirus-associated (VA) RNAs, namely VAI and VAII [60,61]. Moreover, EBERs can completely substitute the critical role of VA RNAs during adenovirus replication [62,63]. Intriguingly, two research groups recently demonstrated that VA RNAs can be further processed into a small functional interference RNA [64,65].

Our preliminary work also identified EBER1 derived small RNA fragments (18–25 nt) from NPC small RNA library cloning [23]. The expressions of these fragments in NPC have been verified by Northern blot analysis (Figure S1A). Nonetheless, EBER1 is unlikely to undergo typical miRNA processing due to the following reasons. Firstly, the predicted stem-loop structure on EBER1 is too small for pre-miRNA processing by Drosha/DGCR8 complex [66]. Secondly, Dicer is unable to cleave EBER1 in an *in vitro* experiment [65]. Thirdly, EBER1 is mainly confined in the nucleus and does not associate with exportin-5, a protein that facilitates nucleus export of pre-mature miRNAs [67]. We further demonstrated that EBER1 and its derived small fragments do not bind to Argonaute2 (Ago2), an integral part in the functional RNA-induced silencing complex (RISC) complex (Figure S1B).

3.2. Bam HI A Rightward Transcripts (BARTs)

BARTs are multi-spliced transcripts originally discovered by cDNA library analysis in C15 NPC xenograft, and their expressions have later been reported in other EBV-infected individuals [68,69]. The contribution of BARTs is dispensable in B-cell transformation since mutant recombinant EBV carrying a null BART region maintains its transforming activity in primary B lymphocyte [70]. However, it has long been speculated that BARTs have important functional roles in EBV-associated epithelial malignancies as it is abundantly expressed in NPC and EBVaGC. In the early study of BARTs, several small ORFs in the spliced cDNA transcripts were the focuses of investigation. RPMS1 and A73 were revealed to be translated into oncogenic proteins in an *in vitro* study and were shown to be negative regulators of Notch and RACK1 signaling pathways, respectively. However, expressions of these two viral proteins have not been demonstrated in natural EBV-infected samples [71,72]. Increasing interest in the functional role of BARTs, has eventually led to the discovery of 22 viral miRNAs (miR-BARTs) and a small nucleolar RNA (snoRNA) in this region [23,73–76]. The details of miR-BARTs' function will be discussed later in this review.

3.3. Discovery of BART-Encoded miRNAs

MicroRNAs (miRNAs), a group of 18–24 nucleotide non protein-coding RNAs, were produced by two endogenous enzymatic digestions (Drosha/DGCR8 complex and Dicer) from the hairpin structures of long RNA transcripts. A mature miRNA recruits a group of cellular proteins, including Argonaute2 (Ago2), to form a stable complex called RNA-induced silencing complex (RISC) [77], in which miRNA acts as a guide strand to negatively regulate gene expression through imperfect complementary sequence pairing to the 3' untranslated region (3'UTR) of the target gene [78]. One mature miRNA is expected to regulate expression of over 100 cellular genes by either repressing protein translation or inducing mRNA degradation [78]. It is estimated that about 30% of cellular protein-coding genes are regulated by miRNA [79].

EBV was the first virus identified to encode miRNA. In 2004, Thomas Tuschl and colleagues used molecular cloning methods to identify a total of 5 miRNAs (ebv-miR-BHRF1-1, -2, -3, ebv-miR-BART1 and 2) in human B-cells infected with laboratory isolated EBV strain B95.8, which carries a 12-kb deletion in the BART region of the EBV genome [73] (Figure 1). Eventually, subsequent work on other latently infected cell lines and biopsy materials identified a total of 25 EBV encoded miRNAs [23,24,74,75,80]. Among them, three miRNAs arising from the UTR of the distinct

Bcl2 homolog gene *BHRF1* (ebv-miR-BHRF1-1, -2 and -3) are predicted to be expressed in the EBV latency III program. The rest of the remaining 22 miRNAs (miR-BART1 to miR-BART22), which are located mainly in two clusters within the non-coding BART region, are generally highly expressed in most of the EBV infected epithelial cells (latency I and II) [20,39].

Figure 1. Expression of miR-BARTs in EBV positive epithelial carcinoma. The fraction of individual miRNA from all tested miR-BARTs is plotted in the upper panel. The middle panel shows the copy number of each miR-BART per nanogram of total RNA. The value represents mean + SD is plotted.

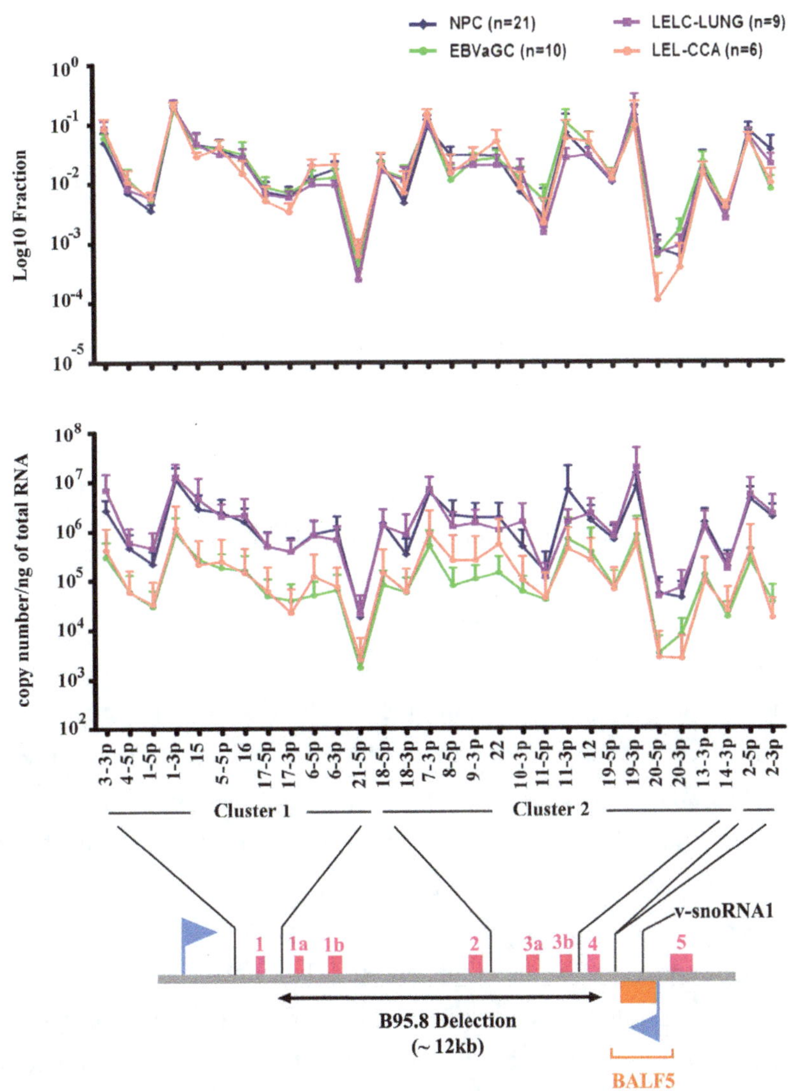

3.4. Discovery of Viral-Encoded snoRNA

Nucleoli contain a large population of conserved small stable RNAs known as small nucleolar RNA (snoRNA). They can mediate nucleotides modification of ribosomal RNAs and facilitate their production by forming a functional snoRNA:protein complex (snoRNPs) with a group of snoRNA core proteins [81]. Based on deep sequencing analysis of cDNA clones from B lymphocytes, Hutzinger *et al.* identified a snoRNA gene within the EBV genome. The viral snoRNA, mapped ~100 bp downstream to the miR-BART2 transcripts, was 65 nucleotides (nt) in length and designated as v-snoRNA1. The authors also confirmed the expression of v-snoRNA1 in a panel of EBV positive BL and LCL cell lines. Given that v-snoRNA1 could associate with three canonical snoRNA proteins (fibrilarin, Nop56 and Nop58), it follows that v-snoRNA1 could be assembled into functional snoRNPs. On the other hand, a 24 nt RNA fragment from the v-snoRNA1 3' terminus (v-snoRNA^{24pp}) had been identified in cDNA library screening, but its expression was only detected during lytic induction on EBV infected 293 cells [76]. In view of the fact that some snoRNA species were reported as precursors for miRNA biogenesis [82,83], v-snoRNA1 might also be further processed into miRNA-like molecule, v-snoRNA^{24pp}. However, the experimental evidence to prove this hypothesis is not yet provided.

The miRNAs possess properties to modify the intracellular environment, a characteristic that is particularly useful and convenient for viral infection. Viral miRNAs were processed from short RNA transcripts (~200 nt) by using cellular transcriptional machinery. Thus, a number of miRNA precursors can be tightly packed into the relatively small viral genome. In addition, miRNAs are non-protein coding molecules that are presumably immunogenically inert, yet they can modulate several target genes that can enhance infected cell survival.

4. Expression of Viral miRNAs in Epithelial Cancers

4.1. Ebv-miR-BHRF1s

All three ebv-miR-BHRF1s are suggested to be derived from two latency III specific Cp/Wp-initiated transcripts, the primary EBNA transcript and the latent BHRF1 transcript. For this reason, BHRF1 miRNAs are mainly expressed in this specific latency program [38,73,74,84–86]. However, several research teams using QRT-PCR technique indicated the presence of BHRF1 miRNAs in latency type II EBV positive epithelial cancer cell lines [87,88]. The inconsistent observation may be the result of active EBV replication in a small portion of infected cells. During EBV replication, miR-BHRF1-2 and -3 may be generated from the lytic form of the BHRF1 transcripts that were produced from the alternative lytic promoter BHRF1p [86]. As a gold standard, Northern blot analysis is used for the detection of miRNA expression. Surprisingly, we could detect the expression of the lytic form of the BHRF1 transcript in NPC samples including cell lines, xenografts and half of the actual patient samples. However, no expression of BHRF1 miRNAs was detected in any of the tested samples (Figure S2A). In contrast to previous studies that reported that induction of lytic EBV replication could facilitate miR-BHRF1s expression in B-cells [74,84], induction of EBV lytic cycle in the native EBV positive NPC cell lines C666-1 did not activate miR-BHRF1 expression at detectable levels (Figure S2B). This finding aligned with the previous report that BHRF1 transcription does not correlate with the expression of mature miR-BHRF1s [86]. Failure of miR-BHRF1 biogenesis may be the result of the absence of their target transcript in NPCs, leading to the sharp degradation of the miRNAs [89]. In addition, while the level of miR-BHRF1s was negligible, miR-BARTs were detected from the biopsies of EBV positive lymphoepithelioma-like cholangiocarcinoma and LELC of the lung (Figure S2C). This is in line with the previous data that BHRF1-derived miRNAs are not processed in epithelial cancers when EBV infection is mainly in latency I/II programs [24,38,80,85].

4.2. Ebv-miR-BARTs

Viral BARTs are initiated from two constitutively active promoters P1 and P2 [86,90]. And miR-BARTs are the only known functional molecules produced from these transcripts. By using RT-PCR and small RNA library sequencing methods, expression profiles of EBV miRNAs in cancer have been extensively studied, albeit most of the studies focused on NPC [24,39,85–88,91,92]. It is obvious in NPC that miR-BARTs are abundantly expressed and constitute up to 23% of the total miRNAs [24]. However the expression of individual miR-BART in the cells is considerably different. The comprehensive miRNA profiling study by Qiu *et al.* revealed that several miR-BARTs that were not detected in normal infected cells (germinal center B cells: latency II and memory B cells: latency 0/I) were upregulated in various EBV-associated malignancies such as BL, EBVaGC, Hodgkin Disease and NPC. These critical findings designated the importance of miR-BARTs in the development of EBV-associated cancer [87].

Qiu *et al.* further concluded that the expression patterns of EBV miRNAs from two epithelial cancers (5 NPCs and 6 EBVaGCs) were extremely similar [87]. To gain insight into the expression pattern in other epithelial cancers, our team employed home-designed QRT-PCR assays to profile the expression of EBV miRNAs in two common and two rare EBV-associated epithelial malignancies in our local primary biopsies, including NPC ($n = 23$), EBVaGC ($n = 10$), LELC of the lung ($n = 9$) and lymphoepithelioma-like CCA ($n = 6$). The presence of EBV in the samples was confirmed by EBER *in situ* hybridization (ISH) on the paraffin-embedded tissue section. Since the tumor contents and transcription activities among individual samples might be different, we measured both the absolute miRNA copy number per nanogram of total RNA and the fraction of viral miRNAs. To obtain an absolute quantification of each miRNA in the sample, an in parallel experiment using synthetic DNA-RNA chimeric oligonucleotides of each mature miRNA sequence as a template for a standard curve was included (Table S1). On the whole, all miR-BARTs were favorably expressed in all the tested EBV associated malignancies except for miR-BART20 and miR-BART21, which showed low expression levels of 10^3 copies per nanogram of total RNA. The low expression levels of these two miRNAs aligned with our previous findings in a small-scale cloning analysis on NPC samples [23]. Whereas the average copy numbers of BART miRNAs expressed in NPC and LELC of the lung was around 5–10 fold higher than EBVaGC and CCA, the miR-BARTs fraction profile of four tumor types were vastly similar (Figure 1).

5. Recent Research Development of Viral Small Nucleolar RNA (v-snoRNA1)

5.1. Expression of v-snoRNA1 in NPC

Apart from the production of miRNAs, abundant BARTs expression in epithelial cancer may have another important role by generating v-snoRNA1 and v-snoRNA1^{24pp}. However, v-snoRNA1 expression is so far only described in EBV infected B-cells. In our recent study, we aimed to characterize the expression of v-snoRNA1 in NPC. We noticed in Northern blot analysis that v-snoRNA1 was highly expressed in almost all NPC samples including cell lines, xenografts and biopsies. Furthermore, v-snoRNA1^{24pp} was detected in a portion of the NPC samples (Figure 2A,B). Previous work by Hutzinger *et al.* demonstrated that inducing EBV lytic replication could upregulate both v-snoRNA1 and v-snoRNA1^{24pp} production in wild type EBV infected 293 cells [76]. In contrast, we demonstrated that in native EBV infected C666-1 cells, induction of the lytic cycle could only slightly up-regulate v-snoRNA1^{24pp} expression and had no observable effect on v-snoRNA1 expression (Figure 2B).

Figure 2. Expression of v-snoRNA1 in NPC: (**A**) Expression of v-snoRNA1 and v-snoRNA1^{24pp} on NPC biopsies (NPC 1 to NPC 9) were analyzed by Northern blot analysis. Biopsies from noncancerous nasopharyngeal tissues (NP 1 to NP 3), NPC cell lines (C666-1) and B95.8 cells were included as controls. The membranes shown in Figure 2A were used previously in our publication [93]; (**B**) To confirm the result on Figure 2A, Northern blot analysis for samples from NPC 1, NPC 2 (*left panel*) and another 7 new NPC biopsies (*middle panel*) were performed on the newly prepared membranes. The effect of EBV replication on v-snoRNA1 expression in C666-1 cells is shown in the *right panel.*

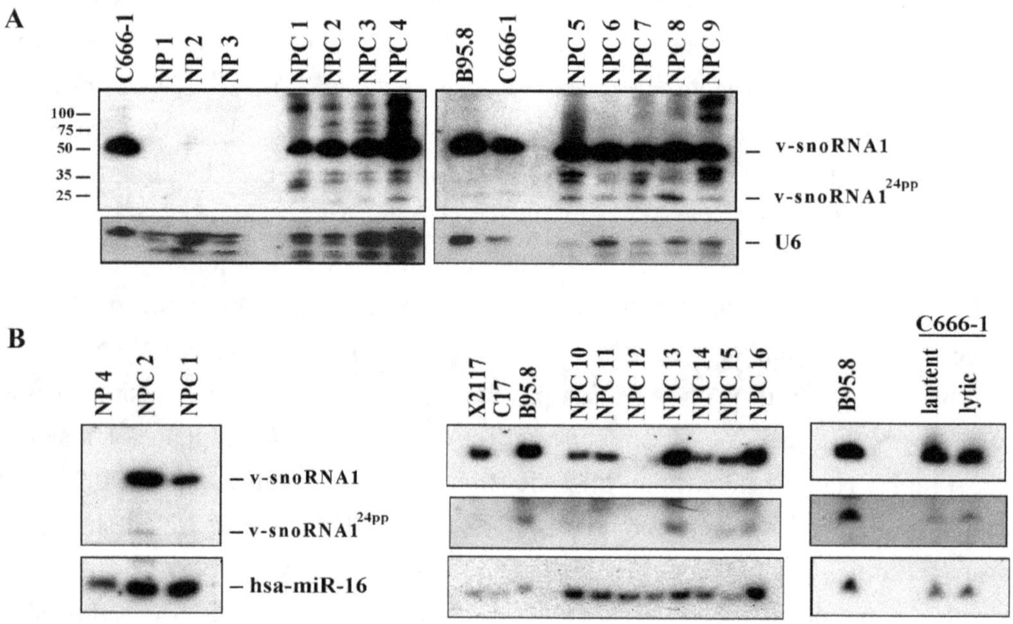

5.2. Nucleotide Polymorphism Is Important for v-snoRNA1 Production

We next compared the v-snoRNA1 sequence in our available cell lines and NPC xenografts to the B95-8-EBVstrain by direct sequencing. Our finding was consistent with the previous observation by Hutzinger *et al.* that v-snoRNA1s from different EBV strains are highly conserved. However, several nucleotide polymorphisms at a position distal from the 3' end of v-snoRNA1 were observed. BC-1-EBV and B95-8-EBV strains had the same sequence polymorphism (variant 1), and all the other tested samples carried another sequence pattern as the C666-1-EBV strain (variant 2) (Figure 3A). To further analyze the variation pattern of EBV strains from 14 normal controls and 31 NPC patients in our population, we showed that the variant distribution of EBV strain was highly similar between NPC patients and normal controls, in which variant 2 accounted for ~70% in our population. Another variant (variant 3) with a three-nucleotide polymorphism to variant 2 accounted for 24% in our locality (Figure 3A). EBV nucleotide variation positioned distal from the hairpin structure was previously reported to affect miR-BART processing [23].

Figure 3. Nucleotide polymorphism of vsoRNA1 flanking sequence: (**A**) The three nucleotide polymorphism patterns (V1–V3) positioned distal from the 3' end of v-snoRNA1 were identified. The sequence shown is corresponded to 152925:152985 (AJ507799.2); (**B**) Expression level of v-snoRNA1 was analyzed by Northern blot using RNA from 293FT cells transiently transfected with expression vectors containing different v-snoRNA1 variants as indicated (AJ507799.2, 152761:153032); (**C**) *In vitro* Dicer and Drosha/DGCR8 processing on the long v-snoRNA1 containing fragment. The digestion products were visualized on 10% PAGE (*left panel*) and analyzed by Northern blot with v-snoRNA24pp complementary oligonucleotide probe (*right panel*). The signal of v-snoRNA24pp is showed as "*". The representative result is shown.

A

```
v-snoRNA1-3'CUGAUGAG--  ----------  ----------  ----------  ----------  ----------
B95.8 (V1)  CUGAUGAGAG  GCUUGGGGGG  GUCUUUGCCC  UCAAGGCGAG  GCUCCUUCUC  CUAGGAAUGC
C666-1 (V2) CUGAUGAGAG  GCUUGGGGGG  -UCUUUGCCC  UCAAGGUGAG  GCUCCUUCU-  --AGGAAUGC
NPC (V3)    CUGAUGAGAG  GCUUAAGGGG  -UCUUUGCCC  UCAAGGUGAG  GCUCCUCCU-  --AGGAATGC
```

	EBV Strain			
	Normal Control	NPC Patient	Overall	Tested EBV positive sample
Variant 1	1 (7%)	2 (6%)	3 (7%)	Cell lines: B95.8 (Prototype EBV), BC-1 (B-cell)
Variant 2	9 (64%)	22 (71%)	31 (69%)	Cell lines: Akata (B-cell), C666-1 (NPC), CB14022 (LCL), Jiyoye (B-cell); Local NPC xenograft: X666; X1915; X2117; X99186
Variant 3	4 (29%)	7 (23%)	11 (24%)	
Total	14	31	45	

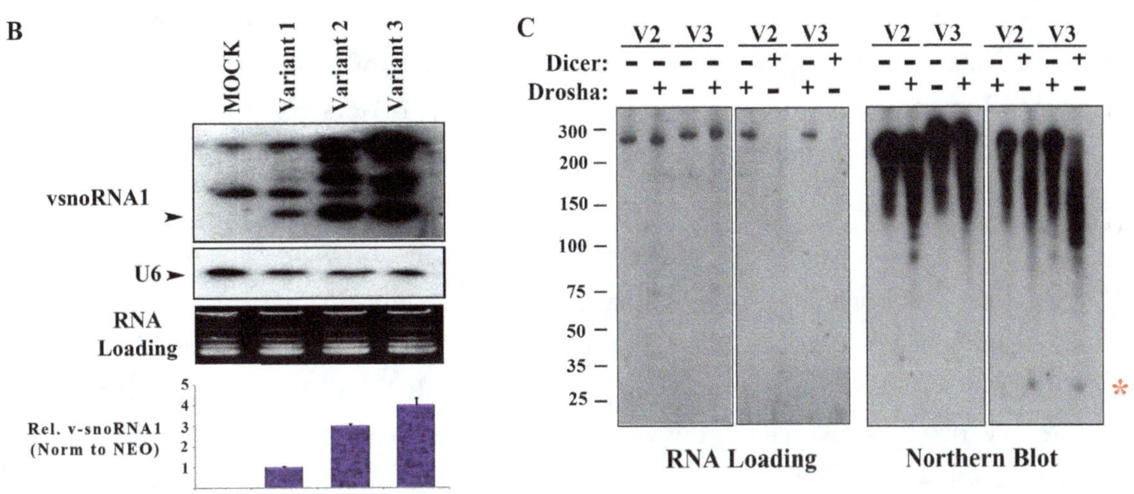

To investigate whether the common variant would affect v-snoRNA1 production, we cloned different variants of v-snoRNA1 containing fragment (~270 nt) into the pcDNA 3.1 (+) expression vector and transiently expressed them into 293 cells. Subsequent Northern blot analysis on the RNA extracts revealed that both fragments from variants 2 and 3 displayed higher expression levels of v-snoRNA1 than the prototype EBV variant 1 (Figure 3B). Surprisingly, the predicted proximal RNA hairpin structures among these three variants are extremely similar to each other (Figure S3). In this sense, the Drosha/DGCR8 complex, which chopped off the hairpin structure (premature miRNA) from the long RNA transcripts, might not be responsible for viral v-snoRNA1 biogenesis. We further substantiated this hypothesis by performing additional *in vitro* enzymatic digestion on T7 transcribed RNA fragments from expression vectors that disabled Drosha/DGCR8 complex's activity to generate v-snoRNA1 [23,94] (Figure 3C). Thus, the v-snoRNA1 or v-snoRNA^{24pp} biogenesis pathway might be dependent on other microprocessor complexes such as the Integrator complex, which was previously proved to take part in the viral pre-miRNA processing in primate herpesvirus [95]. On the other hand, we demonstrated that Dicer activity alone was sufficient to process the *in vitro* transcribed long RNA fragment into v-snoRNA1^{24pp}. Hence, v-snoRNA^{24pp} may be directly produced from v-snoRNA1 or it longer precursor with the help of Dicer activity (Figure 3C).

6. Functional Roles of miR-BARTs in Cancer Development

6.1. Methods for miRNA Targets Identification

Viral miRNA is expected to contribute to cancer development by silencing both the viral and cellular target transcripts in the post-transcriptional level. Thus, identification and validation of the downstream targets of miR-BARTs are critical to understand how EBV is involved in tumorigenesis. It is generally accepted that a perfect match between the miRNA seed sequence (nucleotides 2–8) and the target 3'UTRs is important in the interaction. However, the exact parameter to determine miRNA target recognition is not fully characterized. As a result, a huge amount of false positive targets are predicted by using *in silico* miRNA target prediction method. To avoid considering the elusive universal rule for target site binding, high throughput sequencing of RNAs isolated by cross-linking and immunoprecipitaion (HITS-CLIP) [96] and photoactivatable ribonucleoside enhanced CLIP (PAR-CLIP) [97,98] methods have recently been employed to identify the endogenous miR-BARTs associated targets [99–101]. These biochemical methods have been reported to identify a hundred of miR-BART cellular targets, indicating the heavy involvement of EBV in regulating the cellular transcriptome. Remarkably, most of the known targets of miR-BARTs are mainly responsible for the control of viral latency, host cell immunity and apoptosis [100,101].

6.2. Regulation of Viral Gene Expression

Although target identification is the most challenging part in miRNA study, the targets of some viral miRNAs are easy to identify because they are transcribed as an antisense sequence of the viral genes. The first EBV miRNA identified to target viral product was miR-BART2-5p, which was transcribed in the opposite direction to the 3'UTRs of the viral DNA polymerase gene, *BALF5* [73]. With the perfect complementarity between miR-BART2-5p and BALF5 transcript, miR-BART2-5p was finally proved to down-regulate BALF5 expression via siRNA cleavage at the binding site [102]. Similarly, vsnoRNA1^{24pp}, positioned distal from miR-BART2 and fully complementary to the BALF5 3'UTR, was indirectly proved to down-regulate BALF5 expression by cleaving the transcript [76]. Being a critical protein for DNA replication, down-regulation of BALF5 may facilitate EBV to enter latency.

Apart from the direct cleavage mechanism, EBV-miRNAs also regulate viral gene production by associating with their 3'UTRs. It had been revealed by our previous work in NPC that expression of LMP1 and LMP2A was downregulated by three BART cluster 1 miRNAs (miR-BART1-5p, 16 and 17-5p) and BART22, respectively [23,103]. Although both LMP1 and LMP2A can promote cell proliferation, migration and invasion, high expression of these two proteins in the infected cells may have destructive effects. LMP2A is the most immunogenic viral protein in latency I and II infection, and circulating T-cells specific for LMP2A derived epitopes are frequently detected in healthy individuals [104,105]. Limiting LMP2A expression by miR-BART22 would have potential advantages for the cells to escape host immune surveillance. On the other hand, high LMP1 expression can suppress cell growth and guide the cells to commit apoptosis in response to several stimuli like TNF treatment and serum depletion [106,107]. Thus, LMP1 expression in cancer should be tightly regulated in order to maintain the balance between its growth promoting and suppressing effects [103]. Fascinatingly, more miR-BARTs have been identified to regulate LMP1 expression in lymphoma cell lines. Ranakrishnan *et al.* showed that miR-BART9 from the NK/T-cell lymphoma cell line SNK6 could increase viral LMP1 expression. This might be the key mechanism of EBV to promote cell growth [108]. However, the mechanism behind this regulation was not clearly demonstrated. Using Ago HITS-CLIP direct biochemical assay in EBV-transformed B cells (Jiyoye cells), Riley *et al.* further recognized that both miR-BART5 and miR-BART19-5p could down-regulate LMP1 expression [101]. While miR-BART5 and hsa-miR-18-5p share the same seed sequence, hsa-miR-18-5p is unable to target LMP1 [101]. This suggested that additional factors, in addition to the perfect Watson-Crick complementary between target 3'UTR and miRNA seed sequence, should also be considered for miRNA suppressive activity [23]. Riley *et al.* further uncovered that miR-BART10-3p could work together with miR-142-3p and miR-17-5p to co-repress viral anti-apoptotic gene *BHRF1* in latency III EBV infection cells. The means to target *BHRF1* was unknown, but blocking these miRNAs individually in Jiyoye cells definitely increased the cellular apoptotic rate [101]. Table 1 lists the confirmed viral targets of miR-BARTs.

Table 1. Function of EBV miRNAs in viral targets.

miRNA	Target	Function	miRNA effect	References
v-snoRNA^{24pp}	BALF5	DNA polymerase	Maintain latency	[76]
BART2-5p	BALF5	DNA polymerase	Maintain latency	[73,102]
BART1-5p/BART16/BART17-5p	LMP1	Transforming factor	Anti-apoptosis	[103]
BART19-5p/BART5	LMP1	Transforming factor	Immune evasion;	[101]
BART9	LMP1	Transforming factor	Promote cell growth (NK/T cells)	[108]
BART22	LMP2A	Signaling molecule	Immune evasion	[23]
BART10-3p	BHRF1	Homolog of Bcl2	Unknown	[101]

6.3. Regulation of Cellular Gene Expression

As a successful pathogen, EBV can develop life-long latency in the host with subsequent reactivation. For this reason, EBV needs to protect itself by tightly controlling viral gene expression and regulating host signaling cascades. It is not surprising to learn that miR-BARTs are the key regulators for host gene expression because only a few of viral proteins are expressed in carcinomas.

Until now, multiple lines of evidence have suggested that miR-BARTs target cellular genes mainly for preventing apoptosis and escaping the host immune system. The p53-up-regulated modulator of apoptosis (PUMA) is the first recognized cellular target of miR-BARTs and was targeted by miR-BART5 via the interaction on its 3'UTR [93]. The PUMA is one of the six members of the BH3-only group in the Bcl2 family. The BH3-only proteins, the essential initiators of apoptosis, are responsible to control the release of cytochrome C from the mitochondrial inter-membrane [109]. Although PUMA is the immediate downstream target of the well-known tumor suppressor gene *p53*, it functions as a key pro-apoptotic protein under the influence of the p53 tumor suppressor and other apoptotic stimuli. With apoptotic stimulation, PUMA is believed to interact with Bcl-2 family proteins and helps to mediate mitochondrial dysfunction, followed by the activation of the caspase cascade and eventually induces apoptosis. The interaction of miR-BART5 and PUMA on cell apoptotic effects was clearly demonstrated in EBV-positive NPC cell lines, in which depletion of miR-BART5 or ectopic expression of PUMA could make the cells more susceptible to apoptosis-inducing drugs [93].

Another BH3-only group protein, Bcl-2 interacting mediator of cell death (Bim), was also shown to be the target of the multiple miRNAs encoded in the BART cluster 1 [110]. Interestingly, Bim's 3'UTR is not responsive to any of the individual miR-BARTs in stable transfectants. This observation indicates the importance of the co-operation of miR-BARTs in cluster 1 for Bim expression. In another study on EBV positive NK/T cell lymphoma, Lin *et al.* demonstrated by transient transfection and Western blot analysis that miR-BART20-5p could downregulate the T-box transcription factor TBX21 expression via the binding site located on the 3'UTR of the target mRNA [111]. They further revealed that inhibition of the TBX21 activity indirectly suppressed p53 function in NK/T lymphoma cell lines. This finding underlined the significance of miR-BART20-5p on preventing p53 dependent apoptosis. However, this miRNA is generally weakly expressed in epithelial cancers (Figure 1).

By using direct co-immunoprecipitated miRNA/mRNA containing RISC method, several studies identified a number of putative anti-apoptotic targets regulated by miR-BARTs. TOMM22 was identified and validated as a potential target of miR-BART16 [112]. This protein is a part of the mitochondrial membrane receptor involved in the association of the pro-apoptotic protein, Bcl-2-associated X (Bax). Treatment with antibodies against TOMM22 or siRNA knockdown has been shown to inhibit the association of Bax protein on mitochondria, thus preventing Bax-dependent apoptosis [113]. In Burkitt's lymphoma, BART miRNAs were found to exert anti-apoptotic effects by repressing CASP3 expression, in which the co-suppressive effects of miR-BART1-3p and -16 on CASP3 expression was confirmed by reporter assay [114]. Riley *et al.* also reported in HITS-CLIP result that both viral miR-BART13-3p and host miR-17 could inhibit Caprin-2 translation [101]. Despite Caprin-2's ability to promote activation of the oncogenic Wnt signaling pathway [115], over-expression of this protein would drastically induce apoptosis [116].

Surprisingly, not all miR-BARTs promote cell growth or inhibit apoptosis. Choi *et al.* demonstrated the tumor suppressive effect of miR-BART15 in gastric cancer [117]. The BRUCE protein is a member of the inhibitor of apoptosis (IAP) family that protects cells from apoptosis by ubiquitin-dependent degradation of caspase protein [118,119]. It was indicated in transient transfection assay that miR-BART15 could suppress cell growth and induced early apoptosis in AGS gastric carcinoma cell in part through targeting the *BRUCE* gene. By carrying a similar "seed sequence" with miR-223, miR-BART15 was also reported to repress NLRP3 expression via the miR-223 putative binding sites on the 3'UTR of NLRP3 [120]. NLRP3 is a NOD-like receptor protein that forms a multi-protein complex called the inflammasome. In response to pathogen invasion, NLRP3-inflammasome inside the cells can be activated to trigger the secretion of interleukin-1β (IL-1β), finally resulting in the stimulation of host inflammatory response. Thus, the presence of miR-BART15 may protect the infected cells by functioning as a NLRP3 blocker. Moreover, miR-223 has been proved as a tumor suppressive miRNA in other epithelial cancers. For example, it can suppress oncogenic protein (stathmin1) expression in liver and gastric carcinomas [121,122]. Yet, the effect of miR-BART15 on the same oncogenic proteins has not been evaluated. Further studies may be needed to uncover the complexity of the miRNA network in carcinogenesis.

Recently, several miR-BARTs in infected cells were selectively secreted as an exosome and transported to the adjacent cells to exert their suppressive function [120,123,124]. In addition, the exosomal miR-BART15 from EBV-infected B cells was found to be sufficient for the inhibition activity of the NLRP3 inflammasome in the PMA-differentiated macrophage cell line THP-1 [120]. In this sense, miR-BART15 might indirectly contribute to cancer development by exercising its "anti-cancer" activities in the adjacent immune cells [117].

Besides BART15, two more miR-BARTs were also reported to augment cancer progression by evading host immunity system. For example, EBV miR-BART2-5p has been shown to regulate the stress-induced *MICB* gene [125]. Since MICB is a key cell surface ligand for natural killer (NK) cells recognition, down-regulation of MICB expression may allow them to escape direct killing by NK cells. The innate immunity related gene *importin 7* (*IPO7*) was a putative target recognized by miR-BART1-3p and BART3 [112,114,126]. The *importin 7* gene encodes a critical receptor protein responsible for importing transcription factors into the nucleus [127]. It is believed that IPO7 is involved in the innate immunity system because knocking down of this gene in macrophage reduces pro-inflammatory cytokine IL-6 secretion [112]. The function of IPO7 in epithelial cells is largely unknown. Therefore, miR-BART3 may mainly exert its function in neighboring immune related cells via exosome.

Aside from genes implicated in apoptosis and immune evasion, a few studies independently reported the function of miR-BARTs on regulating tumor suppressor genes and the gene responsible for latency maintenance. Although the direct interaction between miRNAs and the corresponding target sites were not assessed, Wong *et al.* identified in a transient transfection assay that miR-BARTs could extensively down-regulate the expression of several well-known Wnt inhibitory molecules, such as *WIF1* (miR-BART19-3p), *APC* (miR-BART19-3p, 7 and 17-5p) and *NLK* (miR-BART19-3p, 14 and 18-5p) [92]. On the other hand, miR-BART3* has been showed to down-regulate the expression of tumor suppressive gene *DICE1* in NPC cells [128]. Since inactivation of *DICE1* occurs frequently in non-EBV associated solid tumors by either allelic deletion or CpG promoter hypermethylation [129–131], this finding has highlighted the importance of miR-BART3 to downregulate DICE1 activity in EBV related carcinogenesis.

It has been shown in primary B cells that EBV infection could globally suppress the expression of cellular miRNAs, indicating the presence of strong miRNA repressors within the EBV genome [132]. The miR-BART6-5p may be one of such EBV-derived repressor since it can inhibit the expression of human Dicer, a critical enzyme for miRNA biogenesis, via four binding sites within the 3'UTR [133]. In addition, knocking down miR-BART6-5p activity in EBV positive cell lines with latency I or II program can enhance Dicer expression and indirectly re-activate the expression of both lytic and highly immunogenic viral proteins such as Zta, Rta and EBNA2 [133]. These findings suggested that the key function of miR-BART6-5p might assist infected cells to maintain latency.

7. Conclusions and Future Perspectives

In NPC and EBVaGC, the clonal EBV genome is consistently detected in all invasive carcinoma and high-grade dysplastic lesions. This observation underscores the importance of viral infection in the development of epithelial cancers. In this context, EBV should have the ability to amend several cellular signaling pathways for both host cell survival and to facilitate tumor progression. In contrast to the low viral protein levels, functional BART miRNAs are generally highly expressed in a number of epithelial cancers (Figure 1). Although the regulatory mechanism by which miRNAs affect tumor progression is not fully understood, some viral miRNAs have already been recognized to regulate expression of several key cancer-related proteins, including those responsible for facilitating latency maintenance, immune suppression and tumor promotion (Tables 1 and 2). For this reason, suppressing oncogenic viral miRNAs (viral onco-miR) activity in the cancer cell may provide a therapeutic benefit for patients.

Table 2. Function of EBV miRNAs in cellular target.

miRNA	Target	Function	miRNA effect	References
BART5-5p	*PUMA*	Proapoptotic protein	Anti-apoptosis	[93]
BARTs cluster 1	*Bim*	Proapoptotic protein	Anti-apoptosis	[110]
BART20-5p	*T-bet* (*TBX21*)	T-box transcription factor	Anti-apoptosis	[111]
BART16	*TOMM22*	Mitochondrial receptor for proapoptotic protein BAX	Anti-apoptosis	[112]
BART1-3p + BART16	*CASP3*	Proapoptotic protein	Anti-apoptosis	[114]
BART13-3p	*CAPRIN2*	Wnt-signaling enhancer	Anti-apoptosis	[101]
BART15	*BRUCE*	Anti-apoptotic protein		[117]
BART15	*NLPR3*	Regulation of inflammation	Immune evasion	[120]
BART2-5p	*MICB*	NK cell ligand	Immune evasion	[125]
BART3	*IPO7*	Nuclear importer protein	Immune evasion	[112]
BART3 + BART16	*IPO7*	Nuclear importer protein	Immune evasion	[114]
BART19-3p	*WIF1*	Wnt inhibitor	Oncogenic properties	[92]
BART19-3p, 7 and 17-5p	*APC*	Wnt inhibitor	Oncogenic properties	[92]
BART19-3p, 14 and 18-5p	*NLK*	Wnt inhibitor	Oncogenic properties	[92]
BART3*	*DICE1*	Tumor suppressor	Oncogenic properties	[128]
BART6-5p	*DICER*	miRNA biogenesis	Unknown	[133]

After the discovery of the first human miRNA 10 year ago, only the antagomir against miR-122 (SPC3649, Miravirsen) can enter into Phase II clinical trials for Hepatitis C treatment [134,135]. However, this therapeutic strategy is being continuously developed and improved, especially on the issues of delivery system and inhibitory efficiency of the therapeutic particles [136,137]. Currently, the schemes for targeting miR-BARTs are reviewed in great detail by Lo AK *et al.* [138]. In theory, EBV miRNAs are exogenous genes solely expressed in cancer cells, and display low sequence homology to known human miRNAs [139]. Advantage would be given to the development of miR-BART targeting therapeutic approach because of the tissue specific delivery and the potential off-target effect of the designed bullet. However, selecting suitable viral onco-miR candidates for targeting therapy is challenging. In this review, we revealed that the miR-BARTs expression patterns of four common

EBV-associated epithelial cancers are highly similar (Figure 1). This observation has suggested that most EBV-positive epithelial cancers employ similar regulatory mechanisms for controlling ebv-miRNA expression, thus a single formula of miRNA inhibitor may be possible for the treatment of those malignancies. Nevertheless, our understanding of viral miRNA function is still far from complete. Despite that several important cellular viral miRNA targets have been identified, their interactions were solely substantiated in cell culture experiments in which non-physiological levels of miRNAs were used. In addition, a single miRNA molecule might be possible to regulate numerous signaling cascades in response to different cellular environments. Therefore, all the identified miRNA targets should be tested in *in vivo* experiments. We believe that a comprehensive understanding of miR-BART function and expression pattern in epithelial cancers is crucial in designing effective therapeutic regimens to combat EBV-associated malignancies.

The involvement of EBV miRNAs on cellular interactions is even more complicated. Currently, certain EBV miRNAs were shown to be actively secreted from infected cells and transported to neighboring cells through a group of small endosomal-derived vesicles called "exosomes" [124,140–142]. It had been demonstrated in *in vitro* and *in vivo* experiments that secretory exosomal ebv-miRNAs can be internalized and exhibit biological functions in the surrounding uninfected recipient cells such as primary monocyte-derived dendritic cells (MoDC) [123,124,143,144]. Thus, EBV seems to utilize the exosome as a vehicle to extend their miRNA function on the neighboring cells for immune evasion. Being protected from enzymatic degradation by membrane bound microvesicles, exosomal miRNAs are efficiently recovered from biological fluids such as plasma, pleural fluid or sputum. This characteristic has lead scientists to consider exosomal viral miRNA as a non-invasive biomarker. However, the study of exosomal miRNAs is only at the initial stage, and the underlying mechanism to regulate its secretion remains largely unclear. For example, circulating miR-BART17 was consistently detected in plasma and serum samples from NPC patients [92,141,145], yet, the reason why the same miRNA was not enriched in the exosomal fraction of the same patients was unknown [145]. In addition, the exosomal miRNA content in hepatocytes has been reported to vary between different pathological and physiological conditions [146]. Whether these conditions may affect the composition of exosomal viral miRNAs in EBV positive cancer patients is not characterized. To this end, the present information is not sufficient to support the use of exosomal content as a biomarker. Nonetheless, Arroyo *et al.* found that the majority of circulating plasma miRNAs was protected by Ago2 complexes [147]; it would be interesting to know if the HITS-CLIP method can be used for biomarker identification.

Despite the uncertainty of the nature of exosomal miRNAs, some EBV miRNAs exhibit a higher expression in the blood from NPC patients compared to non-NPC controls [92,145,148]. The preliminary work from Courzones and colleagues further revealed that the plasma concentration of miR-BART17 in NPC patients is partly dependent on the tumor mass [145]. Surprisingly, there seemed to be no correlation between the concentration of plasma viral miRNAs and EBV DNA load in NPC patients [145,148]. While cell-free EBV DNA represents an effective biomarker for NPC diagnosis and prognosis, incorporating miR-BARTs in the analysis may provide additional information to enhance the assessment of the tumor status. We envisage that additional experimental data would warrant a more solid conclusion on this issue.

Acknowledgments

This research was supported by the Collaborative Research Fund from the Hong Kong Research Grants Council (CUHK04/CRF/08) and RFCID (11100162). The authors thank Elaine S. Chan (CUHK) and Y.K. Tong (CUHK) for critical reading of the manuscript.

Appendix

Materials and Methods

Patient Samples

All specimens in this study were recruited at the Prince of Wales Hospital, Hong Kong. They are included frozen biopsy tissues of 21 NPCs, one EBVaGC and two LELC of lungs, formalin fixed paraffin embedded (FFPE) specimens from 9 cases of EBVaGC, 7 cases of LELC of lung and 7 cases of LELC of CCA. The present of EBV in the samples were confirmed by EBV-encoded small RNA (EBER) *in situ* hybridization (ISH) on the paraffin-embedded tissue sectioned. All participants provided written informed consent for the collection of samples and subsequent analysis. Ethics approval was obtained from the Joint Chinese University of Hong Kong-New Territories East Cluster Clinical Research Ethics Committee, Hong Kong.

RNA Extraction and Quantitative Reverse Transcription PCR (QRT-PCR)

Total RNA from biopsies and FFPE samples were prepared by TRIzol reagent (Invitrogen, Carlsbad, CA, USA) and Recover All Total nucleic Acid Isolation kit (Ambion Inc, Austin, TX, USA) according to the manufacturer's recommendations respectively. The concentration and quality of all RNA was determined by using a NanoDrop-1000 UV-VIS Spectrophotometer (NanoDrop Company, Wilmington, DE, USA). One microgram of total RNA was reverse-transcribed in 20 μL of reaction with miScript Reversed Transcription Kit (Qiagen, Hilden, Germany) and diluted 5-fold for individual QRT-PCR assay. MiScript SYBR green PCR kit (Qiagen, Hilden, Germany) with kit provided universal primer and in-house designed miRNA-specific forward primers were used for QRT-PCR reaction with the following thermal cycling condition: 95 °C for 15 min, followed by 40 cycles at 94 °C for 15 s, annealing with specific temperature for 30 s and 70 °C for 30 s. The PCR was carried out in LightCycler® 480 Real-Time PCR System (Roche Diagnostic) and the crossing point (Cp) was analyzed by second derivative method. To eliminate the non-specific SYBR green signal, PCR result will be normalized with the signal generated by EBV negative epithelial cell line NP69. Standard curves were included in each experiment for absolute quantification of each miRNA per nanogram of total RNA using serially-diluted synthetic DNA-RNA chimeric oligonucleotides (Integrated DNA Technologies, Coralville, IA, USA) with mature miRNA sequences as RNA template (from 10^7 to 10^3 copies; 10 μM = 6.02×10^{12} copies per micro-liter). It had been demonstrated by other team that standard curve generated from single-stranded oligonucleotides is similar to that obtained by using T7-transcribed RNA [149]. Analysis of each sample was performed in triplicate. The oligonucleotide sequences and specific annealing temperature used for QRT-PCR are shown in Table S1.

Expression of BHRF1 and Northern Blot

Expression of BHRF1 mRNA was analyzed by an RT-PCR based assay as previous described [150]. In brief, 1 µg of RNA was reversed transcribed and used to perform PCR reaction with BHRF1 specific primers (5'-GTC AAG GTT TCG TCT GTG TG-3' and 5'-TTC TCT TGC TGC TAG CTC CA-3') and actin specific primers (5'-TAA GGA GAA GCT GTG CTA CGT C-3' and 5'-GGA GTT GAA GGT AGT TTC GTG G-3). PCR products were analyzed in 2% agarose gel and transferred onto Nytran SPC membrane (Whatman Inc, Piscataway, NJ) for Southern blot analysis using γ-32P ATP end-labeled synthetic oligonucleotides complementary to the lytic form of BHRF1 (5'-ATG CAC ACG ACT GTC CCG TAT ACA C-3'). Northern blot was performed as described previously [23,122]. The sequence of synthetic oligonucleotides used as probes for EBER1-5p and 3p were EBER1-5p: 5'-TAG GGC AGC GTA GGT CCT-3' and EBER1-3p: 5'-AAA CAT GCG GAC CAC CAG CTG G-3'. As a positive control, 5 µg of RNA from EBV-positive marmoset cells line B95-8 was included for Northern blot analysis.

AGO2 Co-Immunoprecipitation

The AGO2 co-immunoprecipitation method was described previously [94].

Induction of Viral Lytic Cycle in NPC Cell Lines

C666-1 cells were treated with 100 ng/mL of Gemcitabine (GEM, Eli Lilly and Company, IN, USA) and 100 ng/mL Trichostatin A (TSA, Sigma-Aldrich, Saint Louis, MO, USA) for 48 days. The induction of viral lytic cycle in C666-1 cells was confirmed by Western blot analysis using antibodies targeting early viral lytic genes, anti-EA-D (BMRF1, Millipore Corp, Billerica, MA, USA), anti-EBV transcription factor R (BRLF1, Argene, Verniolle, France) and anti-EBV-ZEBRA (Argene, Verniolle, France).

Plasmid Constructs

The v-snoRNA1 expression vector were made by inserting the PCR fragments that contains v-snoRNA1 flanking sequence (AJ507799.2, 152761:153032) into pcDNA3.1 expression vector through *Hind III* and *Xba I* sites. The PCR products were amplified using genomic DNA from B95.8 (variant 1), C666-1 (variant 2) and one of the NPC biopsy (variant 3) as template. The sequences of the primes used for the PCR are F: GCC AAG CTT GCC CTT GCG TGT C and R: GGC TCT AGA AGG CTG GCA AAG ATC.

In Vitro Processing Assay

In vitro Drosha/DGCR8 and Dicer digestion assay was described previously [23].

Table S1. Oligonucleotide sequences and the specific annealing temperature used for QRT-PCR.

miRNA	Specific forward primer	Chimeric miRNA mimic	Annealing temp.
BART1-3p	AGCACCGCTATCCACTATGT	TAGCACCGCTATCCACTATrGrUrC	55
BART1-5p	TCTTAGTGGAAGTGACGTGCT	TCTTAGTGGAAGTGACGTGCTrGrUrG	60
BART2-3p	AGGAGCGATTTGGAGAAAATAA	AAGGAGCGATTTGGAGAAAATrArArA	60
BART2-5p	TATTTTCTGCATTCGCCCTTGC	TATTTTCTGCATTCGCCCTrUrGrC	60
BART3	CACCACTAGTCACCAGGTGT	CGCACCACTAGTCACCAGGrUrGrU	60
BART4	ACCTGATGCTGCTGGTGTGC	GACCTGATGCTGCTGGTGTrGrCrU	64
BART5	AAGGTGAATATAGCTGCCCAT	CAAGGTGAATATAGCTGCCCArUrCrG	55
BART6-3p	GGGATCGGACTAGCCTTAGA	CGGGGATCGGACTAGCCTTrArGrA	55
BART6-5p	TAAGGTTGGTCCAATCCATAGG	TAAGGTTGGTCCAATCCATrArGrG	55
BART7	CATCATAGTCCAGTGTCCAGGG	CATCATAGTCCAGTGTCCArGrGrG	60
BART8	TACGGTTTCCTAGATTGTACAG	TACGGTTTCCTAGATTGTArCrArG	55
BART9	TAACACTTCATGGGTCCCGTAGT	TAACACTTCATGGGTCCCGTrArGrU	55
BART10	TACATAACCATGGAGTTGGCTGT	TACATAACCATGGAGTTGGCrUrGrU	60
BART11-3p	ACGCACACCAGGCTGACTG	ACGCACACCAGGCTGACTrGrCrC	62
BART11-5p	AGACAGTTTGGTGCGCTAGT	TCAGACAGTTTGGTGCGCTAGrUrUrG	55
BART12	CTGTGGTGTTTGGTGTGGTT	TCCTGTGGTGTTTGGTGTGrGrUrU	62
BART13	ACACTCCAGCTGGGTGTAACTTGC CAGGGA	TGTAACTTGCCAGGGACGGCrUrGrA	62
BART14	TAAATGCTGCAGTAGTAGGGA	TAAATGCTGCAGTAGTAGGrGrArU	55
BART15	TCAGTGGTTTTGTTTCCTTGA	GTCAGTGGTTTTGTTTCCTrUrGrA	60
BART16	TAGATAGAGTGGGTGTGTGCTC	TTAGATAGAGTGGGTGTGTGCrUrCrU	64
BART17-3p	GTATGCCTGGTGTTCCCCTTA	TGTATGCCTGGTGTCCCCTTrArGrU	60
BART17-5p	TAAGAGGACGCAGGCATACA	TAAGAGGACGCAGCATACrArArG	55
BART18-3p	TATCGGAAGTTTGGGCTTCGT	TATCGGAAGTTTGGGCTTCrGrUrC	60
BART18-5p	TCAAGTTCGCACTTCCTATAC	TCAAGTTCGCACTTCCTATrArCrA	55
BART19-3p	ACACTCCAGCTGGGUUUUGUUUG CUUGGGA	TTTTGTTTGCTTGGGAATrGrCrU	55
BART19-5p	ACATTCCCCGCAAACATGACAT	ACATTCCCCGCAAACATGACrArUrG	55
BART20-3p	CATGAAGGCACAGCCTGTTAC	CATGAAGGCACAGCCTGTTrArCrC	60
BART20-5p	TAGCAGGCATGTCTTCATTCC	TAGCAGGCATGTCTTCATrUrCrC	60
BART21-5p	CACTAGTGAAGGCAACTAAC	TCACTAGTGAAGGCAACTrArArC	55
BART22	TTACAAAGTCATGGTCTAGTAGT	TTACAAAGTCATGGTCTAGTrArGrU	55

Figure S1. Lack of association between novel EBER end terminal fragments and AGO2: (**A**) Expression of small EBER fragments in NPCs was validated by Northern blot analysis. Locations of the EBER1-5p and -3p fragments are indicated with arrow; (**B**) Western blot analysis of hAGO2 protein from C666-1 whole cell lysate before (Input) and after immunoprecipitation (IP) with anti-AGO2 antibody (*left panel*). Monoclonal antibody against FLAG tag (lane 2) and against hAGO2 4 (lane 3) were used for IP from C666-1 cell lysate (lane 1). AGO-2 associated miRNAs were analyzed by Northern blot analysis with synthetic oligonucleotide probe for the specific small EBV fragments listed below the figure. Locations of the probed fragments are indicated with arrows. RNA loading was visualized by SYBR Gold stained PAGE. BART22 IP was included as positive control. The information of EBER1-5p and 3p were described in a previous publication [23].

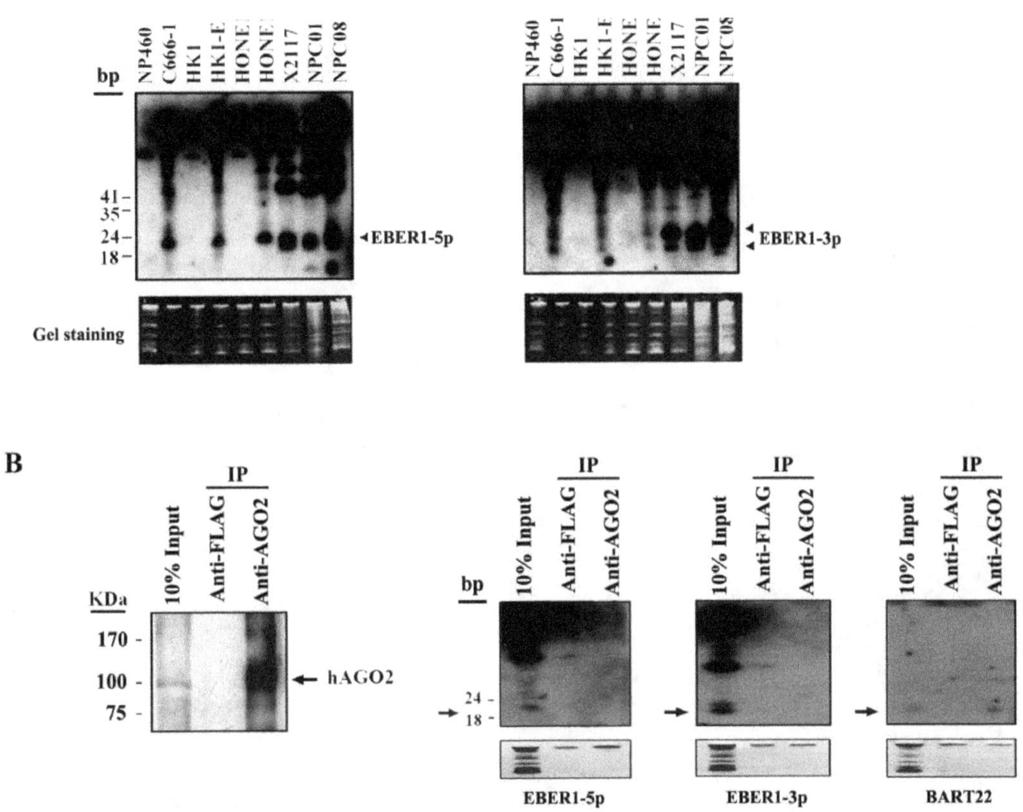

Figure S2. Absence of miR-BHRF1s expression in epithelial cancers: (**A**) Expression of BHRF1 mRNA transcripts in NPC cell lines (lanes 2–6), B-cell cell lines (lanes 7–10), NPC xenografts (lanes 11–12) and NPC biopsies (lanes 13–16) was analyzed by RT-PCR analysis. The RT-PCR products were separated on agarose gel (*left upper panel*) and transferred to a membrane for detection by Southern blot (*left middle panel*). Expression of actin in RT-PCR was used as a control (*left lower panel*). Representative Northern blot results showed the expression of ebv-miRNAs in different EBV positive samples (*right panel*); (**B**) Induction of EBV lytic cycles in C666-1 cells was confirmed by Western blot analysis (*left panel*). Expression of miR-BHRF1 and hsa-miR-16 was demonstrated by Northern blot analysis (*right panel*); (**C**) Representative Northern blot analysis on RNA samples from different EBV positive epithelial carcinomas was performed. The biopsies included NPC, Lymophoepithelial-like cholangiocarcinoma (LEL-CCA) and Lymphoepithelial-like carcinoma of the lung (LELC-Lung).

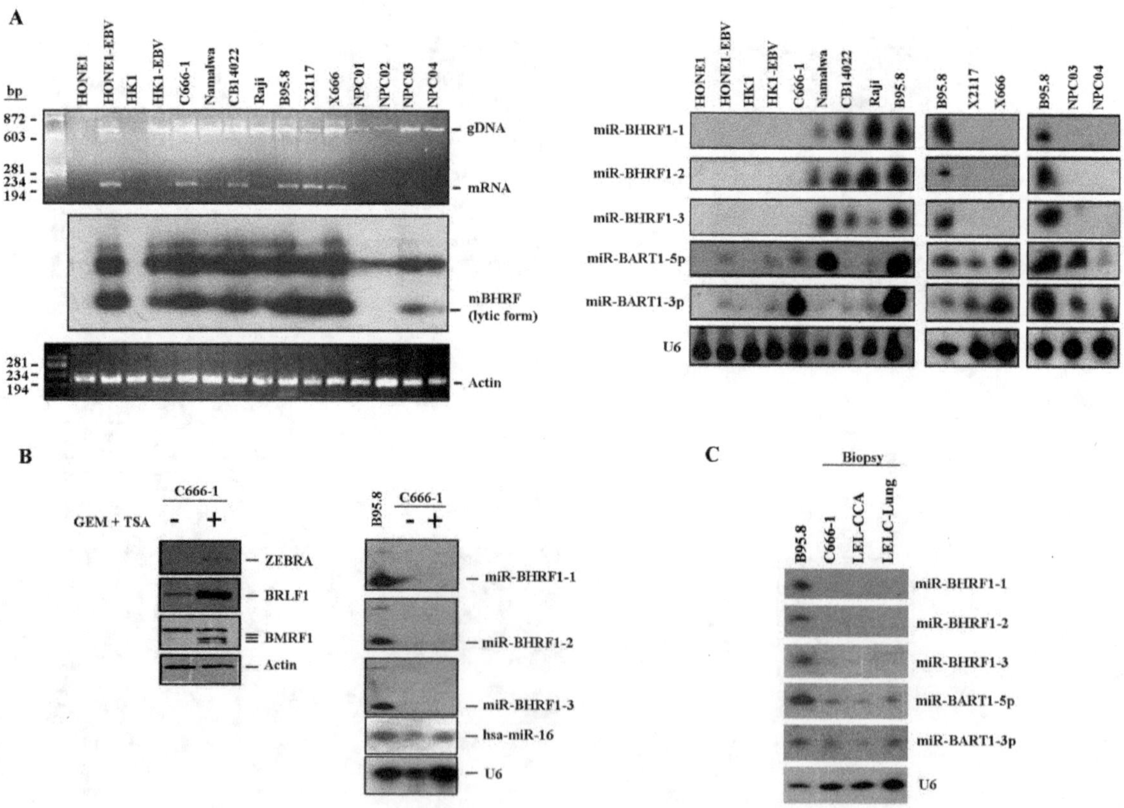

Figure S3. The predicted secondary structures of the long v-snoRNA1 containing fragment. The RNA folding of EBV fragment (AJ507799.2, 152761:153032) predicted by MFOLD are shown in B95.8-EBV (Variant 1) (*left panel*); C666-1-EBV (Variant 2) (*middle panel*) and the common EBV strain in our locality (Variant 3) (*right panel*). The folding energy (dG) with units (kcal/mol) is indicated. The location of v-snoRNA1 is indicated by sequences between red arrows.

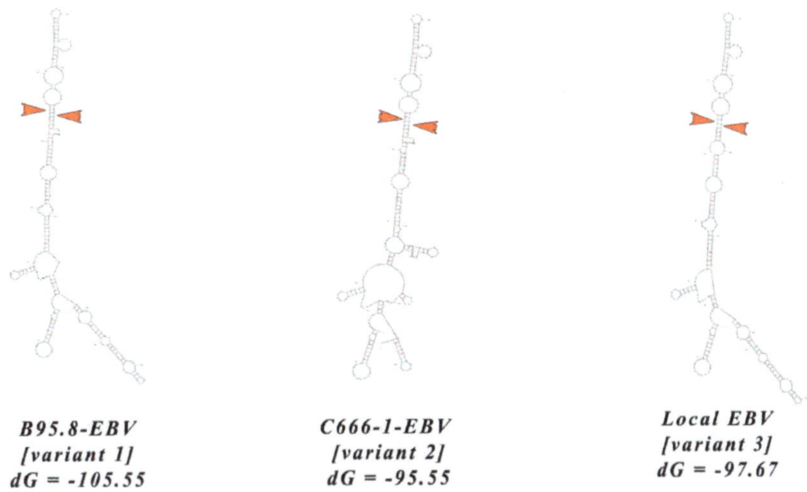

References

1. Epstein, M.A.; Barr, Y.M.; Achong, B.G. A second virus-carrying tissue culture strain (Eb2) of lymphoblasts from burkitt's lymphoma. *Pathol. Biol.* **1964**, *12*, 1233–1234.
2. Zur Hausen, H.; Schulte-Holthausen, H. Presence of EB virus nucleic acid homology in a "virus-free" line of Burkitt tumour cells. *Nature* **1970**, *227*, 245–248.
3. Sitki-Green, D.L.; Edwards, R.H.; Covington, M.M.; Raab-Traub, N. Biology of Epstein-Barr virus during infectious mononucleosis. *J. Infect. Dis.* **2004**, *189*, 483–492.
4. Kutok, J.L.; Wang, F. Spectrum of Epstein-Barr virus-associated diseases. *Annu. Rev. Pathol.* **2006**, *1*, 375–404.
5. Babcock, G.J.; Decker, L.L.; Volk, M.; Thorley-Lawson, D.A. EBV persistence in memory B cells *in vivo*. *Immunity* **1998**, *9*, 395–404.
6. Young, L.S.; Rickinson, A.B. Epstein-Barr virus: 40 years on. *Nat. Rev. Cancer* **2004**, *4*, 757–768.
7. Gregory, C.D.; Rowe, M.; Rickinson, A.B. Different Epstein-Barr virus-B cell interactions in phenotypically distinct clones of a Burkitt's lymphoma cell line. *J. Gen. Virol.* **1990**, *71*, 1481–1495.
8. Timms, J.M.; Bell, A.; Flavell, J.R.; Murray, P.G.; Rickinson, A.B.; Traverse-Glehen, A.; Berger, F.; Delecluse, H.J. Target cells of Epstein-Barr-virus (EBV)-positive post-transplant lymphoproliferative disease: Similarities to EBV-positive Hodgkin's lymphoma. *Lancet* **2003**, *361*, 217–223.
9. Brooks, L.; Yao, Q.Y.; Rickinson, A.B.; Young, L.S. Epstein-Barr virus latent gene transcription in nasopharyngeal carcinoma cells: Coexpression of EBNA1, LMP1, and LMP2 transcripts. *J. Virol.* **1992**, *66*, 2689–2697.

10. Qu, L.; Rowe, D.T. Epstein-Barr virus latent gene expression in uncultured peripheral blood lymphocytes. *J. Virol.* **1992**, *66*, 3715–3724.
11. Shaknovich, R.; Basso, K.; Bhagat, G.; Mansukhani, M.; Hatzivassiliou, G.; Murty, V.V.; Buettner, M.; Niedobitek, G.; Alobeid, B.; Cattoretti, G. Identification of rare Epstein-Barr virus infected memory B cells and plasma cells in non-monomorphic post-transplant lymphoproliferative disorders and the signature of viral signaling. *Haematologica* **2006**, *91*, 1313–1320.
12. Kieff, E.; Johannsen, E.; Calderwood, M.A. *Epstein-Barr Virus: Latency and Transformation*; Robertson, E.S., Ed.;. Caister Academic Press: Norfolk, UK, 2010; pp. 1–24.
13. Lo, K.W.; To, K.F.; Huang, D.P. Focus on nasopharyngeal carcinoma. *Cancer Cell* **2004**, *5*, 423–428.
14. Hong Kong Cancer Registry. Available online: http://www3.ha.org.hk/cancereg/Statistics.html accessed on 30 June 2013, Hong Kong Cancer Registry 2010 (accessed on 20 August 2013).
15. Marks, J.E.; Phillips, J.L.; Menck, H.R. The national cancer data base report on the relationship of race and national origin to the histology of nasopharyngeal carcinoma. *Cancer* **1998**, *83*, 582–588.
16. Imai, S.; Nishikawa, J.; Takada, K., Cell-to-cell contact as an efficient mode of Epstein-Barr virus infection of diverse human epithelial cells. *J. Virol.* **1998**, *72*, 4371–4378.
17. Iizasa, H.; Nanbo, A.; Nishikawa, J.; Jinushi, M.; Yoshiyama, H. Epstein-Barr Virus (EBV)-associated gastric carcinoma. *Viruses* **2012**, *4*, 3420–3439.
18. Raab-Traub, N. Epstein-Barr virus in the pathogenesis of NPC. *Semin. Cancer Biol.* **2002**, *12*, 431–441.
19. Dawson, C.W.; Port, R.J.; Young, L.S. The role of the EBV-encoded latent membrane proteins LMP1 and LMP2 in the pathogenesis of nasopharyngeal carcinoma (NPC). *Semin. Cancer Biol.* **2012**, *22*, 144–153.
20. Lo, A.K.; Lo, K.W.; Ko, C.W.; Young, L.S.; Dawson, C.W. Inhibition of the LKB1-AMPK pathway by the Epstein-Barr virus-encoded LMP1 promotes proliferation and transformation of human nasopharyngeal epithelial cells. *J. Pathol.* **2013**, *230*, 336–346.
21. Lee, S.P. Nasopharyngeal carcinoma and the EBV-specific T cell response: Prospects for immunotherapy. *Semin. Cancer Biol.* **2002**, *12*, 463–471.
22. Gottschalk, S.; Heslop, H.E.; Rooney, C.M. Adoptive immunotherapy for EBV-associated malignancies. *Leuk. Lymphoma* **2005**, *46*, 1–10.
23. Lung, R.W.; Tong, J.H.; Sung, Y.M.; Leung, P.S.; Ng, D.C.; Chau, S.L.; Chan, A.W.; Ng, E.K.; Lo, K.W.; To, K.F. Modulation of LMP2A expression by a newly identified Epstein-Barr virus-encoded microRNA miR-BART22. *Neoplasia* **2009**, *11*, 1174–1184.
24. Chen, S.J.; Chen, G.H.; Chen, Y.H.; Liu, C.Y.; Chang, K.P.; Chang, Y.S.; Chen, H.C. Characterization of Epstein-Barr virus miRNAome in nasopharyngeal carcinoma by deep sequencing. *PLoS One* **2010**, *5*, e12745.
25. Burke, A.P.; Yen, T.S.; Shekitka, K.M.; Sobin, L.H. Lymphoepithelial carcinoma of the stomach with Epstein-Barr virus demonstrated by polymerase chain reaction. *Mod. Pathol.* **1990**, *3*, 377–380.
26. Shibata, D.; Tokunaga, M.; Uemura, Y.; Sato, E.; Tanaka, S.; Weiss, L.M. Association of Epstein-Barr virus with undifferentiated gastric carcinomas with intense lymphoid infiltration. Lymphoepithelioma-like carcinoma. *Am. J. Pathol.* **1991**, *139*, 469–474.

27. Osato, T.; Imai, S. Epstein-Barr virus and gastric carcinoma. *Semin. Cancer Biol.* **1996**, *7*, 175–182.
28. Shibata, D.; Weiss, L.M. Epstein-Barr virus-associated gastric adenocarcinoma. *Am. J. Pathol.* **1992**, *140*, 769–774.
29. Takada, K. Epstein-Barr virus and gastric carcinoma. *Mol. Pathol.* **2000**, *53*, 255–261.
30. Lee, J.H.; Kim, S.H.; Han, S.H.; An, J.S.; Lee, E.S.; Kim, Y.S. Clinicopathological and molecular characteristics of Epstein-Barr virus-associated gastric carcinoma: A meta-analysis. *J. Gastroenterol. Hepatol.* **2009**, *24*, 354–365.
31. Murphy, G.; Pfeiffer, R.; Camargo, M.C.; Rabkin, C.S. Meta-analysis shows that prevalence of Epstein-Barr virus-positive gastric cancer differs based on sex and anatomic location. *Gastroenterology* **2009**, *137*, 824–833.
32. Chen, J.N.; He, D.; Tang, F.; Shao, C.K. Epstein-Barr virus-associated gastric carcinoma: A newly defined entity. *J. Clin. Gastroenterol.* **2012**, *46*, 262–271.
33. Imai, S.; Koizumi, S.; Sugiura, M.; Tokunaga, M.; Uemura, Y.; Yamamoto, N.; Tanaka, S.; Sato, E.; Osato, T. Gastric carcinoma: Monoclonal epithelial malignant cells expressing Epstein-Barr virus latent infection protein. *Proc. Natl. Acad. Sci. USA* **1994**, *91*, 9131–9135.
34. Luo, B.; Wang, Y.; Wang, X.F.; Liang, H.; Yan, L.P.; Huang, B.H.; Zhao, P. Expression of Epstein-Barr virus genes in EBV-associated gastric carcinomas. *World J. Gastroenterol.* **2005**, *11*, 629–633.
35. Zhao, J.; Liang, Q.; Cheung, K.F.; Kang, W.; Lung, R.W.; Tong, J.H.; To, K.F.; Sung, J.J.; Yu, J. Genome-wide identification of Epstein-Barr virus-driven promoter methylation profiles of human genes in gastric cancer cells. *Cancer* **2013**, *119*, 304–312.
36. Zhao, J.; Jin, H.; Cheung, K.F.; Tong, J.H.; Zhang, S.; Go, M.Y.; Tian, L.; Kang, W.; Leung, P.P.; Zeng, Z.; *et al.* Zinc finger E-box binding factor 1 plays a central role in regulating Epstein-Barr virus (EBV) latent-lytic switch and acts as a therapeutic target in EBV-associated gastric cancer. *Cancer* **2012**, *118*, 924–936.
37. Zhao, J.; Liang, Q.; Cheung, K.F.; Kang, W.; Dong, Y.; Lung, R.W.; Tong, J.H.; To, K.F.; Sung, J.J.; Yu, J. Somatostatin receptor 1, a novel EBV-associated CpG hypermethylated gene, contributes to the pathogenesis of EBV-associated gastric cancer. *Br. J. Cancer* **2013**, *108*, 2557–2564.
38. Kim do, N.; Chae, H.S.; Oh, S.T.; Kang, J.H.; Park, C.H.; Park, W.S.; Takada, K.; Lee, J.M.; Lee, W.K.; Lee, S.K. Expression of viral microRNAs in Epstein-Barr virus-associated gastric carcinoma. *J. Virol.* **2007**, *81*, 1033–1036.
39. Marquitz, A.R.; Mathur, A.; Shair, K.H.; Raab-Traub, N. Infection of Epstein-Barr virus in a gastric carcinoma cell line induces anchorage independence and global changes in gene expression. *Proc. Natl. Acad. Sci. USA* **2012**, *109*, 9593–9598.
40. Iezzoni, J.C.; Gaffey, M.J.; Weiss, L.M. The role of Epstein-Barr virus in lymphoepithelioma-like carcinomas. *Am. J. Clin. Pathol.* **1995**, *103*, 308–315.
41. Xiao, P.; Shi, H.; Zhang, H.; Meng, F.; Peng, J.; Ke, Z.; Wang, K.; Liu, Y.; Han, A. Epstein-Barr virus-associated intrahepatic cholangiocarcinoma bearing an intense lymphoplasmacytic infiltration. *J. Clin. Pathol.* **2012**, *65*, 570–573.
42. Hsu, J.L.; Glaser, S.L. Epstein-Barr virus-associated malignancies: Epidemiologic patterns and etiologic implications. *Crit. Rev. Oncol. Hematol.* **2000**, *34*, 27–53.
43. Hippocrate, A.; Oussaief, L.; Joab, I. Possible role of EBV in breast cancer and other unusually EBV-associated cancers. *Cancer Lett.* **2011**, *305*, 144–149.

44. Begin, L.R.; Eskandari, J.; Joncas, J.; Panasci, L. Epstein-Barr virus related lymphoepithelioma-like carcinoma of lung. *J. Surg. Oncol.* **1987**, *36*, 280–283.
45. Han, A.J.; Xiong, M.; Zong, Y.S. Association of Epstein-Barr virus with lymphoepithelioma-like carcinoma of the lung in southern China. *Am. J. Clin. Pathol.* **2000**, *114*, 220–226.
46. Lerner, M.R.; Andrews, N.C.; Miller, G.; Steitz, J.A. Two small RNAs encoded by Epstein-Barr virus and complexed with protein are precipitated by antibodies from patients with systemic lupus erythematosus. *Proc. Natl. Acad. Sci. USA* **1981**, *78*, 805–809.
47. Takada, K. Role of EBER and BARF1 in nasopharyngeal carcinoma (NPC) tumorigenesis. *Semin. Cancer Biol.* **2012**, *22*, 162–165.
48. Sample, J.; Kieff, E. Transcription of the Epstein-Barr virus genome during latency in growth-transformed lymphocytes. *J. Virol.* **1990**, *64*, 1667–1674.
49. Howe, J.G.; Shu, M.D. Epstein-Barr virus small RNA (EBER) genes: Unique transcription units that combine RNA polymerase II and III promoter elements. *Cell* **1989**, *57*, 825–834.
50. Schwemmle, M.; Clemens, M.J.; Hilse, K.; Pfeifer, K.; Troster, H.; Muller, W.E.; Bachmann, M. Localization of Epstein-Barr virus-encoded RNAs EBER-1 and EBER-2 in interphase and mitotic Burkitt lymphoma cells. *Proc. Natl. Acad. Sci. USA* **1992**, *89*, 10292–10296.
51. Gourzones, C.; Busson, P.; Raab-Traub, N. *Nasopharyngeal Carcinoma: Keys for Translational Medicine and Biology*; Busson, P., Ed.; Landes Bioscience and Springer Science: Villejuif, France, 2012; pp. 42–60.
52. Fok, V.; Mitton-Fry, R.M.; Grech, A.; Steitz, J.A. Multiple domains of EBER 1, an Epstein-Barr virus noncoding RNA, recruit human ribosomal protein L22. *RNA* **2006**, *12*, 872–882.
53. Toczyski, D.P.; Steitz, J.A. EAP, a highly conserved cellular protein associated with Epstein-Barr virus small RNAs (EBERs). *EMBO J.* **1991**, *10*, 459–466.
54. Toczyski, D.P.; Matera, A.G.; Ward, D.C.; Steitz, J.A. The Epstein-Barr virus (EBV) small RNA EBER1 binds and relocalizes ribosomal protein L22 in EBV-infected human B lymphocytes. *Proc. Natl. Acad. Sci. USA* **1994**, *91*, 3463–3467.
55. Clarke, P.A.; Schwemmle, M.; Schickinger, J.; Hilse, K.; Clemens, M.J. Binding of Epstein-Barr virus small RNA EBER-1 to the double-stranded RNA-activated protein kinase DAI. *Nucleic Acids Res.* **1991**, *19*, 243–248.
56. Samanta, M.; Iwakiri, D.; Kanda, T.; Imaizumi, T.; Takada, K. EB virus-encoded RNAs are recognized by RIG-I and activate signaling to induce type I IFN. *EMBO J.* **2006**, *25*, 4207–4214.
57. Wong, H.L.; Wang, X.; Chang, R.C.; Jin, D.Y.; Feng, H.; Wang, Q.; Lo, K.W.; Huang, D.P.; Yuen, P.W.; Takada, K.; *et al.* Stable expression of EBERs in immortalized nasopharyngeal epithelial cells confers resistance to apoptotic stress. *Mol. Carcinog.* **2005**, *44*, 92–101.
58. Iwakiri, D.; Sheen, T.S.; Chen, J.Y.; Huang, D.P.; Takada, K. Epstein-Barr virus-encoded small RNA induces insulin-like growth factor 1 and supports growth of nasopharyngeal carcinoma-derived cell lines. *Oncogene* **2005**, *24*, 1767–1773.
59. Iwakiri, D.; Eizuru, Y.; Tokunaga, M.; Takada, K. Autocrine growth of Epstein-Barr virus-positive gastric carcinoma cells mediated by an Epstein-Barr virus-encoded small RNA. *Cancer Res.* **2003**, *63*, 7062–7067.

60. Rosa, M.D.; Gottlieb, E.; Lerner, M.R.; Steitz, J.A. Striking similarities are exhibited by two small Epstein-Barr virus-encoded ribonucleic acids and the adenovirus-associated ribonucleic acids VAI and VAII. *Mol. Cell Biol.* **1981**, *1*, 785–796.

61. Glickman, J.N.; Howe, J.G.; Steitz, J.A. Structural analyses of EBER1 and EBER2 ribonucleoprotein particles present in Epstein-Barr virus-infected cells. *J. Virol.* **1988**, *62*, 902–911.

62. Bhat, R.A.; Thimmappaya, B. Two small RNAs encoded by Epstein-Barr virus can functionally substitute for the virus-associated RNAs in the lytic growth of adenovirus 5. *Proc. Natl. Acad. Sci. USA* **1983**, *80*, 4789–4793.

63. Wang, Y.; Xue, S.A.; Hallden, G.; Francis, J.; Yuan, M.; Griffin, B.E.; Lemoine, N.R. Virus-associated RNA I-deleted adenovirus, a potential oncolytic agent targeting EBV-associated tumors. *Cancer Res.* **2005**, *65*, 1523–1531.

64. Aparicio, O.; Razquin, N.; Zaratiegui, M.; Narvaiza, I.; Fortes, P. Adenovirus virus-associated RNA is processed to functional interfering RNAs involved in virus production. *J. Virol.* **2006**, *80*, 1376–1384.

65. Sano, M.; Kato, Y.; Taira, K. Sequence-specific interference by small RNAs derived from adenovirus VAI RNA. *FEBS Lett.* **2006**, *580*, 1553–1564.

66. Han, J.; Lee, Y.; Yeom, K.H.; Nam, J.W.; Heo, I.; Rhee, J.K.; Sohn, S.Y.; Cho, Y.; Zhang, B.T.; Kim, V.N. Molecular basis for the recognition of primary microRNAs by the Drosha-DGCR8 complex. *Cell* **2006**, *125*, 887–901.

67. Fok, V.; Friend, K.; Steitz, J.A. Epstein-Barr virus noncoding RNAs are confined to the nucleus, whereas their partner, the human La protein, undergoes nucleocytoplasmic shuttling. *J. Cell Biol.* **2006**, *173*, 319–325.

68. Gilligan, K.; Sato, H.; Rajadurai, P.; Busson, P.; Young, L.; Rickinson, A.; Tursz, T.; Raab-Traub, N. Novel transcription from the Epstein-Barr virus terminal EcoRI fragment, DIJhet, in a nasopharyngeal carcinoma. *J. Virol.* **1990**, *64*, 4948–4956.

69. Gilligan, K.J.; Rajadurai, P.; Lin, J.C.; Busson, P.; Abdel-Hamid, M.; Prasad, U.; Tursz, T.; Raab-Traub, N. Expression of the Epstein-Barr virus BamHI A fragment in nasopharyngeal carcinoma: Evidence for a viral protein expressed *in vivo*. *J. Virol.* **1991**, *65*, 6252–6259.

70. Robertson, E.S.; Tomkinson, B.; Kieff, E. An Epstein-Barr virus with a 58-kilobase-pair deletion that includes BARF0 transforms B lymphocytes *in vitro*. *J. Virol.* **1994**, *68*, 1449–1458.

71. Al-Mozaini, M.; Bodelon, G.; Karstegl, C.E.; Jin, B.; Al-Ahdal, M.; Farrell, P.J. Epstein-Barr virus BART gene expression. *J. Gen. Virol.* **2009**, *90*, 307–316.

72. Smith, P.R.; de Jesus, O.; Turner, D.; Hollyoake, M.; Karstegl, C.E.; Griffin, B.E.; Karran, L.; Wang, Y.; Hayward, S.D.; Farrell, P.J. Structure and coding content of CST (BART) family RNAs of Epstein-Barr virus. *J. Virol.* **2000**, *74*, 3082–3092.

73. Pfeffer, S.; Zavolan, M.; Grasser, F.A.; Chien, M.; Russo, J.J.; Ju, J.; John, B.; Enright, A.J.; Marks, D.; Sander, C.; *et al.* Identification of virus-encoded microRNAs. *Science* **2004**, *304*, 734–736.

74. Cai, X.; Schafer, A.; Lu, S.; Bilello, J.P.; Desrosiers, R.C.; Edwards, R.; Raab-Traub, N.; Cullen, B.R. Epstein-Barr virus microRNAs are evolutionarily conserved and differentially expressed. *PLoS Pathog.* **2006**, *2*, e23.

75. Grundhoff, A.; Sullivan, C.S.; Ganem, D. A combined computational and microarray-based approach identifies novel microRNAs encoded by human gamma-herpesviruses. *RNA* **2006**, *12*, 733–750.

76. Hutzinger, R.; Feederle, R.; Mrazek, J.; Schiefermeier, N.; Balwierz, P.J.; Zavolan, M.; Polacek, N.; Delecluse, H.J.; Huttenhofer, A. Expression and processing of a small nucleolar RNA from the Epstein-Barr virus genome. *PLoS Pathog.* **2009**, *5*, e1000547.
77. Hammond, S.M.; Boettcher, S.; Caudy, A.A.; Kobayashi, R.; Hannon, G.J. Argonaute2, a link between genetic and biochemical analyses of RNAi. *Science* **2001**, *293*, 1146–1150.
78. Bartel, D.P. MicroRNAs: Genomics, biogenesis, mechanism, and function. *Cell* **2004**, *116*, 281–297.
79. Carthew, R.W.; Sontheimer, E.J. Origins and mechanisms of miRNAs and siRNAs. *Cell* **2009**, *136*, 642–655.
80. Zhu, J.Y.; Pfuhl, T.; Motsch, N.; Barth, S.; Nicholls, J.; Grasser, F.; Meister, G. Identification of novel Epstein-Barr virus microRNA genes from nasopharyngeal carcinomas. *J. Virol.* **2009**, *83*, 3333–3341.
81. Matera, A.G.; Terns, R.M.; Terns, M.P. Non-coding RNAs: Lessons from the small nuclear and small nucleolar RNAs. *Nat. Rev. Mol. Cell Biol.* **2007**, *8*, 209–220.
82. Ender, C.; Krek, A.; Friedlander, M.R.; Beitzinger, M.; Weinmann, L.; Chen, W.; Pfeffer, S.; Rajewsky, N.; Meister, G. A human snoRNA with microRNA-like functions. *Mol. Cell* **2008**, *32*, 519–528.
83. Saraiya, A.A.; Wang, C.C. snoRNA, a novel precursor of microRNA in Giardia lamblia. *PLoS Pathog.* **2008**, *4*, e1000224.
84. Xing, L.; Kieff, E. Epstein-Barr virus BHRF1 micro- and stable RNAs during latency III and after induction of replication. *J. Virol.* **2007**, *81*, 9967–9975.
85. Imig, J.; Motsch, N.; Zhu, J.Y.; Barth, S.; Okoniewski, M.; Reineke, T.; Tinguely, M.; Faggioni, A.; Trivedi, P.; Meister, G.; *et al.* microRNA profiling in Epstein-Barr virus-associated B-cell lymphoma. *Nucleic Acids Res.* **2011**, *39*, 1880–1893.
86. Amoroso, R.; Fitzsimmons, L.; Thomas, W.A.; Kelly, G.L.; Rowe, M.; Bell, A.I. Quantitative studies of Epstein-Barr virus-encoded microRNAs provide novel insights into their regulation. *J. Virol.* **2011**, *85*, 996–1010.
87. Qiu, J.; Cosmopoulos, K.; Pegtel, M.; Hopmans, E.; Murray, P.; Middeldorp, J.; Shapiro, M.; Thorley-Lawson, D.A. A novel persistence associated EBV miRNA expression profile is disrupted in neoplasia. *PLoS Pathog.* **2011**, *7*, e1002193.
88. Cosmopoulos, K.; Pegtel, M.; Hawkins, J.; Moffett, H.; Novina, C.; Middeldorp, J.; Thorley-Lawson, D.A. Comprehensive profiling of Epstein-Barr virus microRNAs in nasopharyngeal carcinoma. *J. Virol.* **2009**, *83*, 2357–2367.
89. Chatterjee, S.; Fasler, M.; Bussing, I.; Grosshans, H. Target-mediated protection of endogenous microRNAs in C. elegans. *Dev. Cell* **2011**, *20*, 388–396.
90. Chen, H.; Huang, J.; Wu, F.Y.; Liao, G.; Hutt-Fletcher, L.; Hayward, S.D. Regulation of expression of the Epstein-Barr virus BamHI-A rightward transcripts. *J. Virol.* **2005**, *79*, 1724–1733.
91. Gourzones, C.; Jimenez, A.S.; Busson, P. Profiling of Epstein-Barr virus-encoded microRNAs in nasopharyngeal carcinoma reveals potential biomarkers and oncomirs. *Cancer* **2012**, 118, dio:10.1002/cncr.26514.
92. Wong, A.M.; Kong, K.L.; Tsang, J.W.; Kwong, D.L.; Guan, X.Y. Profiling of Epstein-Barr virus-encoded microRNAs in nasopharyngeal carcinoma reveals potential biomarkers and oncomirs. *Cancer* **2012**, *118*, 698–710.

93. Choy, E.Y.; Siu, K.L.; Kok, K.H.; Lung, R.W.; Tsang, C.M.; To, K.F.; Kwong, D.L.; Tsao, S.W.; Jin, D.Y. An Epstein-Barr virus-encoded microRNA targets PUMA to promote host cell survival. *J. Exp. Med.* **2008**, *205*, 2551–2560.
94. Lung, R.W.; Wang, X.; Tong, J.H.; Chau, S.L.; Lau, K.M.; Cheng, S.H.; Woo, J.K.; Woo, J.; Leung, P.C.; Ng, M.H.; *et al.* A single nucleotide polymorphism in microRNA-146a is associated with the risk for nasopharyngeal carcinoma. *Mol. Carcinog.* **2012**, doi:10.1002/mc.21937.
95. Cazalla, D.; Xie, M.; Steitz, J.A. A primate herpesvirus uses the integrator complex to generate viral microRNAs. *Mol. Cell* **2011**, *43*, 982–992.
96. Chi, S.W.; Zang, J.B.; Mele, A.; Darnell, R.B. Argonaute HITS-CLIP decodes microRNA-mRNA interaction maps. *Nature* **2009**, *460*, 479–486.
97. Ascano, M.; Hafner, M.; Cekan, P.; Gerstberger, S.; Tuschl, T. Identification of RNA-protein interaction networks using PAR-CLIP. *Wiley Interdiscip. Rev. RNA* **2012**, *3*, 159–177.
98. Hafner, M.; Lianoglou, S.; Tuschl, T.; Betel, D. Genome-wide identification of miRNA targets by PAR-CLIP. *Methods* **2012**, *58*, 94–105.
99. Gottwein, E.; Corcoran, D.L.; Mukherjee, N.; Skalsky, R.L.; Hafner, M.; Nusbaum, J.D.; Shamulailatpam, P.; Love, C.L.; Dave, S.S.; Tuschl, T.; *et al.* Viral microRNA targetome of KSHV-infected primary effusion lymphoma cell lines. *Cell Host Microbe* **2011**, *10*, 515–526.
100. Skalsky, R.L.; Corcoran, D.L.; Gottwein, E.; Frank, C.L.; Kang, D.; Hafner, M.; Nusbaum, J.D.; Feederle, R.; Delecluse, H.J.; Luftig, M.A.; *et al.* The viral and cellular microRNA targetome in lymphoblastoid cell lines. *PLoS Pathog.* **2012**, *8*, e1002484.
101. Riley, K.J.; Rabinowitz, G.S.; Yario, T.A.; Luna, J.M.; Darnell, R.B.; Steitz, J.A. EBV and human microRNAs co-target oncogenic and apoptotic viral and human genes during latency. *EMBO J.* **2012**, *31*, 2207–2221.
102. Barth, S.; Pfuhl, T.; Mamiani, A.; Ehses, C.; Roemer, K.; Kremmer, E.; Jaker, C.; Hock, J.; Meister, G.; Grasser, F.A. Epstein-Barr virus-encoded microRNA miR-BART2 down-regulates the viral DNA polymerase BALF5. *Nucleic Acids Res.* **2008**, *36*, 666–675.
103. Lo, A.K.; To, K.F.; Lo, K.W.; Lung, R.W.; Hui, J.W.; Liao, G.; Hayward, S.D. Modulation of LMP1 protein expression by EBV-encoded microRNAs. *Proc. Natl. Acad. Sci. USA* **2007**, *104*, 16164–16169.
104. Hislop, A.D.; Taylor, G.S.; Sauce, D.; Rickinson, A.B. Cellular responses to viral infection in humans: Lessons from Epstein-Barr virus. *Annu. Rev. Immunol.* **2007**, *25*, 587–617.
105. Straathof, K.C.; Leen, A.M.; Buza, E.L.; Taylor, G.; Huls, M.H.; Heslop, H.E.; Rooney, C.M.; Bollard, C.M. Characterization of latent membrane protein 2 specificity in CTL lines from patients with EBV-positive nasopharyngeal carcinoma and lymphoma. *J. Immunol.* **2005**, *175*, 4137–4147.
106. Eliopoulos, A.G.; Dawson, C.W.; Mosialos, G.; Floettmann, J.E.; Rowe, M.; Armitage, R.J.; Dawson, J.; Zapata, J.M.; Kerr, D.J.; Wakelam, M.J.; *et al.* CD40-induced growth inhibition in epithelial cells is mimicked by Epstein-Barr virus-encoded LMP1: Involvement of TRAF3 as a common mediator. *Oncogene* **1996**, *13*, 2243–2254.
107. Dirmeier, U.; Hoffmann, R.; Kilger, E.; Schultheiss, U.; Briseno, C.; Gires, O.; Kieser, A.; Eick, D.; Sugden, B.; Hammerschmidt, W. Latent membrane protein 1 of Epstein-Barr virus coordinately regulates proliferation with control of apoptosis. *Oncogene* **2005**, *24*, 1711–1717.

108. Ramakrishnan, R.; Donahue, H.; Garcia, D.; Tan, J.; Shimizu, N.; Rice, A.P.; Ling, P.D. Epstein-Barr virus BART9 miRNA modulates LMP1 levels and affects growth rate of nasal NK T cell lymphomas. *PLoS One* **2011**, *6*, e27271.
109. Lomonosova, E.; Chinnadurai, G. BH3-only proteins in apoptosis and beyond: An overview. *Oncogene* **2008**, *27*, S2–S19.
110. Marquitz, A.R.; Mathur, A.; Nam, C.S.; Raab-Traub, N. The Epstein-Barr virus BART microRNAs target the pro-apoptotic protein Bim. *Virology* **2011**, *412*, 392–400.
111. Lin, T.C.; Liu, T.Y.; Hsu, S.M.; Lin, C.W. Epstein-Barr virus-encoded miR-BART20-5p inhibits T-bet translation with secondary suppression of p53 in invasive nasal NK/T-cell lymphoma. *Am. J. Pathol.* **2013**, *182*, 1865–1875.
112. Dolken, L.; Malterer, G.; Erhard, F.; Kothe, S.; Friedel, C.C.; Suffert, G.; Marcinowski, L.; Motsch, N.; Barth, S.; Beitzinger, M.; *et al.* Systematic analysis of viral and cellular microRNA targets in cells latently infected with human gamma-herpesviruses by RISC immunoprecipitation assay. *Cell Host Microbe* **2010**, *7*, 324–334.
113. Bellot, G.; Cartron, P.F.; Er, E.; Oliver, L.; Juin, P.; Armstrong, L.C.; Bornstein, P.; Mihara, K.; Manon, S.; Vallette, F.M. TOM22, a core component of the mitochondria outer membrane protein translocation pore, is a mitochondrial receptor for the proapoptotic protein Bax. *Cell Death Differ.* **2007**, *14*, 785–794.
114. Vereide, D.T.; Seto, E.; Chiu, Y.F.; Hayes, M.; Tagawa, T.; Grundhoff, A.; Hammerschmidt, W.; Sugden, B. Epstein-Barr virus maintains lymphomas via its miRNAs. *Oncogene* **2013**, doi:10.1038/onc.2013.71.
115. Ding, Y.; Xi, Y.; Chen, T.; Wang, J.Y.; Tao, D.L.; Wu, Z.L.; Li, Y.P.; Li, C.; Zeng, R.; Li, L. Caprin-2 enhances canonical Wnt signaling through regulating LRP5/6 phosphorylation. *J. Cell Biol.* **2008**, *182*, 865–872.
116. Aerbajinai, W.; Lee, Y.T.; Wojda, U.; Barr, V.A.; Miller, J.L. Cloning and characterization of a gene expressed during terminal differentiation that encodes a novel inhibitor of growth. *J. Biol. Chem.* **2004**, *279*, 1916–1921.
117. Choi, H.; Lee, H.; Kim, S.R.; Gho, Y.S.; Lee, S.K. EBV encoded miR-BART15-3p promotes cell apoptosis partially by targeting BRUCE. *J. Virol.* **2013**, doi:10.1128/JVI.03159-12.
118. Chen, Z.; Naito, M.; Hori, S.; Mashima, T.; Yamori, T.; Tsuruo, T. A human IAP-family gene, apollon, expressed in human brain cancer cells. *Biochem. Biophys. Res. Commun.* **1999**, *264*, 847–854.
119. Verhagen, A.M.; Coulson, E.J.; Vaux, D.L. Inhibitor of apoptosis proteins and their relatives: IAPs and other BIRPs. *Genome Biol.* **2001**, *2*, REVIEWS3009.1–REVIEWS3009.10.
120. Haneklaus, M.; Gerlic, M.; Kurowska-Stolarska, M.; Rainey, A.A.; Pich, D.; McInnes, I.B.; Hammerschmidt, W.; O'Neill, L.A.; Masters, S.L. Cutting edge: miR-223 and EBV miR-BART15 regulate the NLRP3 inflammasome and IL-1beta production. *J. Immunol.* **2012**, *189*, 3795–3799.
121. Kang, W.; Tong, J.H.; Chan, A.W.; Lee, T.L.; Lung, R.W.; Leung, P.P.; So, K.K.; Wu, K.; Fan, D.; Yu, J.; *et al.* Yes-associated protein 1 exhibits oncogenic property in gastric cancer and its nuclear accumulation associates with poor prognosis. *Clin. Cancer Res.* **2011**, *17*, 2130–2139.
122. Wong, Q.W.; Lung, R.W.; Law, P.T.; Lai, P.B.; Chan, K.Y.; To, K.F.; Wong, N. MicroRNA-223 is commonly repressed in hepatocellular carcinoma and potentiates expression of Stathmin1. *Gastroenterology* **2008**, *135*, 257–269.

123. Pegtel, D.M.; Cosmopoulos, K.; Thorley-Lawson, D.A.; van Eijndhoven, M.A.; Hopmans, E.S.; Lindenberg, J.L.; de Gruijl, T.D.; Wurdinger, T.; Middeldorp, J.M. Functional delivery of viral miRNAs via exosomes. *Proc. Natl. Acad. Sci. USA* **2010**, *107*, 6328–6333.
124. Meckes, D.G., Jr.; Shair, K.H.; Marquitz, A.R.; Kung, C.P.; Edwards, R.H.; Raab-Traub, N. Human tumor virus utilizes exosomes for intercellular communication. *Proc. Natl. Acad. Sci. USA* **2010**, *107*, 20370–20375.
125. Nachmani, D.; Stern-Ginossar, N.; Sarid, R.; Mandelboim, O. Diverse herpesvirus microRNAs target the stress-induced immune ligand MICB to escape recognition by natural killer cells. *Cell Host Microbe* **2009**, *5*, 376–385.
126. Yang, I.V.; Wade, C.M.; Kang, H.M.; Alper, S.; Rutledge, H.; Lackford, B.; Eskin, E.; Daly, M.J.; Schwartz, D.A. Identification of novel genes that mediate innate immunity using inbred mice. *Genetics* **2009**, *183*, 1535–1544.
127. Ullman, K.S.; Powers, M.A.; Forbes, D.J. Nuclear export receptors: From importin to exportin. *Cell* **1997**, *90*, 967–970.
128. Lei, T.; Yuen, K.S.; Xu, R.; Tsao, S.W.; Chen, H.; Li, M.; Kok, K.H.; Jin, D.Y. Targeting of DICE1 tumor suppressor by Epstein-Barr virus-encoded miR-BART3* microRNA in nasopharyngeal carcinoma. *Int. J. Cancer* **2013**, *133*, 79–87.
129. Ropke, A.; Buhtz, P.; Bohm, M.; Seger, J.; Wieland, I.; Allhoff, E.P.; Wieacker, P.F. Promoter CpG hypermethylation and downregulation of DICE1 expression in prostate cancer. *Oncogene* **2005**, *24*, 6667–6675.
130. Li, W.J.; Hu, N.; Su, H.; Wang, C.; Goldstein, A.M.; Wang, Y.; Emmert-Buck, M.R.; Roth, M.J.; Guo, W.J.; Taylor, P.R. Allelic loss on chromosome 13q14 and mutation in deleted in cancer 1 gene in esophageal squamous cell carcinoma. *Oncogene* **2003**, *22*, 314–318.
131. Wieland, I.; Arden, K.C.; Michels, D.; Klein-Hitpass, L.; Bohm, M.; Viars, C.S.; Weidle, U.H. Isolation of DICE1: A gene frequently affected by LOH and downregulated in lung carcinomas. *Oncogene* **1999**, *18*, 4530–4537.
132. Godshalk, S.E.; Bhaduri-McIntosh, S.; Slack, F.J. Epstein-Barr virus-mediated dysregulation of human microRNA expression. *Cell Cycle* **2008**, *7*, 3595–3600.
133. Iizasa, H.; Wulff, B.E.; Alla, N.R.; Maragkakis, M.; Megraw, M.; Hatzigeorgiou, A.; Iwakiri, D.; Takada, K.; Wiedmer, A.; Showe, L.; *et al.* Editing of Epstein-Barr virus-encoded BART6 microRNAs controls their dicer targeting and consequently affects viral latency. *J. Biol. Chem.* **2010**, *285*, 33358–33370.
134. ClinicalTrials.gov. Available online: http://clinicaltrials.gov/show/NCT01200420 (accessed on 20 August 2013).
135. Janssen, H.L.; Reesink, H.W.; Lawitz, E.J.; Zeuzem, S.; Rodriguez-Torres, M.; Patel, K.; van der Meer, A.J.; Patick, A.K.; Chen, A.; Zhou, Y.; *et al.* Treatment of HCV infection by targeting microRNA. *N. Engl. J. Med.* **2013**, *368*, 1685–1694.
136. Muthiah, M.; Park, I.K.; Cho, C.S. Nanoparticle-mediated delivery of therapeutic genes: Focus on miRNA therapeutics. *Expert. Opin. Drug. Deliv.* **2013**, doi:10.1517/17425247.2013.798640.
137. Garzon, R.; Marcucci, G.; Croce, C.M. Targeting microRNAs in cancer: Rationale, strategies and challenges. *Nat. Rev. Drug Discov.* **2010**, *9*, 775–789.

138. Lo, A.K.; Dawson, C.W.; Jin, D.Y.; Lo, K.W. The pathological roles of BART miRNAs in nasopharyngeal carcinoma. *J. Pathol.* **2012**, *227*, 392–403.
139. Babu, S.G.; Ponia, S.S.; Kumar, D.; Saxena, S. Cellular oncomiR orthologue in EBV oncogenesis. *Comput. Biol. Med.* **2011**, *41*, 891–898.
140. Keryer-Bibens, C.; Pioche-Durieu, C.; Villemant, C.; Souquere, S.; Nishi, N.; Hirashima, M.; Middeldorp, J.; Busson, P. Exosomes released by EBV-infected nasopharyngeal carcinoma cells convey the viral latent membrane protein 1 and the immunomodulatory protein galectin 9. *BMC Cancer* **2006**, *6*, 283.
141. Gourzones, C.; Gelin, A.; Bombik, I.; Klibi, J.; Verillaud, B.; Guigay, J.; Lang, P.; Temam, S.; Schneider, V.; Amiel, C.; *et al.* Extra-cellular release and blood diffusion of BART viral micro-RNAs produced by EBV-infected nasopharyngeal carcinoma cells. *Virol. J.* **2010**, *7*, 271.
142. Rechavi, O.; Erlich, Y.; Amram, H.; Flomenblit, L.; Karginov, F.V.; Goldstein, I.; Hannon, G.J.; Kloog, Y. Cell contact-dependent acquisition of cellular and viral nonautonomously encoded small RNAs. *Genes Dev.* **2009**, *23*, 1971–1979.
143. Verweij, F.J.; van Eijndhoven, M.A.; Hopmans, E.S.; Vendrig, T.; Wurdinger, T.; Cahir-McFarland, E.; Kieff, E.; Geerts, D.; van der Kant, R.; Neefjes, J.; *et al.* LMP1 association with CD63 in endosomes and secretion via exosomes limits constitutive NF-kappaB activation. *EMBO J.* **2011**, *30*, 2115–2129.
144. Pegtel, D.M.; van de Garde, M.D.; Middeldorp, J.M. Viral miRNAs exploiting the endosomal-exosomal pathway for intercellular cross-talk and immune evasion. *Biochim. Biophys. Acta* **2011**, *1809*, 715–721.
145. Gourzones, C.; Ferrand, F.R.; Amiel, C.; Verillaud, B.; Barat, A.; Guerin, M.; Gattolliat, C.H.; Gelin, A.; Klibi, J.; Chaaben, A.B.; *et al.* Consistent high concentration of the viral microRNA BART17 in plasma samples from nasopharyngeal carcinoma patients-evidence of non-exosomal transport. *Virol. J.* **2013**, *10*, 119.
146. Bala, S.; Petrasek, J.; Mundkur, S.; Catalano, D.; Levin, I.; Ward, J.; Alao, H.; Kodys, K.; Szabo, G. Circulating microRNAs in exosomes indicate hepatocyte injury and inflammation in alcoholic, drug-induced, and inflammatory liver diseases. *Hepatology* **2012**, *56*, 1946–1957.
147. Arroyo, J.D.; Chevillet, J.R.; Kroh, E.M.; Ruf, I.K.; Pritchard, C.C.; Gibson, D.F.; Mitchell, P.S.; Bennett, C.F.; Pogosova-Agadjanyan, E.L.; Stirewalt, D.L.; *et al.* Argonaute2 complexes carry a population of circulating microRNAs independent of vesicles in human plasma. *Proc. Natl. Acad. Sci. USA* **2011**, *108*, 5003–5008.
148. Chan, J.Y.; Gao, W.; Ho, W.K.; Wei, W.I.; Wong, T.S. Overexpression of Epstein-Barr virus-encoded microRNA-BART7 in undifferentiated nasopharyngeal carcinoma. *Anticancer Res.* **2012**, *32*, 3201–3210.
149. Bustin, S.A. Absolute quantification of mRNA using real-time reverse transcription polymerase chain reaction assays. *J. Mol. Endocrinol.* **2000**, *25*, 169–193.
150. Oudejans, J.J.; van den Brule, A.J.; Jiwa, N.M.; de Bruin, P.C.; Ossenkoppele, G.J.; van der Valk, P.; Walboomers, J.M.; Meijer, C.J. BHRF1, the Epstein-Barr virus (EBV) homologue of the BCL-2 protooncogene, is transcribed in EBV-associated B-cell lymphomas and in reactive lymphocytes. *Blood* **1995**, *86*, 1893–1902.

5

The Role of microRNA in Gastric Malignancy

Toshihiro Nishizawa [1,2] and Hidekazu Suzuki [1,*]

1 Division of Gastroenterology and Hepatology, Department of Internal Medicine, Keio University School of Medicine, 35 Shinanomachi, Shinjuku-ku, Tokyo 160-8582, Japan; E-Mail: nisizawa@kf7.so-net.ne.jp

2 Division of Gastroenterology, National Hospital Organization Tokyo Medical Center, 2-5-1 Higashigaoka, Meguro-ku, Tokyo 152-8902, Japan

* Author to whom correspondence should be addressed; E-Mail: hsuzuki@a6.keio.jp;

Abstract: *Helicobacter pylori* (*H. pylori*) infection is the main cause of gastritis, gastro-duodenal ulcer, and gastric cancer. MicroRNAs (miRNAs) are small noncoding RNAs that function as endogenous silencers of numerous target genes. Many miRNA genes are expressed in a tissue-specific manner and play important roles in cell proliferation, apoptosis, and differentiation. Recent discoveries have shed new light on the involvement of miRNAs in gastric malignancy. However, at the same time, several miRNAs have been associated with opposing events, leading to reduced inflammation, inhibition of malignancy, and increased apoptosis of transformed cells. The regulation of miRNA expression could be a novel strategy in the chemoprevention of human gastric malignancy. In this article, the biological importance of miRNAs in gastric malignancy is summarized.

Keywords: microRNA; *Helicobacter pylori*; gastric cancer

---

1. Introduction

Helicobacter pylori (*H. pylori*) infection is one of the most prevalent infectious diseases worldwide, and is estimated to affect 40%–50% of the world population [1]. *H. pylori* has been identified as a group 1 carcinogen by the World Health Organization and is associated with the development of gastric cancer [2]. *H. pylori* eradication has been shown to have a prophylactic effect on gastric cancer [3,4].

MicroRNAs (miRNAs) are noncoding RNAs comprising 18–24 nucleotides that can post-transcriptionally downregulate various target genes [5]. It is estimated that the human genome encodes more than one thousand miRNAs, targeting 30%–60% of all protein-coding genes [6]. They are expressed in a tissue-specific manner, and play important roles in cell proliferation, apoptosis, and differentiation [5,7]. Moreover, recent studies have shown a connection between aberrant expression of miRNAs and the development of cancer. In this article, the biological importance of miRNAs in gastric malignancy is summarized.

2. *Helicobacter pylori* and miRNA

CagA of *H. pylori* is a bacterium-derived oncogenic protein closely associated with the development of gastric cancers [8]. After injection into host cells using a type IV secretion system, CagA is phosphorylated at tyrosine residues by the c-Src and Lyn kinases. Phosphorylated CagA then activates the Src homology-2 domain-containing phosphatase 2 (SHP2), which activates the Erk1/2 pathway. CagA translocated into CD44v9-positive gastric cancer stem-like cells is thought to escape from reactive oxygen species-dependent autophagy, resulting in gastric carcinogenesis [9].

Table 1. MicroRNAs (miRNAs) change in response to *Helicobacter pylori.*

miRNAs	Change	Target mRNAs	Biological process targeted
let-7a	↓	**RAB40C** **HMGA2**	**Cell cycle progression** **Invasion**
let-7b/d/e/f	↓	HMGA2	Invasion
miR101	↓	**MCL1**	**Apoptosis**
miR-106b	↓	p21 BIM	Cell cycle progression Apoptosis
miR-125a	↓	ERBB2	Proliferation
miR-141	↓	**FGFR2**	**Proliferation**
miR-200a	↓	ZEB1, ZEB2	Epithelial to mesenchymal transition
miR-200b/c	↓	BCL2, XIAP	Apoptosis
miR-203	↓	**ABL1**	**Proliferation, Invasion**
miR-204	↓	EZR	Proliferation
miR-218	↓	**ROBO1**	**Invasion, Metastasis**
miR-375	↓	**PDK1, 12-3-3** **JAK2**	**Apoptosis** **Proliferation**
miR-429	↓	BCL2, XIAP MYC	Apoptosis Proliferation
miR-17	↑	**p21**	**Cell cycle progression**
miR-20a	↑	**p21**	**Cell cycle progression**
miR-21	↑	**PTEN** **RECK**	**Proliferation** **Metastasis**
miR-146a	↑	**IRAK1, TRAF6** **SMAD4**	**Proliferation, Immune response** **Apoptosis**
miR-155	↑	**IKK-ε, SMAD4** **FADD, PLIα**	**Immune response** **Apoptosis**
miR-223	↑	**EPB41L3**	**Invasion, Metastasis**

↓: miRNA is downregulated in response to *H. pylori*; ↑: miRNA is upregulated in response to *H. pylori*; Bold indicates miRNA changes in the same way in gastric cancer.

miRNA changes in response to *H. pylori* infection are summarized in Table 1. *miR-584* and *miR-1290* expression are upregulated in CagA-transformed cells. The *miR-584* and *miR-1290* target is Foxa1, and knockdown of Foxa1 promotes the epithelial-mesenchymal transition (EMT). Overexpression of *miR-584* and *miR-1290* induces intestinal metaplasia of gastric epithelial cells. These results indicate that *miR-584* and *miR-1290* interfere with cell differentiation and lead to remodeling of the gastric mucosal tissues [10].

CagA enhances *c-myc* and DNA methyltransferase 3B, and attenuates *miR-6a* and *miR-101* expression, which results in the attenuation of *let-7* expression by histone and DNA methylation. CagA induces aberrant epigenetic silencing of *let-7* expression, leading to Ras pathway activation. Thus, the miRNA pathway is a new pathogenic mechanism for CagA.

Shiotani *et al.* [11] reported that the expression of oncogenic miRNAs (*miR-17/92* and the *miR-106b-93-25* cluster, *miR-21, miR-194*, and *miR-196*) is significantly higher in the intestinal metaplasia than in the non-intestinal metaplasia. *H. pylori* eradication improves miRNA deregulation, but not in the intestinal metaplasia. Long-term colonization of *H. pylori* might induce an epigenetic modification of gastric mucosal genes, including the promoter of tumor suppressor miRNAs, but this is not completely reversible by bacterial eradication alone. Epigenetic therapy in severe atrophic or metaplastic gastric mucosa after *H. pylori* eradication might be a possible option for gastric cancer prevention.

3. Cell Cycle Progression and miRNA

The dysregulation of cell cycle progression is a hallmark of malignancy. Cyclin-CDK (cyclin-dependent kinase) complexes regulate this progression through the cell cycle. miRNA dysregulation promotes cell cycle progression by upregulating cyclin expression or downregulating expression of CDK inhibitors (p57, p21, *etc.*) in numerous malignancies (Figure 1). *miR-449* is downregulated in *H. pylori*-infected gastric mucosa and in gastric cancer and targets cyclin E2 and geminin. Both cyclin E2 and geminin are overexpressed in various malignancies and promote G1/S and M/G1 cell cycle progression [6]. Consequently, downregulation of *miR-449*, as occurs following *H. pylori* infection, promotes cell cycle progression and proliferation through the upregulation of cyclin E2 and geminin.

G2/M cell cycle progression and proliferation in gastric cancer cells are regulated by p42.3 [12]; *miR-29a* is significantly downregulated in gastric cancer and targets p42.3 [13] (Figure 1). Thus, the downregulation of *miR-29a* results in a reciprocal increase in p42.3 expression, promoting increased cell cycle progression and proliferation.

Both *miR-93* and *miR-106b* directly target p21, resulting in its transcriptional silencing and impairment of its tumor-suppressing activity [14]. In addition, *miR-25* targets p57, while *miR-221* and *miR-222* target p27 and p57 [15] (Figure 1). These oncogenic miRNA clusters are also significantly upregulated in gastric cancer [16]. Overexpression of most of these miRNAs results in activation of CDK2, thereby promoting G1/S phase progression.

Figure 1. Regulation of cell cycle progression. Cyclins and cyclin-dependent kinases (CDKs) determine a cell's progress through the cell cycle.

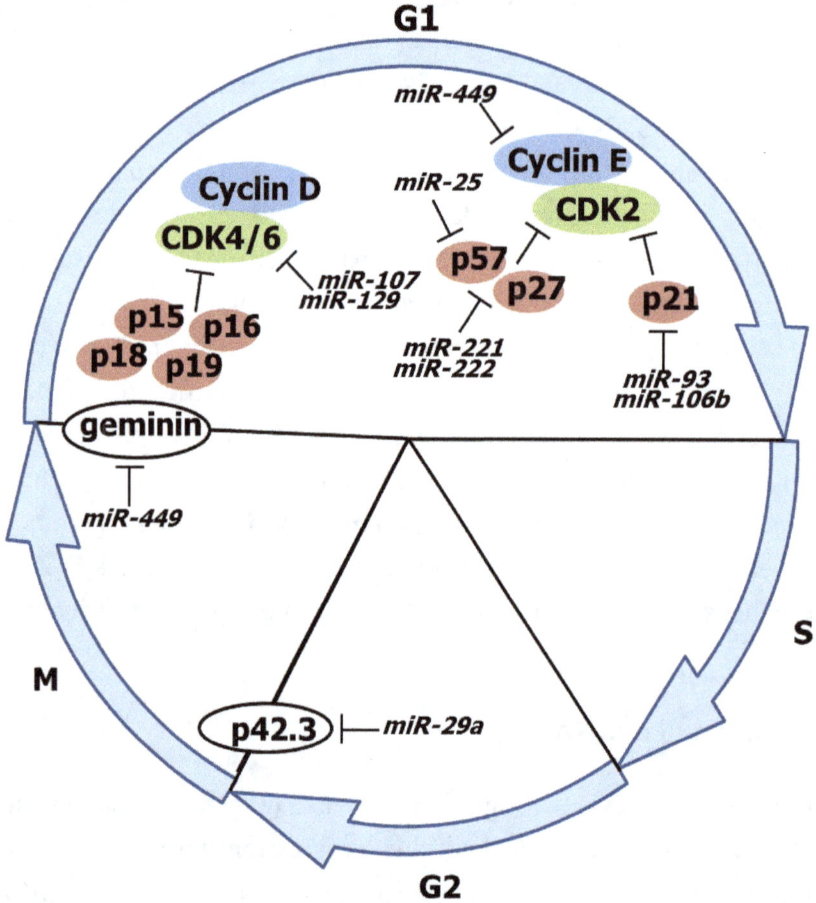

4. Inhibition of Apoptosis and miRNA

Evasion of apoptosis is a common feature of malignancy. Apoptosis is classifiable as either intrinsic or extrinsic pathway-dependent. The extrinsic apoptosis pathway is initiated on the cell surface through the activation of specific pro-apoptotic death receptors. Tumor necrosis factors are cytokines produced mainly by activated macrophages that bind the death receptors as their ligands. Ligand binding induces receptor clustering and the recruitment of the adaptor protein Fas-associated death domain (FADD), leading to induction of caspases and ultimately cell death. *miR-155* targets *FADD*, leading to decreased expression of this key adaptor molecule (Figure 2) [17]. Therefore, the upregulation of *miR-155* by *H. pylori* and during carcinogenesis results in the downregulation of FADD and the inhibition of apoptosis.

The intrinsic pathway is initiated within cells and hinges on the balance of activity between pro-apoptotic (e.g., Bax, Bak, Bim, BNIP3L, and Bid) and anti-apoptotic (e.g., Bcl-2, Bcl-xL, and Mcl-1) proteins from the Bcl-2 (B cell lymphoma 2) superfamily. Some miRNAs overexpressed in gastric cancer function as oncogenic miRNAs by targeting members of the pro-apoptotic proteins. *miR-25*, *miR-93*, *miR-106b*, and *miR-130* inhibit apoptosis by preventing the expression of the pro-apoptotic protein, Bim (Figure 2) [14].

Figure 2. Signaling cascades that regulate the intrinsic and extrinsic pathways of apoptosis. Receptor tyrosine kinase detects survival stimuli such as growth factors and induces phosphoinositide 3-kinase (PI3K)/Akt signaling cascades that ultimately result in the inhibition of the pro-apoptotic protein, Bad. Pro-apoptotic and anti-apoptotic proteins govern the intrinsic cell death pathway, which results in the release of cytochrome *c* from the mitochondria and induction of the caspase cascade. Signaling through death receptors initiates the extrinsic pathway of apoptosis.

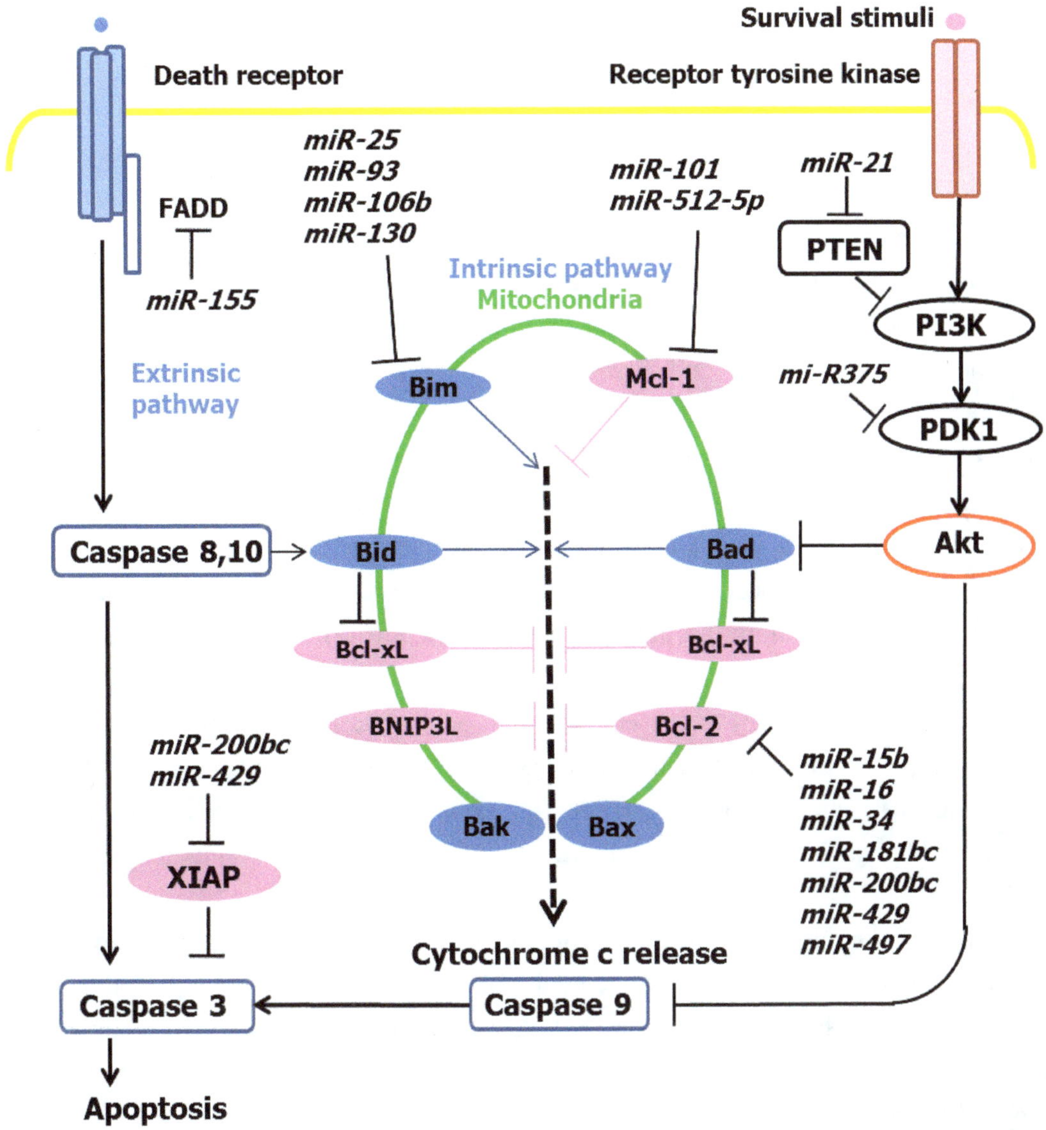

Tumor suppressor miRNAs (*miR-15b*, *miR-16*, *miR-34*, *miR-181b*, *miR-181c,* and *miR-497*) target anti-apoptotic Bcl-2. These miRNA clusters are downregulated in gastric cancer cells, leading to increased expression of Bcl-2 and inhibition of apoptosis. The *miR-200bc/429* cluster is downregulated in *H. pylori*-infected gastric mucosa, and these miRNAs directly target Bcl-2 and XIAP (x-linked inhibitor of apoptosis) [18]. *miR-101* and *miR-515-5p* target Mcl-1, which are downregulated in gastric cancer, lead to increased levels of Mcl-1 and an anti-apoptotic phenotype (Figure 2).

In addition to targeting proteins directly involved in the intrinsic and extrinsic cell death pathways, miRNAs target other factors that ultimately lead to apoptosis inhibition and increased proliferation.

miR-21 targets PTEN (phosphatase and tensin homolog), a tumor suppressor and negative regulator of the PI3K/Akt signaling pathway. *miR-21* is upregulated in gastric cancer, and its overexpression shifts the balance between proliferation and apoptosis, by increasing cellular proliferation and inhibiting apoptosis.

miR-375 targets 3-phophoinositide dependent protein kinase (PDK1), a kinase that directly phosphorylates Akt, thereby regulating the PI3K/Akt signaling pathway. Overexpression of *miR-375* reduces cell viability and *miR-375* is downregulated in gastric cancer (Figure 2) [19].

5. Metastasis and miRNA

Some miRNAs that are known to regulate cell cycle progression and apoptosis pathways are also involved in invasion and metastasis. *miR-181b* is aberrantly overexpressed in *H. pylori* infection and gastric cancer tissues [20]. Cell proliferation, migration, and invasion in the gastric cancer cells were significantly increased after *miR-181b* transfection, and the number of apoptotic cells was also increased. Furthermore, overexpression of *miR-181b* downregulated the protein levels of tissue inhibitor of metalloproteinase 3 (TIMP). The upregulation of *miR-181b* may play an important role in the progression of gastric cancer and *miR-181b* may be a potential molecular target for gastric cancer therapies.

miR-218 is reduced significantly in gastric cancer tissues, *H. pylori*-infected gastric mucosa, and *H. pylori*-infected AGS cells [21]. *miR-218*, a tumor suppressor miRNA, is downregulated in gastric cancer, which correlates with increased metastasis and cancer invasion [22]. This downregulation is thought to occur through the direct targeting of roundabout homolog (ROBO1), which leads to enhanced signaling through the ROBO1 receptor. The SLIT/ROBO signaling pathway is implicated in many biological responses through regulating cell migration. Thus, disruption of this signaling cascade can result in increased invasion and metastasis. *miR-21* also targets RECK (reversion-inducing-cysteine-rich protein with kazal motifs), a tumor and metastasis suppressor that inhibits tumor metastasis and angiogenesis through modulation of matrix metalloproteinases. Recently, Li *et al.* indicated that *miR-21*, *miR-218*, and *miR-223* may be potential biomarkers for gastric cancer detection [23].

miR-148a is downregulated in gastric cancer. The protein interaction network regulated by *miR-148a* is associated with metastasis-related function, such as integrin-mediated signaling, cell-matrix adhesion, and blood coagulation [24]. A single miRNA can provoke a chain reaction and further affect the protein interaction network. This interactive network-based approach could provide insight into carcinogenesis.

6. miRNA and Anticancer Therapy

miRNAs are promising molecular targets for anticancer therapeutics in gastric cancer. Kim *et al.* reported that *miR-10b* was silenced in gastric cancer cells by promoter methylation. *miR-10b* targets the oncogene that encodes microtubule-associated protein, RP/EB family member 1. After 5-aza-2'-deoxycytidine treatment of gastric cancer cells, *miR-10b* methylation is significantly decreased, and the expression of *miR-10b* is restored. The modulation of *miR-10b* may represent a therapeutic approach for treating gastric cancer [25].

Runx3 is an important tumor suppressor that is inactivated in gastric cancer, and promoter hypermethylation of Runx3 is frequent [26]. 5-aza-2'-deoxycytidine treatment reactivates the expression of Runx3. Lai *et al.* reported that *miR-130b* expression is upregulated in gastric cancer, and this is inversely associated with Runx3 hypermethylation. *miR-130b* overexpression increases cell viability, reduces cell death, and decreases the expression of Bim in TGF-beta mediated apoptosis, subsequent to the downregulation of Runx3 protein expression. The attenuation of Runx3 protein levels by miRNA may reduce the growth suppressive potential of Runx3 and contribute to tumorigenesis [27]. Wang *et al.* reported that *miR-301a* is upregulated in gastric cancer, and directly downregulates Runx3 expression [28].

The selective cyclooxygenase-2 (COX-2) inhibitor celecoxib is a potential drug for the treatment of gastrointestinal tumors. We investigated the role of miRNAs in gastric carcinogenesis and the feasibility of a new therapeutic approach for gastric cancer [29]. miRNA microarray analysis revealed that *miR-29c* is significantly downregulated in gastric cancer tissues relative to non-tumor gastric mucosa [29]. *miR-29c* is significantly activated by celecoxib in gastric cancer cells (AGS) [29]. Celecoxib activation of *miR-29c* induces suppression of the oncogene *Mcl-1*, a target of *miR-29c* and apoptosis in gastric cancer cells. These results suggest that the downregulation of the *miR-29c* tumor suppressor plays a critical role in the progression of gastric cancer. As such, selective COX-2 inhibitors may be a clinical option for the treatment of gastric cancer via restoration of *miR-29c*.

7. MALT Lymphoma and miRNA

Gastric B-cell lymphoma of the mucosa-associated lymphoid tissue (MALT lymphoma) develops in the chronically inflamed mucosa of patients infected with the bacterial pathogen. In 60%–80% of these cases, the *H. pylori*-positive gastric MALT lymphoma regresses after *H. pylori* eradication. The t(11;18) (q21;q21) translocation is associated with *API2-MALT1* fusion, and this translocation responds only rarely or not at all to *H. pylori* eradication.

We previously reported that a hematopoietic-specific miRNA, *miR-142*, and an oncogenic miRNA, *miR-155*, are overexpressed in MALT lymphoma lesions [30]. *miR-142-5p* and *miR-155* suppress the proapoptotic gene *TP53INP1* as their target [30]. The expression levels of *miR-142-5p* and *miR-155* are significantly increased in MALT lymphomas that do not respond to *H. pylori* eradication [30]. The expression levels of *miR-142-5p* and *miR-155* are associated with the clinical courses of gastric MALT lymphoma cases and these miRNAs may have a potential application as novel biomarkers for gastric MALT lymphoma.

Craig *et al.* reported the strong downregulation of the putative tumor suppressor miRNA, *miR-203*, in human MALT lymphoma samples, which results from extensive promoter hypermethylation of the *miR-203* locus and coincides with the deregulation of its target, *ABL1*. Treatment of lymphoma B cells with demethylating agents leads to increased *miR-203* expression and concomitant downregulation of ABL1, confirming the effectiveness of epigenetic regulation of this miRNA. These results show that the transformation from gastritis to MALT lymphoma is epigenetically regulated by *miR-203* promoter methylation and identifies ABL1 as a novel target for treatment [31].

Although generally considered an indolent disease, MALT lymphoma may have the ability to transform into gastric diffuse large B-cell lymphoma (gDLBCL). Craig *et al.* reported that Myc overexpression is detected in 80% of gDLBCLs, but only 20% of MALT lymphomas are spotted on a tissue microarray. FoxP1 overexpression is detectible in gDLBCL, but not in gastric MALT lymphoma. *miR-34a* is downregulated in malignant lymphoma, as are the targets of *miR-34a*, Myc and FoxP1 and *miR-34a* shows strong antiproliferative properties when overexpressed in DLBCL cells. *miR-34a* replacement therapy is therefore a promising strategy in lymphoma treatment [32].

8. Conclusions

As we are just beginning to understand the relationship between miRNAs and gastric malignancies and the number of identified miRNA genes is increasing, there is a potential for a large number of therapeutic targets and biomarkers in this area. Further studies are necessary to investigate whether miRNA-oriented therapy is an effective strategy for the chemoprevention of gastric malignancies.

References

1. Suzuki, H.; Iwasaki, E.; Hibi, T. *Helicobacter pylori* and gastric cancer. *Gastric Cancer* **2009**, *12*, 79–87.
2. Suzuki, H.; Nishizawa, T.; Hibi, T. *Helicobacter pylori* eradication therapy. *Future Microbiol.* **2010**, *5*, 639–648.
3. Nishizawa, T.; Suzuki, H.; Nakagawa, I.; Minegishi, Y.; Masaoka, T.; Iwasaki, E.; Hibi, T. Early *Helicobacter pylori* eradication restores sonic hedgehog expression in the gastric mucosa of Mongolian gerbils. *Digestion.* **2009**, *79*, 99–108.
4. Fukase, K.; Kato, M.; Kikuchi, S.; Inoue, K.; Uemura, N.; Okamoto, S.; Terao, S.; Amagai, K.; Hayashi, S.; Asaka, M. Effect of eradication of *Helicobacter pylori* on incidence of metachronous gastric carcinoma after endoscopic resection of early gastric cancer: An open-label, randomised controlled trial. *Lancet* **2008**, *372*, 392–397.
5. Saito, Y.; Suzuki, H.; Hibi, T. The role of microRNAs in gastrointestinal cancers. *J. Gastroenterol.* **2009**, *44*, 18–22.
6. Noto, J.M.; Peek, R.M. The role of microRNAs in *Helicobacter pylori* pathogenesis and gastric carcinogenesis. *Front Cell Infect Microbiol.* **2011**, *1*, 21.
7. Saito, Y.; Suzuki, H.; Tsugawa, H.; Suzuki, S.; Matsuzaki, J.; Hirata, K.; Hibi, T. Dysfunctional gastric emptying with down-regulation of muscle-specific microRNAs in *Helicobacter pylori*-infected mice. *Gastroenterology* **2011**, *140*, 189–198.
8. Suzuki, H.; Nishizawa, T.; Tsugawa, H.; Mogami, S.; Hibi, T. Roles of oxidative stress in stomach disorders. *J. Clin. Biochem. Nutr.* **2012**, *50*, 35–39.
9. Tsugawa, H.; Suzuki, H.; Saya, H.; Hatakeyama, M.; Hirayama, T.; Hirata, K.; Nagano, O.; Matsuzaki, J.; Hibi, T. Reactive oxygen species-induced autophagic degradation of *Helicobacter pylori* caga is specifically suppressed in cancer stem-like cells. *Cell Host Microbe* **2012**, *12*, 764–777.

10. Zhu, Y.; Jiang, Q.; Lou, X.; Ji, X.; Wen, Z.; Wu, J.; Tao, H.; Jiang, T.; He, W.; Wang, C.; *et al.* MicroRNAs up-regulated by CagA of *Helicobacter pylori* induce intestinal metaplasia of gastric epithelial cells. *PLoS One* **2012**, *7*, e35147.
11. Shiotani, A.; Uedo, N.; Iishi, H.; Murao, T.; Kanzaki, T.; Kimura, Y.; Kamada, T.; Kusunoki, H.; Inoue, K.; Haruma, K. *H. pylori* eradication did not improve dysregulation of specific oncogenic miRNAs in intestinal metaplastic glands. *J. Gastroenterol.* **2012**, *47*, 988–998.
12. Xu, X.; Li, W.; Fan, X.; Liang, Y.; Zhao, M.; Zhang, J.; Tong, W.; Wang, J.; Yang, W.; Lu, Y. Identification and characterization of a novel p42.3 gene as tumor-specific and mitosis phase-dependent expression in gastric cancer. *Oncogene* **2007**, *26*, 7371–7379.
13. Cui, Y.; Su, W.Y.; Xing, J.; Wang, Y.C.; Wang, P.; Chen, X.Y.; Shen, Z.Y.; Cao, H.; Lu, Y.Y.; Fang, J.Y. MiR-29a inhibits cell proliferation and induces cell cycle arrest through the downregulation of p42.3 in human gastric cancer. *PLoS One* **2011**, *6*, e25872.
14. Kan, T.; Sato, F.; Ito, T.; Matsumura, N.; David, S.; Cheng, Y.; Agarwal, R.; Paun, BC.; Jin, Z.; Olaru, A.V.; *et al.* The miR-106b-25 polycistron, activated by genomic amplification, functions as an oncogene by suppressing p21 and Bim. *Gastroenterology* **2009**, *136*, 1689–1700.
15. Kim, Y.K.; Yu, J.; Han, T.S.; Park, S.Y.; Namkoong, B.; Kim, D.H.; Hur, K.; Yoo, M.W.; Lee, H.J.; Yang, H.K.; *et al.* Functional links between clustered microRNAs: Suppression of cell-cycle inhibitors by microRNA clusters in gastric cancer. *Nucleic Acids Res.* **2009**, *37*, 1672–1681.
16. Volinia, S.; Calin, G.A.; Liu, C.G.; Ambs, S.; Cimmino, A.; Petrocca, F.; Visone, R.; Iorio, M.; Roldo, C.; Ferracin, M.; *et al.* A microRNA expression signature of human solid tumors defines cancer gene targets. *Proc. Natl. Acad. Sci. USA* **2006**, *103*, 2257–2261.
17. Xiao, B.; Liu, Z.; Li, B.S.; Tang, B.; Li, W.; Guo, G.; Shi, Y.; Wang, F.; Wu, Y.; Tong, W.D.; *et al.* Induction of microRNA-155 during *Helicobacter pylori* infection and its negative regulatory role in the inflammatory response. *J. Infect Dis.* **2009**, *200*, 916–925.
18. Zhu, W.; Xu, H.; Zhu, D.; Zhi, H.; Wang, T.; Wang, J.; Jiang, B.; Shu, Y.; Liu, P. miR-200bc/429 cluster modulates multidrug resistance of human cancer cell lines by targeting BCL2 and XIAP. *Cancer Chemother. Pharmacol.* **2012**, *69*, 723–731.
19. Tsukamoto, Y.; Nakada, C.; Noguchi, T.; Tanigawa, M.; Nguyen, L.T.; Uchida, T.; Hijiya, N.; Matsuura, K.; Fujioka, T.; Seto, M.; *et al.* MicroRNA-375 is downregulated in gastric carcinomas and regulates cell survival by targeting PDK1 and 14-3-3zeta. *Cancer Res.* **2010**, *70*, 2339–2349.
20. Guo, J.X.; Tao, Q.S.; Lou, P.R.; Chen, X.C.; Chen, J.; Yuan, G.B. miR-181b as a potential molecular target for anticancer therapy of gastric neoplasms. *Asian Pac. J. Cancer Prev.* **2012**, *13*, 2263–2267.
21. Gao, C.; Zhang, Z.; Liu, W.; Xiao, S.; Gu, W.; Lu, H. Reduced microRNA-218 expression is associated with high nuclear factor kappa B activation in gastric cancer. *Cancer* **2010**, *116*, 41–49.
22. Tie, J.; Pan, Y.; Zhao, L.; Wu, K.; Liu, J.; Sun, S.; Guo, X.; Wang, B.; Gang, Y.; Zhang, Y.; *et al.* MiR-218 inhibits invasion and metastasis of gastric cancer by targeting the Robo1 receptor. *PLoS Genet.* **2010**, *6*, e1000879.
23. Li, B.S.; Zhao, Y.L.; Guo, G.; Zhu, E.D.; Luo, X.; Mao, X.H.; Zou, Q.M.; Yu, P.W.; Zuo, Q.F.; Li, N.; *et al.* Plasma microRNAs, miR-223, miR-21 and miR-218, as novel potential biomarkers for gastric cancer detection. *PLoS One* **2012**, *7*, e41629.

24. Tseng, C.; Lin, C.C.; Chen, C.N.; Huang, H.C.; Juan, H.F. Integrative network analysis reveals active microRNA and their functions in gastric cancer. *BioMed. Central Syst. Biol.* **2011**, *5*, 1–11.
25. Kim, K.; Lee, H.C.; Park, J.L.; Kim, M.; Kim, S.Y.; Noh, S.M.; Song, K.S.; Kim, J.C.; Kim, Y.S. Epigenetic regulation of microRNA-10b and targeting of oncogenic MAPRE1 in gastric cancer. *Epigenetics* **2011**, *6*, 740–751.
26. Suzuki, M.; Suzuki, H.; Minegishi, Y.; Ito, K.; Nishizawa, T.; Hibi, T. *H. pylori*-eradication therapy increases runx3 expression in the glandular epithelial cells in enlarged-fold gastritis. *J. Clin. Biochem. Nutr.* **2010**, *46*, 259–264.
27. Lai, K.W.; Koh, K.X.; Loh, M.; Tada, K.; Subramaniam, M.M.; Lim, X.Y.; Vaithilingam, A.; Salto-Tellez, M.; Iacopetta, B.; Ito, Y.; *et al.* MicroRNA-130b regulates the tumour suppressor RUNX3 in gastric cancer. *Eur. J. Cancer* **2010**, *46*, 1456–1463.
28. Wang, M.; Li, C.; Yu, B.; Su, L.; Li, J.; Ju, J.; Yu, Y.; Gu, Q.; Zhu, Z.; Liu, B.; Overexpressed miR-301a promotes cell proliferation and invasion by targeting RUNX3 in gastric cancer. *J. Gastroenterol.* **2013**, in press.
29. Saito, Y.; Suzuki, H.; Imaeda, H.; Matsuzaki, J.; Hirata, K.; Tsugawa, H.; Hibino, S.; Kanai, Y.; Saito, H.; Hibi, T. The tumor suppressor microRNA-29c is downregulated and restored by celecoxib in human gastric cancer cells. *Int. J. Cancer* **2013**, *132*, 1751–1760.
30. Saito, Y.; Suzuki, H.; Tsugawa, H.; Imaeda, H.; Matsuzaki, J.; Hirata, K.; Hosoe, N.; Nakamura, M.; Mukai, M.; Saito, H.; *et al.* Overexpression of miR-142-5p and miR-155 in Gastric Mucosa-Associated Lymphoid Tissue (MALT) lymphoma resistant to *Helicobacter pylori* eradication. *PLoS One* **2012**, *7*, e47396.
31. Craig, V.J.; Cogliatti, S.B.; Rehrauer, H.; Wundisch, T.; Muller, A. Epigenetic silencing of microRNA-203 dysregulates ABL1 expression and drives *Helicobacter*-associated gastric lymphomagenesis. *Cancer Res.* **2011**, *71*, 3616–3624.
32. Craig, V.J.; Cogliatti, S.B.; Imig, J.; Renner, C.; Neuenschwander, S.; Rehrauer, H.; Schlapbach, R.; Dirnhofer, S.; Tzankov, A.; Muller, A. Myc-mediated repression of microRNA-34a promotes high-grade transformation of B-cell lymphoma by dysregulation of FoxP1. *Blood* **2011**, *117*, 6227–6236.

6

The Intertwining of Transposable Elements and Non-Coding RNAs

Michael Hadjiargyrou [1] and Nicholas Delihas [2,*]

[1] Department of Life Sciences, Theobald Science Center, Room 420, New York Institute of Technology, Old Westbury, NY 11568, USA; E-Mail: mhadji@nyit.edu

[2] Department of Molecular Genetics and Microbiology, School of Medicine, Stony Brook University, Stony Brook, NY 11794, USA

* Author to whom correspondence should be addressed; E-Mail: Nicholas.delihas@stonybrook.edu;

Abstract: Growing evidence shows a close association of transposable elements (TE) with non-coding RNAs (ncRNA), and a significant number of small ncRNAs originate from TEs. Further, ncRNAs linked with TE sequences participate in a wide-range of regulatory functions. Alu elements in particular are critical players in gene regulation and molecular pathways. Alu sequences embedded in both long non-coding RNAs (lncRNA) and mRNAs form the basis of targeted mRNA decay via short imperfect base-pairing. Imperfect pairing is prominent in most ncRNA/target RNA interactions and found throughout all biological kingdoms. The piRNA-Piwi complex is multifunctional, but plays a major role in protection against invasion by transposons. This is an RNA-based genetic immune system similar to the one found in prokaryotes, the CRISPR system. Thousands of long intergenic non-coding RNAs (lincRNAs) are associated with endogenous retrovirus LTR transposable elements in human cells. These TEs can provide regulatory signals for lincRNA genes. A surprisingly large number of long circular ncRNAs have been discovered in human fibroblasts. These serve as "sponges" for miRNAs. Alu sequences, encoded in introns that flank exons are proposed to participate in RNA circularization via Alu/Alu base-pairing. Diseases are increasingly found to have a TE/ncRNA etiology. A single point mutation in a SINE/Alu sequence in a human long non-coding RNA leads to brainstem atrophy and death. On the other hand, genomic clusters of repeat sequences as well as lncRNAs function in epigenetic regulation. Some clusters are unstable, which can lead to formation of diseases such as facioscapulohumeral muscular dystrophy. The future may hold more surprises regarding diseases associated with ncRNAs andTEs.

Keywords: non-coding RNAs; transposable elements; microRNAs; Alu sequences; endogenous retrovirus LTR; epigenetics; disease formation

---

1. Introduction

The genome is a dynamic entity, ever-changing as a result of endogenous DNA movement or the acquisition of exogenous DNA leading to genomic rearrangements. Such events have contributed to the plasticity and evolution of the genome and all of its complexity, much of which has slowly come to light over the past decades but whose pace has certainly accelerated in the last few years as a result of breakthroughs in genomic technologies, development of newer sequencing techniques, and availability of data in public databases by scientists all over the world. This new and vast genomic knowledge has led to a revolutionary and unparalleled in depth examination of the genome, transcriptome, proteome, interactome, *etc.* Indeed, the concept and definition of a gene may have to be altered [1,2].

However, one key question before us is what new functional loci and regulatory mechanisms have been formed during genomic evolution, especially as they pertain to the genesis of non-coding RNAs (ncRNAs), ncRNA regulatory roles and their association and interaction with transposable elements (TEs). In this review, we outline recent advances in origins of microRNAs (miRNA) and functional properties of ncRNAs as they pertain to their interaction with TEs, especially in humans. What emerges is a fascinating new picture of interconnected molecular interactions and regulatory pathways.

1.1. Non-coding RNAs

The development and use of new sequencing techniques, such as RNA-Seq has greatly increased our discovery of new RNAs [3,4]. The Encyclopedia of DNA Elements (ENCODE), an international project with the intent goal of determining functional elements of the entire human genome, has employed these techniques and found thousands of new RNA transcripts [2,5]. Surprisingly, about 75%–85% of the human genome is transcribed into primary and processed transcripts [2]; yet only 1.2% of the human genome encodes proteins. This suggests that most of the human genome space is devoted to RNA synthesis that is not devoted to protein-coding. Functions for most non-coding RNA (ncRNA) transcripts are unknown, but the future may hold fascinating prospects of finding new roles and molecular pathways. For example, thousands of circular ncRNAs (cirRNA) have recently been identified; they represent scrambled coding sequences that originate from exons (nonrandom products of RNA splicing) and are involved in small ncRNA regulation [6,7]. These cirRNAs are transcripts that do not encode proteins but have a regulatory role in the cell and thus are regulatory ncRNAs. A significant number of ncRNAs stem from non-protein-coding regions of the genome (intergenic regions), but many also originate from protein-coding regions as antisense transcripts, or from intron regions, and as just mentioned, from scrambled coding sequences. Many ncRNAs target mRNAs and induce their degradation. On the other hand, others are associated with regulation of transcription. Indeed the biological significance of regulation by RNA was grossly underestimated in the past.

Given the barrage of published studies on newly discovered ncRNAs, especially with eukaryotes, their classification and subclassification is indeed very challenging. However, Di Leva and Garofalo [8] used a simple classification system and presented three basic categories: (1) Housekeeping RNAs (rRNAs, tRNAs, snRNAs and snoRNAs); (2) short non-coding RNAs that are less than 200 nucleotides that include but are not limited to microRNAs (miRNAs), Piwi-interacting RNAs (piRNAs) and retrotransposon-derived ncRNAs, and (3) long non-coding RNAs (lncRNAs) that are greater than 200 nucleotides. lncRNAs are currently divided into long intergenic ncRNAs (lincRNAs) as they are encoded in intergenic regions, transcripts from introns, long ncRNA that are antisense transcripts in coding regions but do not encode proteins, and circular RNA transcripts from coding regions that have scrambled exon sequences and also do not encode proteins. Recent identification and classification of long ncRNAs lists additional categories [5].

A common theme that prevails in target RNA regulation by ncRNAs is the formation of short intermolecular RNA/RNA base-paired stems that contain Watson–Crick pairs, imperfect pairing (bulged and looped out positions), and non-canonical pairs. Imperfect ncRNA/target RNA pairing was determined experimentally with the first discovered and functionally characterized non-coding RNA in prokaryotes [9–11]. This is a type of ncRNA/target RNA binding that prevails throughout all biological kingdoms, although RNA binding proteins are also key factors in stable binding. Most of the ncRNAs discussed in this paper interact with their target RNAs via imperfect pairing.

1.2. Transposable Elements

TEs are defined as mobile genetic elements (pieces of DNA capable of moving to new locations); they also constitute the "mobilome" in that they can impact cell transcription [12]. TEs are characterized as either Class I retrotransposons or Class II DNA transposons. Retrotransposons are further subdivided based on the presence of their long terminal repeat (LTR) that contains the element's functions for mobility and regulatory sequences. LTRs flank endogenous retroviruses (ERV) and are capable of transposition. ERVs are mostly inactive viruses due to accumulation of mutations, but LTRs are active transposons, encode for all the essential factors for mobility and can multiply within a cell independent of the ERV. They carry promoter and enhancer sequences enabling host genes to be transcribed, as well as lncRNA genomic sequences. There are about 500,000 copies of LTR sequences in Homo sapiens, which make up about 8 percent of the human genome.

The other type of Class I retrotransposons are composed of the Long Interspersed Nuclear Elements (LINEs) as well as the Short Interspersed Nuclear Elements (SINEs). LINEs are mobile, whereas SINEs are non-autonomous DNA transposable elements and require LINEs for their mobility and propagation [13]. Duplicate copies are generated during mobility, with sequences identical to the original element inserted in a new location on the genome. In time, such copies may accumulate mutations independently and therefore will differ in sequence from their original sequence leading to increased divergence.

SINEs do not encode proteins. Alu sequences are classified as SINE elements and are about 300 nucleotides. They originated from the 7SL RNA transcript via retrotranscription into DNA. 7SL RNA forms part of the signal recognition particle. A hallmark of Alu sequences is that Alu RNAs fold into specific stable stem-loop structures, albeit with extensive imperfect base-pairing (Figure 1). Alus are highly abundant in mammalian cells, e.g., there ~10^6 copies of Alu in the human genome that make up ~11 percent of the genome [14], but most of these cannot be mobilized due to accumulation of mutations. Alu sequences are also found embedded in lncRNAs, where they are found to directly participate in base-pairing to target mRNAs (see Section 3.1). Additionally they are also found in piRNA genetic clusters (see Section 3.9).

Figure 1. Secondary structural model of Alu RNA. Modified from [15] with permission from Dr. Jennifer Doudna.

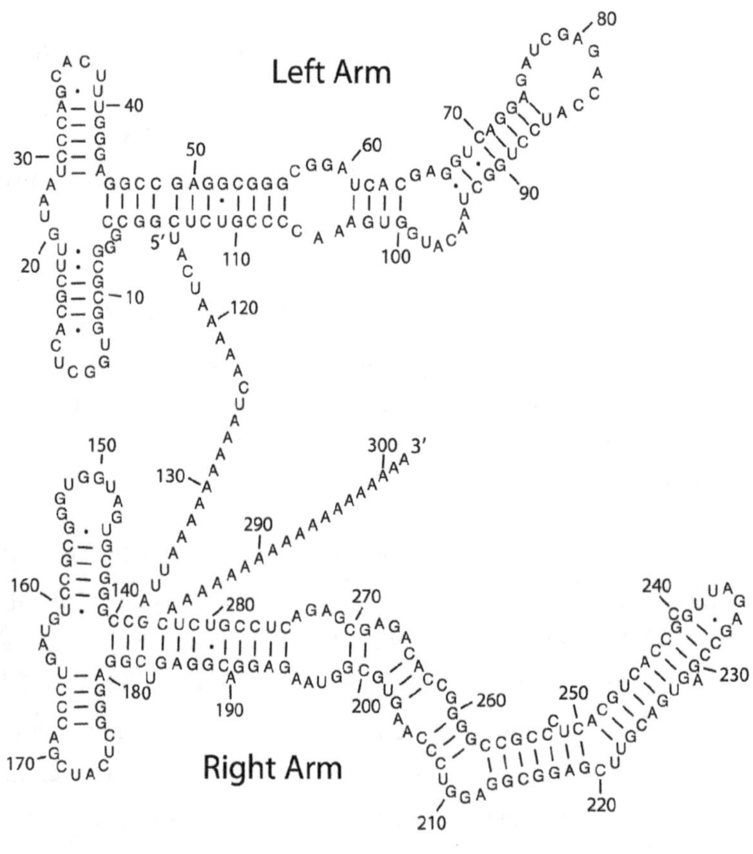

All-in-all, repeat sequences comprise 50%–75% of the human genome [14,16]. Repeats are broadly classified either as TE repeats or tandem repeats [17], but the major fraction of repeats represent transposable elements, either active or inactive. Tandem repeats represent a rather heterogeneous group, but some may have originated from TEs [18]. Repeats are regions where there is high recombination and this may sometimes result in genetic abnormalities.

In this review we focus on the origins of ncRNAs, the microRNAs from TEs and the interaction of ncRNA with TEs, primarily as found in mammalian tissues. Evidence is rapidly accumulating to show that this intimate association plays a central role in molecular and genetic mechanisms, such as RNA-based immunity. Furthermore, Cowley and Oakley have already described some of the impact of TEs in the promotion of human transcript diversity [12].

2. TE Origins of miRNAs

microRNAs (miRNA) are small non-coding regulatory RNAs molecules that function post-transcriptionally by binding to the 3'UTR of target mRNAs and ultimately inducing inhibition of target mRNA function. While the majority of miRNAs originate from intergenic genomic sequences, some arise from genes and TEs. The molecular origins of many miRNAs support the hypothesis that miRNA hairpin generation is based on the insertion of two related TEs flanking a single genomic locus (see below). As such, transcription that occurs across this locus leads to the biogenesis of functional miRNAs. One of the earliest studies to indicate that a number of mammalian miRNAs are derived from TEs utilized a bioinformatics approach, where the authors analyzed the Sanger miRNAi database using a software program that specifically detects well characterized repeats [19]. Specifically, 11 different miRNA precursors contained repeat sequences (4 derived from LINE-2 repeats and others with SINEs, LTRs and simple repeats). The majority of these miRNAs are highly conserved across human, mouse and rat, but some are confined to only one or two species.

In a subsequent and more in depth study, Piriyapongsa and Jordan investigated the relationship between human miRNAs and TEs by comparing the genomic locations of experimentally characterized human miRNA genes with the locations of annotated genomic TE sequences [20]. A correlation was observed, and nt sequence comparisons showed a high identity between seven members of the family of miRNAs hsa-mir-548 and the miniature inverted repeat transposable element (MITE), Made1. By use of human genome tilling arrays that visualize genomic expression, one Made1 element was found to be inserted into a transcriptionally active intergenic site. Made1 and other MITEs have palindromic sequences, and when transcribed, show a segment that has an imperfect stem-loop RNA structure. As RNAi-related enzymes can recognize this type of imperfect stem-loop and process it into the 22 bp mature miRNA sequences, the authors proposed that Made1 TE transcripts are processed into hsa-mir-548 miRNAs. The expression date and high sequence identities strongly support the proposed TE origin of several hsa-mir-548 family members.

In a related study, Piriyapongsa *et al.* used comparative genomic sequence data from the UCSC Genome Browser and evaluated the evolution of TE-derived human miRNAs [21]. They found 55 experimentally characterized human miRNA genes that were derived from TEs (LINE and SINE, LTRs and DNA transposons). Sequence comparisons showed that on average, TE-derived miRNAs are less conserved than non-TE-derived miRNAs. Further, a subset of these, are related to the ancient L2 and MIR families. Results also predicted an additional 85 novel TE-derived miRNA genes. Lastly, for some of the TE-derived miRNAs and their putative target genes, a comparison of expression patterns (miRNA *vs.* mRNA) was performed and revealed a number of them to have anti-correlated expression, consistent with regulation via mRNA degradation and thus supporting their regulatory function.

Examination of fourteen previously identified marsupial (Monodelphis domestica) specific miRNAs and their flanking sequences revealed that half of these miRNAs evolved from marsupial-specific TEs [22]. More specifically, six of these TE sequences were identified as LINEs and one as a Mariner DNA transposon. In a subsequent study, Yuan and colleagues also investigated another placental-specific miRNA gene family (miR-1302) that at the time of the analysis had 11 members that were distributed in the human genome (present in the miRBase) [23]. They demonstrated that all members of this family were derived from a single transposon (MER53 element). MER53 is a TE with a 193-bp consensus sequence and is characterized by the presence of terminal inverted repeats and TA target site duplications that can form palindromic structures [24]. Further analysis of the phylogenetic distribution and evolution dynamics of the miR-1302 family identified 36 potential paralogs of

MER53-derived miR-1302 genes in the human genome and another 58 potential orthologs in placental mammals and showed that these members of the hsa-mir-1302 family emerged within the last 180 million years since placental mammals diverged from marsupials. Lastly, the authors also explored the targets of the mature human miR-1302 and found 1835 genes with predicted function in transportation, localization, system development processes and their regulation, as well as in binding and in transcription regulation [23].

Genome-wide studies were performed using a comparative genomics approach in order to identify human miRNA paralogs (in mouse and rhesus) in segmental duplication pair data [25]. Of ~1000 miRNA genes and ~1000 mature sequences from human, ~700 miRNA genes and ~1000 mature sequences from mouse, and ~500 miRNA genes and ~500 mature sequences from rhesus, they identified 228 novel miRNA homologs in the rhesus genome and 22 novel miRNA homologs in the mouse genome (by using miRBase 16). Further, they also found 12 and 2 novel miRNA paralogs in the human and mouse genome, respectively, but none were found in the rhesus genome. In a separate analysis, the authors also examined the coverage density of repetitive elements, and if it was at least 50% in a miRNA gene or 100% in one of the associated mature miRNA sequences, then the miRNA gene was considered to be a RdmiR. Using this rule, the study identified a large number of miRNAs genes that overlap with repeats (TEs: LINEs, SINEs and LTRs) and other types of repetitive elements (DNA transposons, specifically MADE1 elements) within the three genomes; 226 (human), 115 (rhesus) and 141 (mouse). The study also identified a smaller number of possible repeat derived miRNAs, which they termed RrmiRs. Lastly, a computational analysis was conducted to investigate the functions of 19 of the conserved RrmiR families (between the three genomes), by identifying their target genes and it was found that the most significant targets are involved in transcriptional regulation, central nervous system development, and negative regulation of biological process. Collectively, the results of this study suggest that repetitive elements contribute to the de novo origin of miRNAs, and that large segmentation duplication events most likely accelerate the expansion of miRNA families (including RdmiRs).

A more recent study involved a comprehensive analysis of the genomic events responsible for the formation of ~15,000 annotated miRNAs against the principle datasets for TEs and ncRNAs and found 2392 (~15%) TE-based miRNAs [26]. The majority of these TE-based miRNAs may have originated via the proposed mechanism depicted in Figure 2.

The authors further investigated the exact TE origins of these 2392 miRNAs and showed that DNA transposons comprise the TE most frequently responsible for miRNA generation (891); others were: LTR Retrotransposon (414), Non-LTR Retrotransposon (814), LINE (312), SINE (353), Satellite (137) and others (136). This last category ("Other") had significant sequence identity to known noncoding RNA sequences (e.g., snoRNAs, scaRNAs, tRNAs). Lastly, a hypothetical scheme proposes that the regulatory miRNAs may have arisen via selective subfunctionalization created by the associated benefit of regulating host genes containing portions of TEs.

Figure 2. Schematic of proposed origin of TE-based miRNAs. When two related but not identical LINE1elements insert themselves near each other, but on opposite strands of the DNA (in blue), they can create a precursor miRNA containing an imperfect stem-loop upon transcription (red line). The pre-miRNA is shown above the arrow and transcription is indicated from the positive strand LINE1. The stem is potentially recognized and processed by the endogenous RNAi machinery. The pre-miRNA stem-loop depicted is representational. Modified from [26].

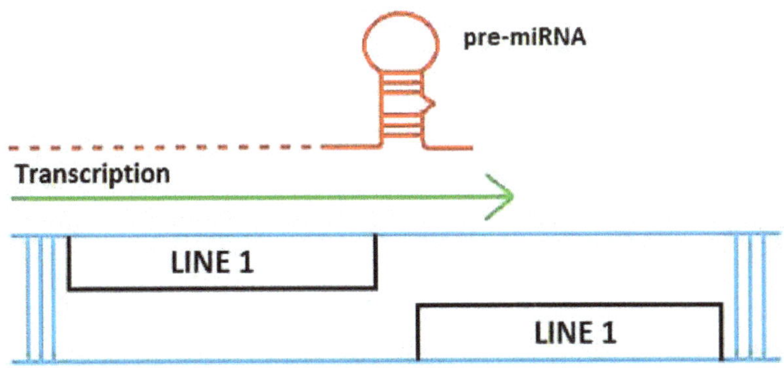

Using the miRBASE database, a more recent study sought to map all miRNA precursors to several genomes and to determine the repetition and dispersion of the corresponding loci, as well as the repetitive elements overlapping these loci. To facilitate this analysis an automatic method called ncRNAclassifier was used in order to classify the relationship of TEs with pre-ncRNAs [27]. By applying this method, a correlation between the number of pre-ncRNA candidates and the presence of TEs was determined using six genomes (frog, human, mouse, nematode, rat and sea squirt). The results indicate that 235 and 68 mis-annotated pre-miRNAs correspond completely to TEs out of 1426 human and 721 mouse pre-miRNAs of miRBase (10.0 release), respectively. Further, the various types of TEs involved were also identified and include (MADE1 and other MITEs, DNA transposons, LTR/ERV, CR1/RTE, L1, SINE, other non-LTR). Lastly, the authors suggest that the ncRNAclassifier can be openly used to determine if a given ncRNA hairpin sequence corresponded to a TE sequence.

An investigation of the TE origins of miRNAs focused on the MER (MEdium Reiteration frequency), interspersed repeats in the genomes of primates, rodentia, and lagomorpha) transposon-derived miRNAs in human genome. Once again, a bioinformatics approach was undertaken to identify the specific miRNAs that are derived from palindromic MERs, by analyzing MER paralogs in human genome. Results from this study identified three miRNAs derived from MER96 located on chromosome 3, and MER91C paralogs located on chromosome 8 and chromosome 17 [28]. More importantly, this study also experimentally validated the interactions between these MER-derived miRNAs with AGO1, AGO2, and AGO3 proteins (involved in gene silencing and act as the catalytic component of the RNA induced silencing complex [RISC]).

Lastly, there are additional classes of small ncRNAs that originated from TEs and/or consist of TE sequences, e.g., certain piRNAs and the specialized SINE and Alu transcripts that function as small ncRNAs; these are discussed below with respect to their functional roles.

In summary, a sizable proportion of miRNAs appear to be derived from TEs. It is highly probably that future bioinformatic analyses will increase the number of miRNA-transposable element relationship as not all miRNAs have been discovered and most likely, all consensus repetitive elements have not yet been described. As ncRNAs serve such a critical regulatory role, TE colonization of the genome has given rise to a number of regulatory processes, several of which we discuss here.

3. Interaction of TEs with ncRNAs—Functional and Disease-Related Significance

3.1. Alu Element Embedded in Long ncRNAs and mRNAs—Crucial Role in Target mRNA Decay

Gong and Maguat revealed the importance of Alu intermolecular base-pairing to lncRNA-induced degradation of mRNA [29,30]. By computational analysis, Alu sequences are found present in ~380 lncRNAs in HeLa cells [30]. In addition, mRNAs were identified that contain an Alu sequence in their 3'UTR regions. Certain mRNAs are targets of the double-stranded RNA binding protein Staufen1 (Stau1), which can induce degradation of the mRNAs. Co-immunoprecipitation experiments show an Alu-containing lncRNA, which originates from chromosome 11, binds to and decreases the abundance of target messages, *i.e.*, plasminogen activator inhibitor type 1 (SERPINE 1) mRNA and an mRNA that encodes an unknown protein termed FLJ21870. Both these mRNAs have an Alu sequence in their 3'UTRs. By secondary structure modeling, it was shown that Alu sequences in lncRNA can base-pair to Alu sequences in the 3'UTR of target mRNAs by intermolecular base-pairing with a stable $-\Delta$ G (Figure 3). This pairing is imperfect and contains bulged positions. The intermolecular stem structure, formed by the interaction between the Alu sequences present in the lncRNA and in the 3' UTR of the mRNA serves as the binding site for Stau1, which subsequently recruits UPF1, a protein required to initiate mRNA decay. Several hundred other lncRNAs contain Alu sequences and have the potential to base-pair with Alu-containing mRNAs, but possible functions of the majority of the several hundred Alu-containing lncRNAs are unknown.

Alu/Alu sequence pairing is not the only interaction seen with Stau1-binding mRNAs, but it represents an important "variation on a theme". In previous experiments with another Stau1-binding mRNA that contains no Alu sequences but a perfect 19 bp stem in the 3'UTR, it was shown that Stau1 binds the perfect 19 bp stem that is formed intramolecularly between distal sequences in the 3'UTR of the mRNA [31] (Figure 3, left). ARF1 is an ADP-ribosylation factor 1 protein, and the *arf1* mRNA transcript has a Stau1-binding site. The 19 bp stem is phylogenetically conserved in different mammalian species. The question remains, how many Stau1-binding mRNAs are targets for decay via intramolecular bp within a message and how many by intermolecular Alu/Alu pairing. As mentioned, there are several hundred lncRNAs that have Alu sequences. In addition, in a related topic, it should also be pointed out that inverted Alu elements are found in many human mRNA 3'UTR sequences. These can form double-stranded intramolecular stems, and they appear to affect mRNA translation efficiency [32].

Figure 3. Schematic of binding of Staul to 3' UTR mRNA intramolecular stem (*arf1* mRNA) (**left**) and to intermolecular stem formed by Alu sequence in lncRNA with Alu sequence within 3'UTR mRNA. Upf1is an RNA helicase. The two RNA duplex stems shown are not drawn to scale. Modified from Gong and Maquat [30].

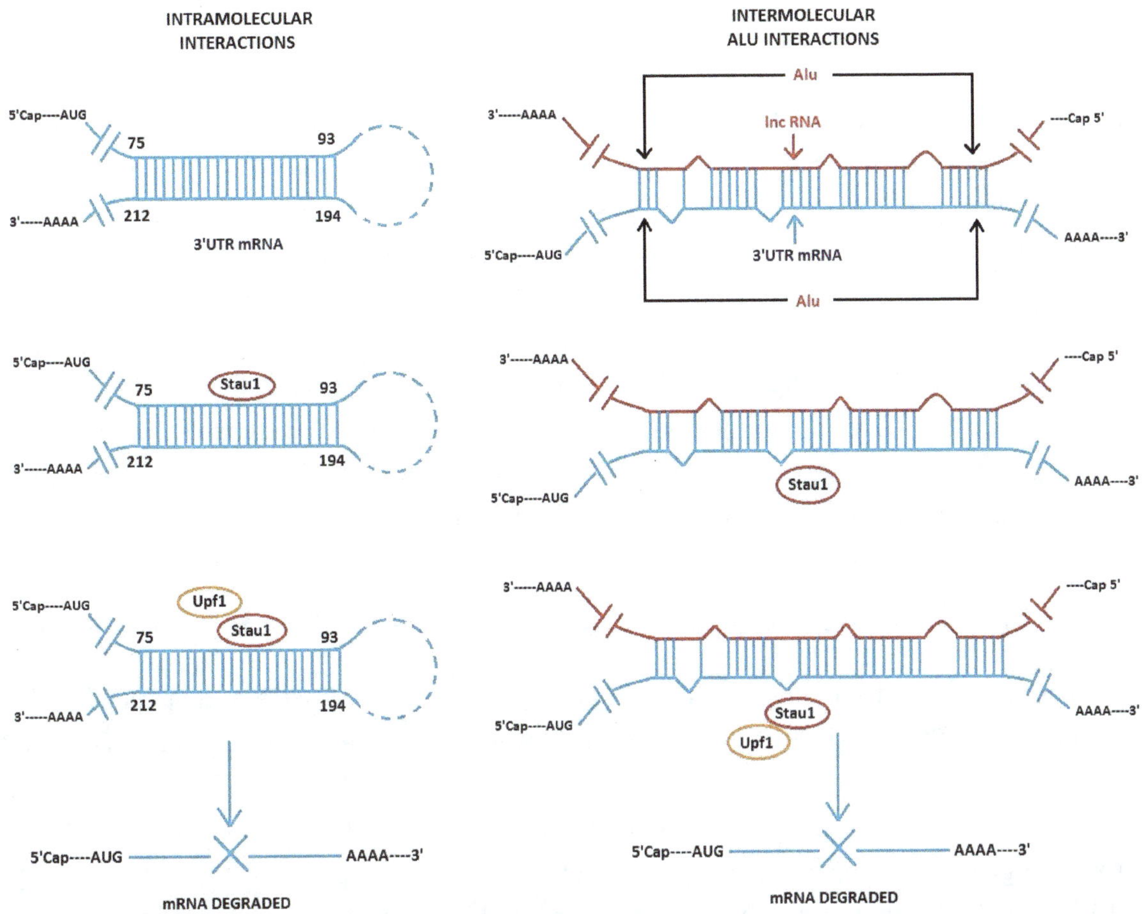

Thus, mRNAs that have a Staul binding site, whether it is a formed intermolecularly via Alu/Alu lncRNA/mRNA imperfect base-pairing or by perfect Watson–Crick intramolecular pairing within the mRNA can be targeted for degradation. This raises an interesting question concerning the specificity of Staul and its recognition sites on duplex RNA stems and imperfect *vs.* perfect double stranded stems. The probability of mistakes in recognition must be very low, yet two or more types of RNA tertiary structures are recognized with great accuracy. Crystal structures of perfect and imperfect double-stranded RNA/Staul-protein complexes would be of major interest. Gleghorn *et al.* have already determined crystal structures of Staul and showed that dimerization of Staul occurs by a degenerate dsRNA-binding domain on Staul [33]. And in another recent study, it was revealed that dimerization can also involve Stau2 [34].

In another study from this laboratory, rodents appear to use the same mechanism of mRNA regulation involving intermolecular imperfect base-pairing between lncRNAs and mRNAs, only this occurs via SINE elements B1, B2 or B4 found at 3'UTRs and in lncRNAs [35].

3.2. Long Non-Coding Antisense RNA Controls mRNA Translation—Importance of Embedded SINE/Alu Repeats

Carrieri *et al.* determined that a long ncRNA transcript, which is partially antisense to ubiquitin carboxy-terminal hydrolase L1 (Uchl1), is essential for increased translation of uch1 mRNA [36]. Uch1 is a neuron associated protein in mammals and may be involved in neurological disease formation [37,38]. Two segments of the antisense lncRNA are crucial for function: the 5' end of the lncRNA transcript that overlaps the uch1 sense transcript, and SINEB2 and Alu repeat segments located downstream on the antisense lncRNA. By using a bioinformatics approach, 31 antisense lncRNAs have been pinpointed that contain SINE/Alu sequences in their 3' end half regions. These RNAs can potentially base-pair to sense transcripts via their 5' ends to the 3'UTR of mRNAs, in a similar manner as uch1 sense RNA/antisense RNA transcript pairing occurs. However the mechanism by which the SINE sequences act to control translation of uch1 mRNA has not been determined, but a hint comes from data showing the orientation of the SINE in the antisense lncRNA is important in rescue experiments [36]. This implies a possible RNA/RNA base-pairing mechanism. This work may define a separate class of regulatory lncRNAs in mammals that appears to differ from the Alu-containing lncRNAs described by Gong and Maquat [29].

3.3. Point Mutation in LINE-1/Alu Element Embedded in a lncRNA Results in Lethal Brain Disease

A primate conserved LINE-1 sequence is found embedded in a lncRNA that maps to human chromosome 8p22 [39]. This LINE sequence also overlaps with an Alu sequence. The lncRNA constitutes a unique transcript originating from an intron. This RNA most likely has regulatory functions. A rare single point mutation, A to G in the LINE-1/Alu sequence is associated with brainstem cell atrophy, a genetic abnormality that results in lethal infantile encephalopathy in humans [39]. The LINE-1 is a degenerate retrotransposon and assumed not to be mobile. It was experimentally determined that in patient brain tissues, the expression of the mutant lncRNA was reduced nearly 10-fold relative to unmutated RNA control levels. mRNAs of two genes that map in the same locus as the lnRNA were found to be unchanged. In addition, knockdown experiments against wild-type lncRNA using siRNA showed a significant increase in apoptotic cells.

Several hypotheses have been presented to explain the drastic phenotypic effects of the single base-pair change in the LINE-1/Alu sequence [39]. One is that piRNAs accidentally target the lncRNA transcript via base-pairing with the mutated LINE-1/Alu sequence and induce silencing of the lncRNA. Another involves inadvertent SRP protein recognition of the mutated sequence, which resides in a conserved internal loop of the Alu secondary structure embedded in the lncRNA. This loop is also present in the 7SL RNA secondary structure.

Thus, this is an example of a point mutation in an embedded TE in a lncRNA sequence that produces human disease. What is not known is what exact role the unique lncRNA plays in normal cell functions and the normal function of the embedded LINE-1/Alu sequence, although there may be involvement in regulatory networks during brain development [39].

3.4. Facioscapulohumeral Muscular Dystrophy (FSHD)-Involvement of Tandem Repeats and Long ncRNA

Repeat elements in primate genomes are primarily of two types: repeats of transposable element integration into the genome and tandem repeats (microsatellites, minisatellites and macrosatellites). There is evidence that ~25 percent of minisatellite tandem repeats are derived from TEs [18]. The evolutionary origin of most macrosatellites is uncertain, however the macrosatellite tandem repeat at human chromosome region 4q35 discussed below has an interesting origin and possibly related to a retrotransposition. It represents a remnant of an unprocessed mRNA from a primate ancestral retrogene. The retrogene was lost but the retrotranscribed unprocessed mRNA (which included introns) was retained. It proceeded through an expansion in the last 25 million years and was greatly multi-copied in Homo sapiens [40–43].

The tandem repeat macrosatellite sequences at chromosome region 4q35 and a chromatin associated long ncRNA are intimately involved in a human genetic disorder termed facioscapulohumeral muscular dystrophy (FSHD) [17,44]. This is a process perhaps best described as a "cause and effect" mechanism involving loss of repeat sequences, the activation of a lncRNA and resultant epigenetic changes in FAHD patients.

Normal individuals carry a certain number of repeats (11–100) in the facioscpulohumeral (FSHD) locus. FSHD patients have a decreased number of repeats [17]. The FSHD locus maps to the chromosomal region 4q35. The region containing the repeat sequences is termed D4Z4 (Figure 4). A lncRNA termed DBE-T is partly encoded by the D4Z4 repeat locus at its 3' end. The genomic repeat sequences serve as binding sites for Polycomb proteins (PcG). PcG proteins have multiple functions but are epigenetic suppressors and are needed to suppress the *FSHD* genes that are silent in normal individuals. However, when there is a loss of genomic repeats resulting in less than 11 repeats, as is found in FSHD patients, DBE-T lncRNA transcription is activated. The RNA is normally silent and is only detected in FSHD patients. DBE-T lncRNA recruits Ash1L, a histone methytransferase to the FHDS locus, which subsequently results in transcription of locus genes due to epigenetic changes in chromatin. Thus these are changes involving the loss of repeat sequences in the D4Z4 region, transcription of a lncRNA with resultant chromatin remodeling and transcription of normally silent genes from the FSHD locus (Figure 4).

Figure 4. Involvement of repeat units (in orange) and lncRNA (in red) in FSHD muscular dystrophy. DBE-T, lncRNA; TRG, Trithorax proteins; other terms are defined in text. From Casa and Gabellini [17]. Reprinted with permission of publisher.

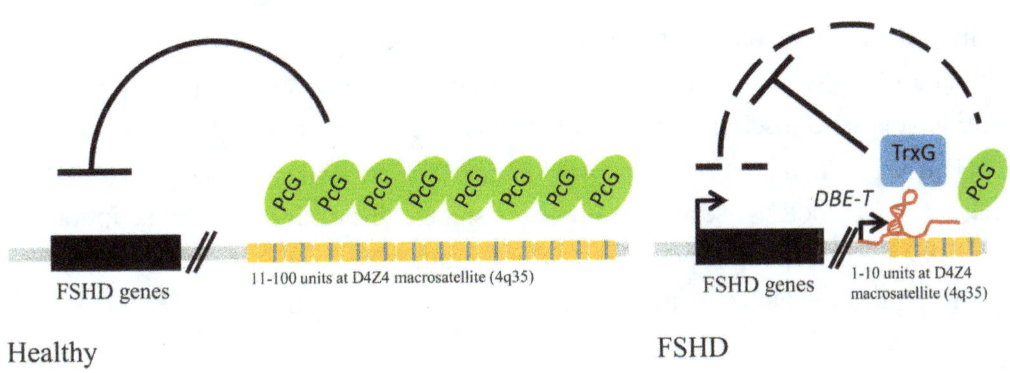

This fascinating study raises questions regarding the cell's reliance of repeat sequences for crucial epigenetic regulation. The D4Z4 region is highly variable in normal individuals, again, 11–100 copies of repeats. Genetic rearrangements appear to have take place in the 4q35 region in FSHD patients [45]. It seems that the cell's reliance on variable number of repeat sequences, as in the D4Z4 region for regulation appears to be a flawed mechanism used for epigenetic silencing. There are also other genetic abnormalities involving deletions in unstable chromosomal regions that have repeat sequences, as in chromosome locus 22q11.2 in DiGeorge Syndrome [46]. Although these repeat regions widely differ in structure and properties, e.g., DiGeorge syndrome can involve rearrangements between palindromic AT-rich repeats, it appears that genomic repeat sequences may constitute weak points in terms of maintenance of genomic fidelity, even though they have their important functions in the cell.

3.5. Embedded Alu Sequences Can Take Part in Alternative Splicing and A to I Editing in Human mRNAs

As opposed to other embedded Alu sequences described in this manuscript, the association of Alus with alternate splicing and RNA editing represent nuclear processes that Alu sequences participate in. Alu sequences are found in exons in about 5% of alternatively spliced mRNAs [47]. The presence of an Alu in exons of pre-messenger RNA transcripts can provide alternative splice sites and parts of the embedded Alu sequences can be incorporated into the processed mRNA [48–51].

Approximately 45 percent of Alu sequences are found in introns in both 5' to 3' and reverse orientations and are present in multiple copies [47,52]. These Alu sequences can potentially form double-stranded stems within a transcript when two Alu RNAs are in antiparallel orientation [53]. This enables RNA editing to take place [54] and can lead to premature stop codons or changes in codon reading [47]. Alus are prominent targets for RNA editing [54]. This is an example of how Alu RNA secondary structure can participate in altering molecular processes.

3.6. Human Endogenous Retrovirus (HERV) LTR Transcripts

About 8% of the human genome consists of human endogenous human retroviruses (HERV) [55]. HERVs cannot produce a viable virus due to mutations, but its associated LTR transposons serve a vital role in cell transcription. In addition, the human genome contains several thousand copies of single long terminal repeats (sLTRs), which originally stem from HERV [55]. These sLTRs carry no viral genes but can function as promoters and enhancers when found upstream of genes. However some sLTRs are situated in introns and are transcribed into RNA. Xu and co-workers, while studying the expression of HERV-9 U3 sLTR show that sLTR RNA transcripts are both sense and antisense RNAs, but the U3 sLTR antisense transcript can bind key transcription factors involved in cell proliferation. The sense sLTR RNA does not bind transcription factors. Importantly, malignant cells express lower levels of antisense sLTR RNA relative to sense transcripts than normal cells. The antisense sLTR RNA, which is ~550 nt appears to be a novel sLTR RNA species. The authors propose that the antisense sLTR lncRNA serves as a trap for some cell proliferation transcription factors. This may have significance in terms of a possible lack of inhibition of growth in cancer cells. Thus this is an example of a regulatory lncRNA that is encoded by a transposon (sLTR), but binds to and inactivates proteins and not other RNAs. Importantly, this ncRNA may play a crucial role in cell proliferation [55].

3.7. Interrelatedness between HERV LTRs and Intergenic Long Non-Coding RNAs

In a different study concerning HERV LTRs, other TEs and lncRNAs, Kelly and Rinn provide a comprehensive analysis of human TE sequences in long intergenic non-coding RNAs (lincRNAs) and conversely, the presence of lincRNAs sequences in transposons [56]. About 7700 lincRNAs overlap TEs and about 1530 lincRNAs are devoid of TEs; thus about 80 percent of human lincRNAs are associated with TEs. lincRNAs display a strikingly non-random association with transposable elements; the majority overlap human endogenous retrovirus (HERV) LTRs and a small minority are associated with LINE or SINE elements.

Interesting observations were made on the orientation of HERV transposons relative to lincRNAs and expression in specific cells. A large number of HERV LTRs are situated at the transcriptional start sites (TSS) of lincRNAs and in the sense orientation. This suggests that HERV LTRs provide regulatory signals for lincRNAs. lincRNAs display a marked stem cell specificity in expression, but lincRNAs that have no LTR associations are expressed highest in testes. On the other hand, lincRNAs that contain Alu sequences are expressed in all cell lines but testes. lincRNAs most likely function in specific tissues, but Alu-containing lincRNAs may be deleterious in testes, as they are not expressed in these tissues. Thus, there is a tissue-specificity in expression [56]. TEs may work "hand-in-hand" with lincRNAs as functional units in particular cells.

3.8. Regulatory Non-Coding Circular RNAs

Circular RNAs were first characterized in human and other mammalian cells about 20 years ago [57,58], however they were initially detected in electron micrographs over 30 years ago [59]. These RNAs consist of scrambled protein coding exons, *i.e.*, the order of exons is not the same as in the genomic sequence of protein coding regions. Scrambled exon sequences were discovered in RNA transcripts in rodents and humans [60]. Subsequently, additional cirRNAs were found [61–63]. Recently, by using deep sequencing of RNA techniques and a bioinformatics approach, Saltzman *et al.* [64] discovered several hundred circular RNAs in human cells and surprisingly, Jeck *et al.* [65] using circular enrichment techniques as well as bioinformatics determined that greater than 14% of human fibroblast gene transcripts are cirRNAs (over 25,000 circular transcripts).

Although circular RNAs arise from protein-coding regions, they do not encode proteins. They are thus a separate class of long non-coding RNAs. Functions of cirRNAs were not elucidated for over 20 years since their discovery. However, the field has now moved dramatically, with two laboratories determining that some circular RNAs serve as "sponges" that can bind approximately 70 microRNAs and thus inactivate the microRNAs [6,7]. This shows that circular RNAs have regulatory functions, *i.e.*, they "regulate the regulator", the microRNAs.

Via bioinformatics analyses, it was shown that Alu elements are found in upstream and downstream introns that straddle the exons that are circularized, and that Alu sequences tended to be inverted and thus complementary [65]. Intron pairing may contribute and be essentially to circularization of exons by complementary base-pairing between Alu elements in the upstream and downstream introns. If this is so, then Alu elements play a major role in formation of circularized RNAs. Related to this, there is precedent for Alu pairing in intons during alternative splicing [53].

3.9. piRNAs—Known Regulators of TEs

piRNAs are a class of small non-coding RNAs that are 26–31 nt. They interact with Piwi proteins, hence their name. The Piwi family is regulatory proteins that were originally defined in Drosophila as P-element induced wimpy testis [66]. piRNAs are abundantly found in germ line cells, especially in mammals, e.g., several million piRNAs are found in mammalian testes. Genetic regions that encode piRNAs consist of clusters. These clusters have repeats of piRNA sequences and there can be as many as 1000 copies of piRNAs in a cluster. piRNAs are processed from long precursors transcripts but little is known of the biogenesis of piRNAs and the number and functions of the associated proteins.

Some piRNA clusters consist of transposon or remnants of transposon sequences. Thus piRNAs can have sequences complementary to transposon sequences and can recognize their targets by base-pairing, either by perfect or imperfect base-pairing. A major role of the piRNA/Piwi protein complex in germ line cells is to protect cells from invading transposons. This is a type of "genetic immune system" that is found in both eukaryotes and prokaryotes. For example, the CRISPR complex in bacteria and archae functions by a comparable mechanism (albeit with significant variations on a theme) to protect cells from invasion by plasmids and viruses [67–69]. Both the piRNA/Piwi and CRISPR immune complexes function by an RNA-based mechanism.

piRNA functions have been studied in detail in *Drosophila*, *C. elegans* and mammalian cells. Functions are complex and may differ in different species, but a large fraction of piRNAs represent antisense transcripts to transposon transcripts [70–72]. A basic mechanism of action of piRNAs, first deduced in *Drosophila*, has the following scenario: when the cell is previously exposed to a TE but now experiences an overload of this transposon, piRNAs containing complementary sequences to the TE will base pair with and induce degradation of the TE RNA via the Piwi proteins. When the cell encounters a transposon that it has not been exposed to before, the TE by chance, may incorporate into the DNA in a piRNA-encoding cluster and thus its sequence can become part of the piRNA cluster (however, we are not aware that the probability of incorporation has been experimentally determined). Via the same mechanism mentioned above, piRNA transcripts that are antisense to the new transposon RNA induce degradation of the TE RNA via the piRNA–Piwi complex [71]. Thus, this is an immune system that helps keep transposons in check [70]. It is of interest that cell survival depends in part, on the probability of incorporation of the TE into a piRNA cluster, *vs.* the probability of insertion into and inactivation of an essential gene. However, other protective mechanisms also operate to limit TE activity.

Additional processes of piRNAs have been determined. In nematodes, piRNAs detect a TE sequence via imperfect base-pairing and then induce another small RNA class, termed 22G-RNAs to silence a transposon [73]. Some processes involve epigenetic mechanisms. For example, in Drosophila, nuclear piRNAs can target a transposon and thus direct Piwi proteins to repression chromatin and thus transcription of the TE [74]. Additionally, piRNAs may also induce the methylation of TE LINE-1 DNA in humans. This can prevent transcription of the transposon and thus assure that the TE DNA will remain dormant and not be expressed [75].

The piRNA/Piwi complex is also essential in genetic imprinting in the case involving DNA methylation of the imprinted locus Ras protein-specific guanine nucleotide-releasing factor 1 (Rasgfr1) locus in mouse germ line cells [76]. Mutants affecting piRNA expression correlate with defects in DNA methylation of Rasgrfl. The differentially methylated region (DMR) associated with Rasgfrl contains a LINE1 retrotransposon and sequences consisting of 23–31 nt small RNAs; these correspond to piRNAs. Yet, a different locus on chromosome 7 also has a region that produces piRNAs that have a good match to the Rasgfrl DMR piRNA sequences. The authors propose that piRNAs generated from chromosome 7 target the retrotransposon in the DMR of the imprinted Rasgrfl locus and that chromosome 7 piRNAs may direct methylation of the Rasgrfl locus [76]. Thus piRNA, in addition to being a post-transcriptional regulator may also be involved in epigenetic regulation of chromosomal genes and genetic imprinting.

3.10. SINE/Alu Transcripts Function as ncRNAs in Gene Regulation at the Transcription Level

The non-autonomous retrotransposon SINE sequences are transcribed into small ncRNAs, but like some piRNAs, they can function at the level of transcription. A SINE transcript termed B2, which is found in the nucleus of mice was shown to be 177 nt. This B2 RNA transcript is conserved in rodents. This RNA binds polymerase II during the heat shock response, disrupts the polymerase/promotor interaction and represses transcription from protein gene promoters [77–79]. It is unclear whether the ncRNAs also interact with the promoter sequence. The human counter part, an Alu RNA, functions in a similar manner during heat shock, even though it's nucleotide sequence and secondary structure differs from B2 RNA [47,77]. Expression of both these small ncRNAs is increased during the heat shock response.

In other studies, a processed human Alu RNA has a sequence that is identical to that of a piRNA present in mammalian testes [80]. The processed transcript, termed piAluRNA is found in the nucleus of human adult stem cells, appears to interact with several nuclear proteins and may be involved in several processes. These include transcription, chromatin organization, organelle organization, DNA repair and cell cycle control [80]. An RNA affinity assay with synthetic oligonucleotides representing a segment of the piAluRNA and high-resolution mass spectrometry-LC-MS were used to identify interacting proteins. Functional studies are needed, but the current binding data strongly suggest the involvement of piAluRNA in several of these nuclear functions. These studies may greatly extend the roles of small ncRNAs in cells.

It is important to point out that binding and repression of proteins by ncRNAs also occurs in prokaryotes. For example, the bacterial ncRNA 6S RNA regulates RNA polymerase by binding sigma factor 70 factor and subsequently repressing RNA polymerase activity from sigma 70 promoters [81–83]. Thus, this is another example of the basic principles of molecular regulation that encompass all biological kingdoms.

4. Conclusions

We presented examples of ncRNAs originating from TEs, such as miRNAs derived from MADE1 TE. In addition, there are piRNAs that consist of TE sequences and processed SINE and/or Alu transcripts that function as small ncRNAs. Some findings show TE-derived miRNAs to be less conserved than non-TE-derived miRNAs [21], which may imply a species-specific function of TE-derived ncRNAs. There is growing evidence for the meshing of TE sequences with ncRNAs involving both structure and function, and this association has resulted in formation of new regulatory pathways. It is obvious that TE transcripts are of an enormous asset to organisms, either as embedded sequences in ncRNAs or as individual RNAs. The interaction between TEs and ncRNAs could be looked at as a "symbiotic partnership" between the cell and transposable elements involving structure and function.

Cells offer TEs the means to multiple and maintain stability. Nevertheless, when active, TEs become overabundant and can become a threat to survival. The cell then elicits mechanisms to limit their replication. In this process, cells often use the TE sequences themselves to limit proliferation, as in the case with piRNAs involved in the genetic immune system [71].

Of special significance is that ~80% of long intergenic non-coding RNAs are associated with TEs, and in a nonrandom fashion [56]. They most likely serve functional roles, e.g., retrotransposon LTRs may provide regulatory signals for associated lincRNAs. This may be the tip of the iceberg in terms of TE/lincRNA functions as there are thousands of transcripts found in humans.

Most known ncRNA/target RNA interactions consist of short imperfect base-paired stems, and many miRNAs can bind to and regulate multiple target sequences. But this raises the question of the probability of making mistakes and targeting the wrong RNA. Other factors such as RNA-binding proteins also contribute to RNA recognition and stable formation, but the probability of mistakes must be very low, as imperfect RNA/RNA interactions appear to be highly specific. For example, the Alu-ncRNA/Alu-target RNA recognition must be as stable and specific as that of the intramolecular Watson–Crick base-paired stem recognized by Stau-1 in mRNA-induced degradation in human cells [29,30]. These stems are an interesting example of divergent RNA/RNA structures that bind the same RNA-binding protein. Three-dimensional RNP structures are needed to understand duplex conformations and protein binding sites on the two types of RNA stems.

Multiple ncRNA/target RNA pairings by the same ncRNA are also found in prokaryotic interactions with binding of ncRNAs to different target sequences and with different predicted base-pairings, e.g., see [84,85]. Thus stable short ncRNA/target RNA sequence pairings and multiple targeting are found in prokaryotes and eukaryotes. There is a beauty in the specificity and stability of small imperfect RNA/RNA pairings, and with employment of transposable elements in this binding process, and at least in eukaryotes, the cell appears to also use TE sequences to a significant extent in this form of binding.

It is of interest that some TE ncRNAs have been shown to bind proteins. For example, the lncRNA from an LTR retrotransposon situated in an intron binds transcription factors involved in cell proliferation [55]. The B2 ncRNA from an Alu transcript binds polymerase II [47]. The piAlu RNA binds nuclear proteins [80]. Thus transposon RNA transcripts show versatility in function in that these can also repress nuclear protein functions. The 6S RNA in bacteria, although not a TE transcript, is an example of a prokaryotic ncRNA that binds and inhibits the bacterial polymerase enzyme. This adds to the universality of ncRNA-related regulatory mechanisms in biological species. The 6S RNA was the first ncRNA to be sequenced [86], albeit its function was not determined until ~30 years later [81]!

The LINE-1/Alu element in a human lncRNA plays a pivotal role in formation of disease. A single mutation in the embedded TE causes human brainstem atrophy [39]. Whether the embedded

LINE-1/Alu element is more prone to mutation than the rest of the lncRNA is not known, but this shows that a point mutation in a TE is the cause of atrophy and death and not a mutation in a protein gene. This adds to the spectrum of mutations that cause human disease, *i.e.*, non-protein-coding genomic sequences can be important factors in human disease. Related to this, the deregulation of lncRNA transcription in diseases such as cancer has recently been highlighted [87,88]. As there are thousands of ncRNAs associated with TEs whose functions have not been determined, the future may possibly hold some interesting surprises with respect to diseases that may have an aberrant ncRNA etiology.

Acknowledgements

We thank Kiran Kumar Govindaiah for aid with the figure drawings, and Drs. Stefanie Mortimer and Jennifer Doudna for providing the Alu secondary structure drawing. We also are grateful to Dr. Davide Gabellini for clarification of the evolution of and molecular mechanisms pertaining to the FSDH repeat region. ND acknowledges support from the Department of Molecular Genetics and Microbiology, Stony Brook University.

References

1. Mattick, J.S. Rocking the foundations of molecular genetics. *Proc. Natl. Acad. Sci. USA* **2012**, *109*, 16400–16401.
2. Djebali, S.; Davis, C.A.; Merkel, A.; Dobin, A.; Lassmann, T.; Mortazavi, A.; Tanzer, A.; Lagarde, J.; Lin, W.; Schlesinger, F.; *et al.* Landscape of transcription in human cells. *Nature* **2012**, *489*, 101–108.
3. Wang, Z.; Gerstein, M.; Snyder, M. RNA-Seq: A revolutionary tool for transcriptomics. *Nat. Rev. Genet.* **2009**, *10*, 57–63.
4. Habegger, L.; Sboner, A.; Gianoulis, T.A.; Rozowsky, J.; Agarwal, A.; Snyder, M.; Gerstein, M. RSEQtools: A modular framework to analyze RNA-Seq data using compact, anonymized data summaries. *Bioinformatics* **2011**, *27*, 281–283.
5. Derrien, T.; Johnson, R.; Bussotti, G.; Tanzer, A.; Djebali, S.; Tilgner, H.; Guernec, G.; Martin, D.; Merkel, A.; Knowles, D.G. The GENCODE v7 catalog of human long noncoding RNAs: Analysis of their gene structure, evolution, and expression. *Genome Res.* **2012**, *22*, 1775–1789.
6. Memczak, S.; Jens, M.; Elefsinioti, A.; Torti, F.; Krueger, J.; Rybak, A.; Maier, L.; Mackowiak, S.D.; Gregersen, L.H.; Munschauer, M.N.; *et al.* Circular RNAs are a large class of animal RNAs with regulatory potency. *Nature* **2013**, *495*, 333–338.
7. Hansen, T.B.; Jensen, T.I.; Clausen, B.H.; Bramsen, J.B.; Finsen, B.; Damgaard, C.K.; Kjems, J. Natural RNA circles function as efficient microRNA sponges. *Nature* **2013**, *495*, 384–388.
8. Di Leva, G.; Garofalo, M. Non-Coding RNAs and Cancer. In *Oncogene and Cancer—From Bench to Clinic*; Siregar, Y., Ed.; InTech: New York, NY, USA, 2013; Chapter 14, pp. 317–358.

9. Mizuno, T.; Chou, M.Y.; Inouye, M. A unique mechanism regulating gene expression: Translational inhibition by a complementary RNA transcript (micRNA). *Proc. Natl. Acad. Sci. USA* **1984**, *81*, 1966–1970.
10. Andersen, J.; Forst, S.A.; Zhao, K.; Inouye, M.; Delihas, N. The function of micF RNA: micF RNA is a major factor in the thermal regulation of OmpF protein in *Escherichia coli*. *J. Biol. Chem.* **1989**, *264*, 17961–17970.
11. Schmidt, M. Zheng, P.; Delihas, N. Secondary structures of *Escherichia coli* antisense *micF* RNA, the 5'-end of the target ompF mRNA, and the RNA/RNA Duplex. *Biochemistry* **1995**, *34*, 3621–3631.
12. Cowley, M.; Oakey, R.J. Transposable elements re-wire and fine-tune the transcriptome. *PLoS Genet.* **2013**, *9*, e1003234.
13. Goodier, J.L.; Kazazian, H.H., Jr. Retrotransposons revisited: The restraint and rehabilitation of parasites. *Cell* **2008**, *135*, 23–35.
14. Lander, E.S.; Linton, L.M.; Birren, B.; Nusbaum, C.; Zody, M.C.; Baldwin, J.; Devon, K.; Dewar, K.; Doyle, M.; FitzHugh, W.; *et al.* Initial sequencing and analysis of the human genome. *Nature* **2001**, *409*, 860–921.
15. Doudna, J.A. The Doudna Lab. Translational control by mRNA secondary structure—Alu-element regulated miRNA interactions. Available online: http://doudna.berkeley.edu/ (accessed on 24 June 2013).
16. De Koning, A.P.; Gu, W.; Castoe, T.A.; Batzer, M.A.; Pollock, D.D. Repetitive elements may comprise over two-thirds of the human genome. *PLoS Genet.* **2011**, *7*, e1002384.
17. Casa, V.; Gabellini, D. A repetitive elements perspective in Polycomb epigenetics. *Front. Genet.* **2012**, *3*, 199.
18. Ahmed, M.; Liang, P. Transposable elements are a significant contributor to tandem repeats in the human genome. *Comp. Funct. Genomics* **2012**, *12*, 947089.
19. Smalheiser, N.R.; Torvik, V.I. Mammalian microRNAs derived from genomic repeats. *Trends Genet.* **2005**, *21*, 322–326.
20. Piriyapongsa, J.; Jordan, I.K. A family of human microRNA genes from miniature inverted-repeat transposable elements. *PLoS One* **2007**, *2*, e203.
21. Piriyapongsa, J.; Mariño-Ramírez, L.; Jordan, I.K. Origin and evolution of human microRNAs from transposable elements. *Genetics* **2007**, *176*, 1323–1337.
22. Devor, E.J.; Peek, A.S.; Lanier, W.; Samollow, P.B. Marsupial-specific microRNAs evolved from marsupial-specific TEs. *Gene* **2009**, *448*, 187–191.
23. Yuan, Z.; Sun, X.; Jiang, D.; Ding, Y.; Lu, Z.; Gong, L.; Liu, H.; Xie, J. Origin and evolution of a placental-specific microRNA family in the human genome. *BMC Evol. Biol.* **2010**, *10*, 346.
24. Kapitonov, V.V.; Jurka, J. MER53, a non-autonomous DNA transposon associated with a variety of functionally related defense genes in the human genome. *DNA Seq.* **1998**, *8*, 277–288.
25. Yuan, Z.; Sun, X.; Liu, H.; Xie, J. MicroRNA genes derived from repetitive elements and expanded by segmental duplication events in mammalian genomes. *PLoS One* **2011**, *6*, e17666.
26. Borchert, G.M.; Holton, N.W.; Williams, J.D.; Hernan, W.L.; Bishop, I.P.; Dembosky, J.A.; Elste, J.E.; Gregoire, N.S.; Kim, J.A.; Koehler, W.W.; *et al.* Comprehensive analysis of microRNA genomic loci identifies pervasive repetitive-element origins. *Mob. Genet. Element* **2011**, *1*, 8–17.

27. Tempel, S.; Pollet, N.; Tahi, F. ncRNAclassifier: A tool for detection and classification of transposable element sequences in RNA hairpins. *BMC Bioinforma.* **2012**, *13*, 246.
28. Ahn, K.; Gim, J.A.; Ha, H.S.; Han, K.; Kim, H.S. The novel MER transposon-derived miRNAs in human genome. *Gene* **2013**, *512*, 422–428.
29. Gong, C.; Maquat, L.E. lncRNAs transactivate STAU1-mediated mRNA decay by duplexing with 3' UTRs via Alu elements. *Nature* **2011**, *470*, 284–288.
30. Gong, C.; Maquat, L.E. "Alu" strious long ncRNAs and their role in shortening mRNA half-lives. *Cell Cycle* **2011**, *10*, 1882–1883.
31. Kim, Y.K.; Furic, L.; Parisien, M.; Major, F.; DesGroseiller, L.; Maquat, L.E. Staufen1 regulates diverse classes of mammalian transcripts. *EMBO J.* **2007**, *26*, 2670–2681.
32. Capshew, C.R.; Dusenbury, K.L.; Hundley, H.A. Inverted Alu dsRNA structures do not affect localization but can alter translation efficiency of human mRNAs independent of RNA editing. *Nucleic Acids Res.* **2012**, *40*, 8637–8645.
33. Gleghorn, M.L.; Gong, C.; Kielkopf, C.L.; Maquat, L.E. Staufen1 dimerizes through a conserved motif and a degenerate dsRNA-binding domain to promote mRNA decay. *Nat. Struct. Mol. Biol.* **2013**, *20*, 515–524.
34. Park, E.; Gleghorn, M.L.; Maquat, L.E. Staufen2 functions in Staufen1-mediated mRNA decay by binding to itself and its paralog and promoting UPF1 helicase but not ATPase activity. *Proc. Natl. Acad. Sci. USA* **2013**, *110*, 405–412.
35. Wang, J.; Gong, C.; Maquat, L.E. Control of myogenesis by rodent SINE-containing lncRNAs. *Genes Dev.* **2013**, *27*, 793–804.
36. Carrieri, C.; Cimatti, L.; Biagioli, M.; Beugnet, A.; Zucchelli, S.; Fedele, S.; Pesce, E.; Ferrer, I.; Collavin, L.; Santoro, C. Long non-coding antisense RNA controls Uchl1 translation through an embedded SINEB2 repeat. *Nature* **2012**, *491*, 454–457.
37. Choi, J.; Levey, A.I.; Weintraub, S.T.; Rees, H.D.; Gearing, M.; Chin, L.S.; Li, L. Oxidative modifications and down-regulation of ubiquitin carboxyl-terminal hydrolase L1 associated with idiopathic Parkinson's and Alzheimer's diseases. *J. Biol. Chem.* **2004**, *279*, 13256–13264.
38. Barrachina, M.; Castaño, E.; Dalfó, E.; Maes, T.; Buesa, C.; Ferrer, I. Reduced ubiquitin C-terminal hydrolase-1 expression levels in dementia with Lewy bodies. *Neurobiol. Dis.* **2006**, *22*, 265–273.
39. Cartault, F.; Munier, P.; Benko, E.; Desguerre, I.; Hanein, S.; Boddaert, N.; Bandiera, S.; Vellayoudom, J.; Krejbich-Trotot, P.; Bintner, M. Mutation in a primate-conserved retrotransposon reveals a noncoding RNA as a mediator of infantile encephalopathy. *Proc. Natl. Acad. Sci. USA* **2012**, *109*, 4980–4985.
40. Hewitt, J.E.; Lyle, R.; Clark, L.N.; Valleley, E.M.; Wright, T.J.; Wijmenga, C.; van Deutekom, J.C.; Francis, F.; Sharpe, P.T.; Hofker, M.; *et al.* Analysis of the tandem repeat locus D4Z4 associated with facioscapulohumeral muscular dystrophy. *Hum. Mol. Genet.* **1994**, *8*, 1287–1295.
41. Clapp, J.; Mitchell, L.M.; Bolland, D.J.; Fantes, J.; Corcoran, A.E.; Scotting, P.J.; Armour, J.A.; Hewitt, J.E. Evolutionary conservation of a coding function for D4Z4, the tandem DNA repeat mutated in facioscapulohumeral muscular dystrophy. *Am. J. Hum. Genet.* **2007**, *81*, 264–279.

42. Snider, L.; Asawachaicharn, A.; Tyler, A.E.; Geng, L.N.; Petek, L.M.; Maves, L.; Miller, D.G.; Lemmers, R.J.; Winokur, S.T.; Tawil, R. RNAtranscripts, miRNA-sizedfragments and proteinsproduced from D4Z4units: New candidates for the pathophysiology of facioscapulohumeral dystrophy. *Hum. Mol. Genet.* **2009**, *18*, 2414–2430.
43. Cabianca, D.S.; Gabellini, D. FSHD: Copy number variations on the theme of muscular dystrophy *J. Cell Biol.* **2010**, *191*, 1049–1060.
44. Cabianca, D.S.; Casa, V.; Bodega, B.; Xynos, A.; Ginelli, E.; Tanaka, Y.; Gabellini, D. A long ncRNA links copy number variation to a polycomb/trithorax epigenetic switch in FSHD muscular dystrophy. *Cell* **2012**, 819–831.
45. Wijmenga, C.; Hewitt, J.E.; Sandkuijl, L.A.; Clark, L.N.; Wright, T.J.; Dauwerse, H.G.; Gruter, A.M.; Hofker, M.H.; Moerer, P.; Williamson, R.; *et al.* Chromosome 4q DNA rearrangements associated with facioscapulohumeral muscular dystrophy. *Nat. Genet.* **1992**, *2*, 26–30.
46. Emanuel, B.S. Molecular mechanisms and diagnosis of chromosome 22q11.2 rearrangements. *Dev. Disabil. Res. Rev.* **2008**, *14*, 11–18.
47. Ponicsan, S.L.; Kugel, J.F.; Goodrich, J.A. Genomic gems: SINE RNAs regulate mRNA production. *Curr. Opin. Genet. Dev.* **2010**, *20*, 149–155.
48. Sorek, R.; Ast, G.; Graur, D. Alu-containing exons are alternatively spliced. *Genome Res.* **2002**, *12*, 1060–1067.
49. Ram, O.; Schwartz, S.; Ast, G. Multifactorial interplay controls the splicing profile of Alu-derived exons. *Mol. Cell Biol.* **2008**, *28*, 3513–3525.
50. Gal-Mark, N.; Schwartz, S.; Ast, G. Alternative splicing of Alu exons—Two arms are better than one. *Nucleic Acids Res.* **2008**, *36*, 2012–2023.
51. Schwartz, S.; Gal-Mark, N.; Kfir, N.; Oren, R.; Kim, E.; Ast, G. Alu exonization events reveal features required for precise recognition of exons by the splicing machinery. *PLoS Comput. Biol.* **2009**, *5*, e1000300.
52. Dagan, T.; Sorek, R.; Sharon, E.; Ast, G.; Graur, D. AluGene: A database of Alu elements incorporated within protein-coding genes. *Nucleic Acids Res.* **2004**, *32*, D489–D492.
53. Lev-Maor, G.; Ram, O.; Kim, E.; Sela, N.; Goren, A.; Levanon, E.Y.; Ast, G. Intronic Alus influence alternative splicing. *PLoS Genet.* **2008**, *4*, e1000204.
54. Athanasiadis, A.; Rich, A.; Maas, S. Widespread A-to-I RNA editing of alu-containing mRNAs in the human transcriptome. *PLoS Biol.* **2004**, *2*, e391.
55. Xu, L.; Elkahloun, A.G.; Candotti, F.; Grajkowski, A.; Beaucage, S.L.; Petricoin, E.F.; Calvert, V.; Juhl, H.; Mills, F.; Mason, K. A novel function of RNAs arising from the long terminal repeat of human endogenous retrovirus 9 in cell cycle arrest. *J. Virol.* **2013**, *87*, 25–36.
56. Kelley, D.; Rinn, J. Transposable elements reveal a stem cell-specific class of long noncoding RNAs. *Genome Biol.* **2012**, *13*, R107.
57. Cocquerelle, C.; Mascrez, B.; Hetuin, D.; Bailleul, B. Mis-splicing yields circular RNA molecules. *FASEB J.* **1993**, *7*, 155–160.
58. Capel, B.; Swain, A.; Nicolis, S.; Hacker, A.; Walter, M.; Koopman, P.; Goodfellow, P.; Lovell-Badge, R. Circular transcripts of the testis-determining gene Sry in adult mouse testis. *Cell* **1993**, *73*, 1019–1030.

59. Hsu, M.-T.; Coca-Prados, M. Electron microscopic evidence for the circular form of RNA in the cytoplasm of eukaryotic cells. *Nature* **1979**, *280*, 339–340.
60. Nigro, J.M.; Cho, K.R.; Fearon, E.R.; Kern, S.E.; Ruppert, J.M.; Oliner, J.D.; Kinzler, K.W.; Vogelstein, B. Scrambled exons. *Cell* **1991**, *64*, 607–613.
61. Zaphiropoulos, P.G. Circular RNAs from transcripts of the rat cytochrome P450 *2C24* gene: Correlation with exon skipping. *Proc. Natl. Acad. Sci. USA* **1996**, *93*, 6536–6541.
62. Li X.-F.; Lytton, J. A circularized sodium-calcium exchanger exon 2 transcript. *J. Biol. Chem.* **1999**, *274*, 8153–8160.
63. Gualandi, F.; Trabanelli, C.; Rimessi, P.; Calzolari, E.; Toffolatti, L.; Patarnello, T.; Kunz, G.; Muntoni, F.; Ferlini, A. Multiple exon skipping and RNA circularisation contribute to the severe phenotypic expression of exon 5 dystrophin deletion. *J. Med. Genet.* **2003**, *40*, e100.
64. Salzman, J.; Gawad, C.; Wang, P.L.; Lacayo, N.; Brown, P.O. Circular RNAs are the predominant transcript isoform from hundreds of human genes in diverse cell types. *PLoS One* **2012**, *7*, e30733.
65. Jeck, W.R.; Sorrentino, J.A.; Wang, K.; Slevin, M.K.; Burd, C.E.; Liu, J.; Marzluff, W.F.; Sharpless, N.E. Circular RNAs are abundant, conserved, and associated with ALU repeats. *RNA* **2013**, *19*, 141–157.
66. Lin, H.; Spradling, A.C. A novel group of pumilio mutations affects the asymmetric division of germline stem cells in the Drosophila ovary. *Development* **1997**, *124*, 2463–2476.
67. Karginov, F.V.; Hannon, G.J. The CRISPR system: Small RNA-guided defense in bacteria and archaea. *Mol. Cell* **2010**, *37*, 7–19.
68. Wiedenheft, B.; van Duijn, E.; Bultema, J.B.; Waghmare, S.P.; Zhou, K.; Barendregt, A.; Westphal, W.; Heck, A.J.; Boekema, E.J.; Dickman, M.J.; *et al.* RNA-guided complex from a bacterial immune system enhances target recognition through seed sequence interactions. *Proc. Natl. Acad. Sci. USA* **2011**, *108*, 10092–10097.
69. Sashital, D.G.; Jinek, M.; Doudna, J.A. An RNA-induced conformational change required for CRISPR RNA cleavage by the endoribonuclease Cse3. *Nat. Struct. Mol. Biol.* **2011**, *18*, 680–687.
70. Brennecke, J.; Aravin, A.A.; Stark, A.; Dus, M.; Kellis, M.; Sachidanandam, R.; Hannon, G.J. Discrete small RNA-generating loci as master regulators of transposon activity in *Drosophila.* *Cell* **2007**, *128*, 1089–1103.
71. Malone, C.D.; Hannon, G.J. Small RNAs as guardians of the genome. *Cell* **2009**, *136*, 656–668.
72. Simonelig, M. Developmental functions of piRNAs and transposable elements: A *Drosophila* point-of-view. *RNA Biol.* **2011**, *8*, 754–759.
73. Bagijn, M.P.; Goldstein, L.D.; Sapetschnig, A.; Weick, E.M.; Bouasker, S.; Lehrbach, N.J.; Simard, M.J.; Miska, E.A. Function, targets, and evolution of *Caenorhabditis elegans* piRNAs. *Science* **2012**, *337*, 574–578.
74. Le Thomas, A.; Rogers, A.K.; Webster, A.; Marinov, G.K.; Liao, S.E.; Perkins, E.M.; Hur, J.K.; Aravin, A.A.; Tóth, K.F. Piwi induces piRNA-guided transcriptional silencing and establishment of a repressive chromatin state. *Genes Dev.* **2013**, *27*, 390–399.
75. Sigurdsson, M.I.; Smith, A.V.; Bjornsson, H.T.; Jonsson, J.J. The distribution of a germline methylation marker suggests a regional mechanism of LINE-1 silencing by the piRNA-PIWI system. *BMC Genet.* **2012**, *13*, 31.

76. Watanabe, T.; Tomizawa, S.; Mitsuya, K.; Totoki, Y.; Yamamoto, Y.; Kuramochi-Miyagawa, S.; Iida, N.; Hoki, Y.; Murphy, P.J.; Toyoda, A.; *et al.* Role for piRNAs and noncoding RNA in de novo DNA methylation of the imprinted mouse *Rasgrf1* locus. *Science* **2011**, *332*, 848–852.
77. Espinoza, C.A.; Goodrich, J.A.; Kugel, J.F. Characterization of the structure, function, and mechanism of B2 RNA, an ncRNA repressor of RNA polymerase II transcription. *RNA* **2007**, *13*, 583–596.
78. Yakovchuk, P.; Goodrich, J.A.; Kugel, J.F. B2 RNA and Alu RNA repress transcription by disrupting contacts between RNA polymerase II and promoter DNA within assembled complexes. *Proc. Natl. Acad. Sci. USA* **2009**, *106*, 5569–5574.
79. Yakovchuk, P.; Goodrich, J.A.; Kugel, J.F. B2 RNA represses TFIIH phosphorylation of RNA polymerase II. *Transcription* **2011**, *2*, 45–49.
80. Blackwell, B.J.; Lopez, M.F.; Wang, J.; Krastins, B.; Sarracino, D.; Tollervey, J.R.; Dobke, M.; Jordan, I.K.; Lunyak, V.V. Protein interactions with piALU RNA indicates putative participation of retroRNA in the cell cycle, DNA repair and chromatin assembly. *Mob. Genet. Elements* **2012**, *2*, 26–35.
81. Wassarman, K.M.; Storz, G. 6S RNA regulates *E. coli* RNA polymerase activity. *Cell* **2000**, *101*, 613–623.
82. Wassarman, K.M. 6S RNA: A regulator of transcription. *Mol. Microbiol.* **2007**, *65*, 1425–1431.
83. Decker, K.B.; Hinton, D.M. The secret to 6S: Regulating RNA polymerase by ribo-sequestration. *Mol. Microbiol.* **2009**, *73*, 137–140.
84. Holmqvist, E.; Unoson, C.; Reimegård, J.; Wagner, E.G. A mixed double negative feedback loop between the sRNA MicF and the global regulator. *Lrp. Mol. Microbiol.* **2012**, *4*, 414–427.
85. Corcoran, C.P.; Podkaminski, D.; Papenfort, K.; Urban, J.H.; Hinton, J.C.; Vogel, J. Superfolder GFP reporters validate diverse new mRNA targets of the classic porin regulator, MicF RNA. *Mol. Microbiol.* **2012**, *84*, 428–445.
86. Brownlee, G.G. Sequence of 6S RNA of *E. coli*. *Nat. New Biol.* **1971**, *229*, 147–149.
87. Hauptman, N.; Glavač, D. Long non-coding RNA in cancer. *Int. J. Mol. Sci.* **2013**, *14*, 4655–4669.
88. Gutschner, T.; Diederichs, S. The Hallmarks of Cancer: A long non-coding RNA point of view. *RNA Biol.* **2012**, *9*, 703–719.

7

Bicistronic Gene Transfer Tools for Delivery of miRNAs and Protein Coding Sequences

Michelle L. Stoller [1,2], Henry C. Chang [1,2] and Donna M. Fekete [1,2,*]

1 Department of Biological Sciences, Purdue University, 915 W State St, West Lafayette, IN 47907-1392, USA; E-Mails: mlstolle@purdue.edu (M.L.S.); hcchang@purdue.edu (H.C.C.)

2 Purdue University Center for Cancer Research, Purdue University, 201 S University Dr, West Lafayette, IN 47907-2064, USA

* Author to whom correspondence should be addressed; E-Mail: dfekete@purdue.edu;

Abstract: MicroRNAs (miRNAs) are a category of small RNAs that modulate levels of proteins via post-transcriptional inhibition. Currently, a standard strategy to overexpress miRNAs is as mature miRNA duplexes, although this method is cumbersome if multiple miRNAs need to be delivered. Many of these miRNAs are found within introns and processed through the RNA polymerase II pathway. We have designed a vector to exploit this naturally-occurring intronic pathway to deliver the three members of the sensory-specific miR-183 family from an artificial intron. In one version of the vector, the downstream exon encodes the reporter (GFP) while another version encodes a fusion protein created between the transcription factor Atoh1 and the hemaglutinin epitope, to distinguish it from endogenous Atoh1. *In vitro* analysis shows that the miRNAs contained within the artificial intron are processed and bind to their targets with specificity. The genes downstream are successfully translated into protein and identifiable through immunofluorescence. More importantly, Atoh1 is proven functional through *in vitro* assays. These results suggest that this cassette allows expression of miRNAs and proteins simultaneously, which provides the opportunity for joint delivery of specific translational repressors (miRNA) and possibly transcriptional activators (transcription factors). This ability is attractive for future gene therapy use.

Keywords: miRNAs; gene therapy; miR-183 family

---

1. Introduction

MicroRNAs (miRNAs) are small non-coding RNAs that are usually between 22 and 24 nucleotides in length. Each miRNA contains a seed region defined by nucleotides 2–7/8 that is perfectly complementary to a sequence found within its target messenger RNA (mRNA). This bond allows the miRNA to control the levels of its target proteins by downregulating the translation or stability of the target mRNA [1]. Since the discovery of microRNAs (miRNAs), research has focused on identifying conserved miRNA families and determining how these small molecules regulate a multitude of cellular processes that occur during cancer [2,3] and development [4].

In development, a subset of miRNAs has received attention because their expression patterns are relatively specific for distinct cell types or organs. One approach to explore the function of such miRNAs is to either knockdown their levels or to force their overexpression *in vivo* or *in vitro*. Intracellular injection or transfection of miRNA mimics has been successful to overexpress mature miRNAs, although the elevated level of miRNA mimics is transient because they are not stably transduced. As an alternative, exogenous miRNAs can be stably expressed using two distinct transcriptional pathways. Some vectors use the RNA polymerase III (polIII) pathway via the U6 promoter to drive expression of pre-miRNA hairpins [5,6], while others use the RNA polymerase II (polII) pathway, for example to express two pre-miRNA sequences downstream of a tet-responsive PolII promoter [7]. A major drawback of these approaches is that cells overexpressing miRNAs cannot be easily identified, making subsequent phenotypic analysis difficult. To circumvent this problem, it is common to combine the delivery of these miRNA elements with some type of reporter gene using IRES (internal ribosomal entry sites) to make a bicistronic mRNA [2,8] or to deliver miRNAs and a reporter gene using two different promoters. In the latter case, a polII- or polIII-based promoter controls the production of the miRNA and a polII-based promoter drives expression of the reporter gene [9,10]. While the use of two promoters allows production of miRNAs and a protein-coding gene, the production of the two factors is not necessarily coordinated. Such a tenuous link between the relative levels of miRNAs and any associated reporter (such as GFP) could compromise the use of the latter as an estimate of the former in functional studies. An alternative approach to overexpress miRNAs is to generate vectors that resemble the 38% of endogenous miRNA genomic loci where miRNAs are found within the introns of protein-coding genes [11]. When used in this context, both miRNAs and an exogenous gene, such as a *GFP* reporter or cell-surface marker, can be placed under the control of the same polII-dependent promoter [12].

Here we describe the development and functional testing of an intronic cassette to deliver a small family of miRNAs, the miR-183 family, that is specifically expressed in primary sensory cells in variety of vertebrate sensory systems, including vision, hearing, taste, olfaction and somatosensory. The evolutionarily conserved miR-183 family miRNAs has three members (miR-183, -96 and -182) that are transcribed as a single polycistronic pri-miRNA [13]. Although our interest is in the role these miRNAs play in the specification of mechanosensory cells of the inner ear, members of this sensory-specific miRNA family are also upregulated in several different types of cancer [14–16].

MIR96 was the first miRNA locus to be associated with a hereditary human disease when it was linked to the DFNA50 locus in two families with dominant non-syndromic progressive hearing loss [17]. Each family has a point mutation in the seed region of *MIR96*, but at different nucleotides. A third deafness allele of DFN50 maps to a location in the pre-miR-96 transcript that likely interferes with miRNA processing [18]. Further supporting the link between deafness and mutations in miR-96, a semidominant deaf mouse mutant (diminuendo) was found with yet a third point mutation in the seed region [19]. The physiological and anatomical defects, present in either heterozygous or homozygous diminuendo mice, indicate that hair cells (HC) fail to fully mature [20].

In mouse, *Mir183*, *96* and *182* are located within an intronic region on Chromosome 6, and are transcribed as a single polycistronic pri-miRNA [21,22]. This coordinated expression is restricted to HCs as they begin to differentiate in both mice and zebrafish [23–26], suggesting that these miRNAs participate in HC development. Indeed, morpholino-mediated knockdown of each of the three miRNAs in zebrafish caused smaller inner ear sensory organ size and reduced numbers of HCs 2 days after injection [26]. Furthermore, overexpression of miR-96 and miR-182, by injection of double-stranded miRNA mimics into one-celled zebrafish, generated duplicate inner ears and produced supernumerary and ectopic inner ear HCs [26].

In total, data from humans, mice and zebrafish argue that the miR-183 family is essential for proper HC development and maintenance. As such, they should be considered as potential therapeutic agents for treating deafness due to HC loss. The vast majority (90%) of hearing loss is categorized as sensorineural, of which the most common type results from the destruction or malformation of the HCs occupying the organ of Corti, while sparing the associated supporting cells. One therapeutic approach is to deliver the HC-promoting transcription factor, Atonal1 (Atoh1), to the supporting cells of damaged ears. This has met with some success in animal models [27,28], although further studies are needed. Since it has been established that initiation and maturation of HCs require a complex regulatory network to turn off and on certain genes [29], we reasoned that the reprogramming of supporting cells into HCs might be enhanced by combining the delivery of an activating factor (Atoh1) and repressive elements (the miR-183 family). As every miR-183 family member is present during HC formation, we desired a gene transfer strategy that could efficiently and simultaneously deliver all 3 miRNAs along with a known HC-specification gene (*Atoh1*) to the same target cell population.

We produced two vectors containing the entire miR-183 family within a single artificial intron located upstream of a protein-encoding exon. The exon encoded either GFP as a reporter gene or a traceable version of Atoh1. We demonstrated that this design facilitates the coordinated expression of all three mature miRNAs and the associated protein. The data suggest that by simply exploiting one of the natural miRNA production pathways, it is possible to simultaneously deliver multiple negative and positive regulatory elements. Since many cellular processes require the joint activation and repression of downstream pathways, this delivery system provides an opportunity to achieve that dual manipulation efficiently.

2. Results and Discussion

2.1. Construction of Bifunctional Atoh1-HA and miRNA Expression Vector

In order to coordinate the expression of the miRNAs and Atoh1 with high precision within the same cell, both elements are synthesized from the same RNA transcript. To accomplish this, an artificial intron containing the miRNAs is placed downstream of EF1α (human elongation factor 1 alpha;

pEF1X [30]) and upstream of *Atoh1* coding sequence (Figure 1). Within the mouse genome, about 3.5 kb of sequence separates *Mir182* from the nearest other family member *Mir96* [21]. To accommodate the size restrictions of certain delivery vectors planned for the future, such as the RCAS avian retrovirus and adeno-associated virus, we removed this large intervening stretch between *Mir182* and *Mir96* while retaining the natural pre-miRNA sequences for all 3 family members. Thus, all of the endogenous sequence between *Mir183* and *Mir96* (~120 bp) along with ~100 bp of sequence flanking the end of each pre-miRNA sequence was kept. Then, the pre-miR-182 sequence, with 120 bps flanking each end, was fused to the *Mir183/Mir96* fragment by PCR.

Figure 1. Bifunctional vector design and processing of transcripts. The vector consists of the EF1α promoter that will drive expression of the *miR-183* family of genes from the intron designated by the splice donor (SD) and splice acceptor (SA) site, and an exon encoding Atoh1 fused to the hemagluttinin influenza epitope (HA). Once the plasmid is transcribed into RNA, endogenous enzymes present in transfected cells should recognize the SD and SA sites to release the intron containing the primary miRNA transcript (**A**); clip it into the three distinct pre-miRNAs, export them from the nucleus (**B**); and then further process them into mature miRNAs (**C**). As the miRNAs follow their own maturation pathway, the spliced Atoh1-HA-encoding polyA+ transcript is exported from the nucleus and processed as mRNA.

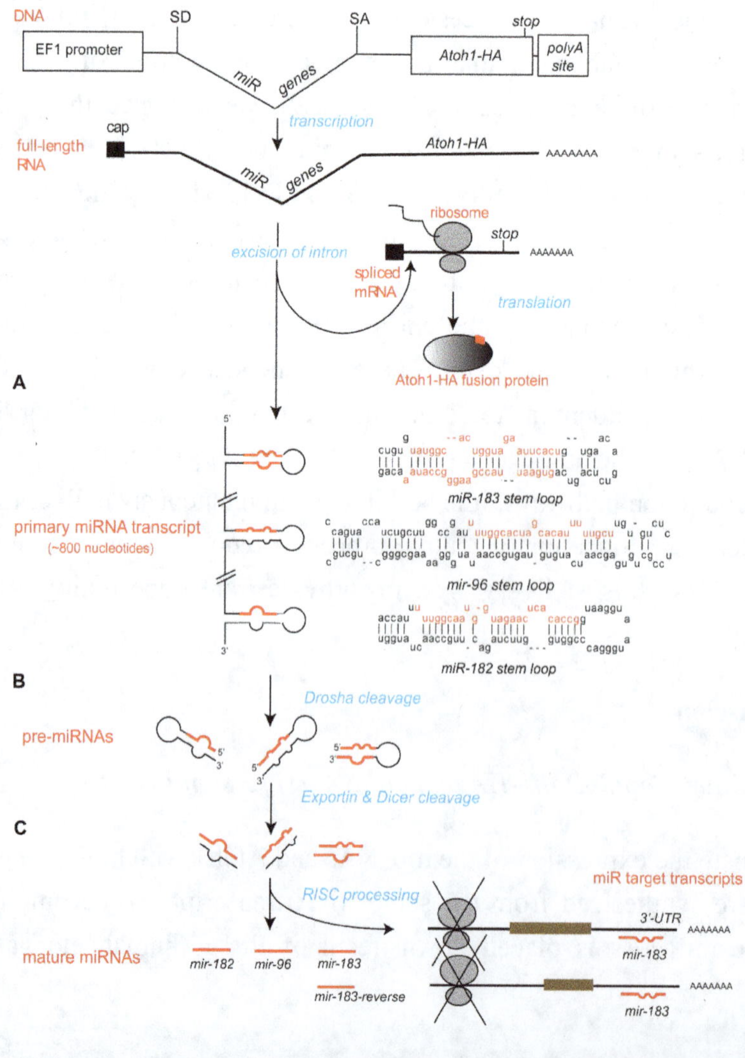

The combined pri-miRNA sequences were inserted into the artificial intron sequence (intron discussed by Lin and colleagues [31,32]). This artificial intron of only ~100 bps contains a splice donor site at the 5' end of the pri-miRNAs. The 3' end flanking the pri-miRNAs houses a branch point domain, polypyrimidine tract, and splice acceptor site. The polypyrimidine tract allows spliceosome assembly, while the branch point is necessary for the cell to recognize that a splicing event should occur to excise the element between the splice donor and acceptor sites.

Downstream of the miRNA intron is the murine *Atoh1* coding region. This *Atoh1* sequence was proven bioactive by its ability to induce ectopic HCs in utero [33]. To facilitate the detection of Atoh1 expression from transfected plasmids, an influenza hemaglutinin (HA) peptide tag (YPYDVPDYA) was fused in-frame to the *Atoh1* coding sequence. Figure 2 displays the overall design of the resulting plasmid, pEF1X-sd-miR183F-sa-Atoh1-HA, hereafter referred to as p183F-Atoh1-HA. Figure 2 also provides details for the introns and exons of the other constructs and their abbreviated names that will be introduced below.

Figure 2. Content and design of overexpression vectors. Each overexpression vector discussed in the paper is listed with its formal name, abbreviated name, and contents. Black boxes represent exons. Intron 1 and exons 1 and 2 were present within the plasmid backbone prior to modification. Checkmarks indicate the presence of artificial intronic flanking sequences. Empty spaces indicate a specific component is not found within that particular vector. TSS: transcription start site.

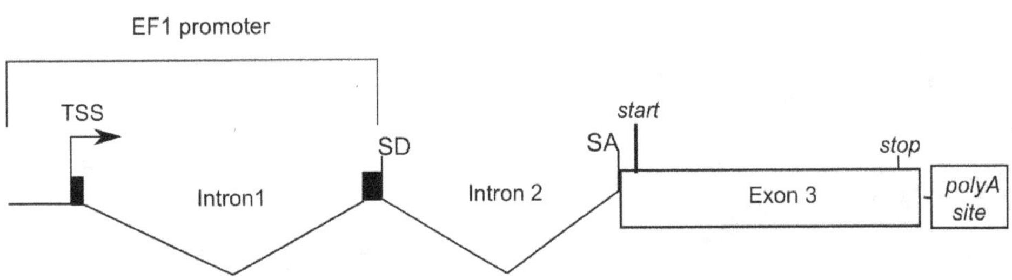

Formal Names	Abbreviated Names	SD	Intron 2	SA	Exon 3
pEF1X-Atoh1-HA	pAtoh1-HA				Atoh1-HA
pEF1X-sda-Atoh1-HA	pSDA-Atoh1-HA	√		√	Atoh1-HA
pEF1X-sd-miR183F-sa-Atoh1-HA	p183F-Atoh1-HA	√	miR-183 family	√	Atoh1-HA
pEF1X-sd-miR183F-sa-Atoh1(N162I)-HA	p183F-Atoh1(N162I)-HA	√	miR-183 family	√	Atoh1(N162I)-HA
pEF1X-sd-miR9-sa-Atoh1-HA	p9-Atoh1-HA	√	miR-9	√	Atoh1-HA
pEF1X-GFP	pGFP				GFP
pEF1X-sd-miR9-sa-GFP	p9-GFP	√	miR-9	√	GFP
pEF1X-sd-miR183F-sa-GFP	p183F-GFP	√	miR-183 family	√	GFP

2.2. Confirmation of Atoh1-HA Production and Function from a Bifunctional Cassette

To ascertain that Atoh1 is expressed from this bicistronic system, HEK293T cells transfected with p183F-Atoh1-HA were stained with anti-HA antibody. In cells 24 h after transfection, HA-positive staining was readily seen in the nuclei using immunofluorescence (Figure 3A), consistent with the fact that Atoh1 is a transcription factor. No HA-positive staining was seen in mock-transfected cells, demonstrating that the signal in p183F-Atoh1-HA-transfected cells is specific (data not shown).

Figure 3. The Atoh1-HA fusion protein is functional and detectable by immunofluorescence. (**A**) Detection of Atoh1-HA fusion protein with HA.11 antibody in cells transfected with p183F-Atoh-HA. Scale bar = 100 microns; (**B**) Illustration of Atoh1 reporter construct; (**C**) Relative luciferase activity of cells transfected with Atoh1 reporter alone or with the indicated versions of the Atoh1-HA overexpression constructs. Luciferase activities are all referenced to cells transfected only with the reporter construct, which is set at 1.0. All constructs showed a significant increase in luminescence compared to the control except p183F-Atoh1(N162I)-HA. Each bar represents mean (±standard error) within each group. Each experiment was replicated at least three times; (**D**) Alignment of conserved Atoh1 segment between fly and mouse. Highlighted is the location of the amino acid mutated to make Atoh1 non-functional while maintaining the HA tag. * $p < 0.05$, ** $p < 0.005$, *** $p < 0.0001$.

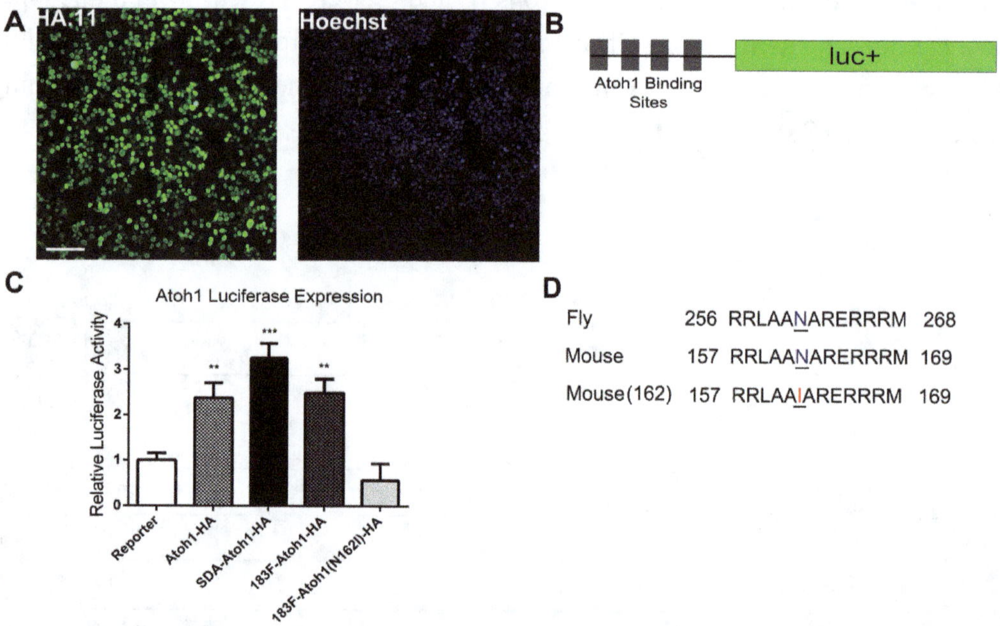

While the immunofluorescence suggests that HA-tagged Atoh1 was expressed and properly localized, it remains possible that the addition of a peptide hinders its bioactivity. To ensure that the HA-tagged Atoh1 is functional, we tested its ability to activate the expression of a luciferase-based reporter gene (4E-box), which has a firefly (FF) luciferase coding sequence placed under the control of four Atoh1-binding sites [34]. In addition, hpRL-SV40 (Promega, Madison, WI, USA), a plasmid with Renilla luciferase driven by a constitutive promoter, was included for normalization. HEK293T cells transfected with pAtoh1-HA showed a 138% increase ($p = 0.0031$) in FF luminescence, compared to those transfected with the pEF1X empty vector. Similarly, cells transfected with pSDA-Atoh1-HA and p183F-Atoh1-HA showed significant increase in FF luciferase luminescence (Figure 3C; pSDA-Atoh1-HA, 225% increase, $p < 0.0001$; p183F-Atoh1-HA, 149% increase, $p = 0.0004$).

To ensure that this increase in FF luciferase expression required a functional Atoh1 protein, we took advantage of the fact that mutating a highly conserved asparagine in the homeobox domain has been shown to disrupt Atoh1 function [35]. We generated p183F-Atoh1(N162I)-HA, which expresses Atoh1-HA with the N162I substitution (this mutation is analogous to the point mutation affecting amino acid 261 in the fly) (Figure 3D). In cells transfected with p183F-Atoh1(N162I)-HA, mutant Atoh1-HA was still detectable by immunofluorescence (data not shown), although its ability to activate FF luciferase expression was diminished (69% decrease in luminescence relative to 3 vectors carrying the wild type *Atoh1* sequence; ANOVA; $p < 0.0001$). Interestingly, the N162I mutation seems to act as a dominant negative, as the luminescence in p183F-Atoh1(N162I)-HA transfected cells decreased by 43% compared to the control (Figure 3C; $p = 0.8213$). Atohl is believed to act as a heterodimer that binds to other bHLH (basic helix loop helix) transcription factors such as E47 [36]. It is likely that expression of N162I prevents the formation of functional Atoh1 heterodimers by depleting the pool of endogenous E47 or other such transcription factors. In any case, our results showed clearly that functional HA-tagged Atoh1 is expressed from these constructs.

2.3. Confirmation of miRNA Production and Function from a Bifunctional Cassette

To assess whether the miRNAs were synthesized from the artificial intron, small RNAs collected from HEK293T cells 30 h after p183F-Atoh1-HA transfection were analyzed by Northern blots. While none was detected in untransfected or pAtoh1-HA transfected cells, bands corresponding to mature miRNA of each 183 family member were seen in p183F-Atoh1-HA-transfected cells (Figure 4A). It is notable that the relative levels of the three miRNAs are distinctly different, with miR-96 most prominent. The observation that these family members are not uniformly expressed has also been reported for murine retina [22] and cochlea [24].

A dual luciferase assay system was used to confirm bioactivity of the miR-183 family miRNAs produced from the cassette. For each miRNA, a reporter construct was created beginning with psiCHECK-2 (Promega, Madison, WI, USA), into which two binding sites complementary to a mature miRNA and separated by a spacer sequence were inserted downstream of the *Renilla* luciferase gene (Figure 4B). The psiCHECK-2 vector also contains the firefly luciferase gene driven off a separate promoter, so that luminescence from the firefly protein serves as an internal transfection control. Reporters containing the miRNA binding sites for miR-182, miR-96 or miR-183 were co-transfected withp183F-Atoh1-HA into HEK293T cells. The luminescence ratio (corrected for transfection efficiency) from the experimental wells was compared to control wells transfected with the relevant miR-183 family reporter and the pAtoh1-HA plasmid lacking the miRNA intron. As shown in Figure 4C–E, each miRNA-reporter construct showed a significant knockdown in luminescence compared to its corresponding control (miR-96: 95% knockdown, $p = 0.0013$; miR-182: 92% knockdown, $p = 0.0008$; miR-183: 89% knockdown, $p < 0.0001$). Thus, all 3 miRNAs are produced from the bifunctional cassette and appear functional.

Figure 4. miRNAs are expressed from the 183F-Atoh1-HA vector and functional. (**A**) p183F-Atoh1-HA transfection leads to production of mature miR-183 family members. Untransfected HEK293T cells and cells transfected with pAtoh1-HA show no detectable expression of 183 family members; whereas miR-183, -96, and -182 are each detected in cells transfected with p183F-Atoh1-HA. U6 levels are provided as the loading control; (**B**) Illustration of miRNA-specific reporter plasmids. PsiCHECK-2 luciferase reporters contain 2 sites complementary to miR-96, -182, or -183 in the 3'UTR; (**C–E**) Knockdown of luciferase activity in reporters specific to each member of the miR-183 family; (**C**) Cells co-transfected with reporter containing miR-183 sites and p183F-Atoh1-HA showed a marked decrease in luciferase activity compared to wells transfected with the miR-183 reporter and pAtoh1-HA. Experiments in (**D**) and (**E**) were conducted in the same manner except with miR-96 or miR-182 complementary binding sites in the luciferase reporter. All showed significant decrease in luminescence. Each bar represents mean (±standard error) for each group. Each experiment was replicated at least three times. * $p < 0.05$, ** $p < 0.005$, *** $p < 0.0001$.

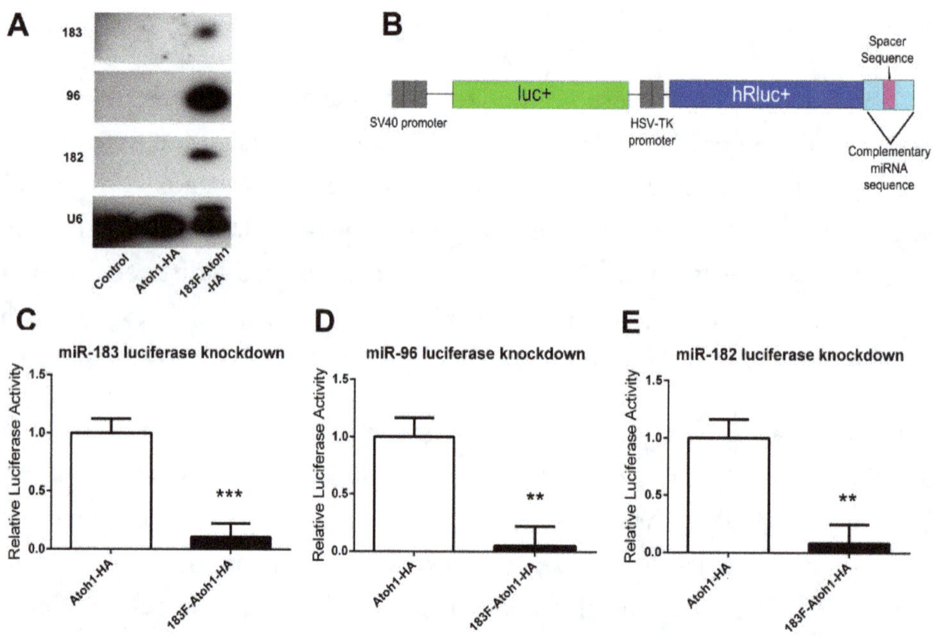

2.4. Overexpression of Functional miRNAs from GFP Expression Vectors

While our primary interest involved the dual expression of the miR-183 family and Atoh1, it was also desirable to express the miR-183 family alone to assess how much of an impact this family can have on HC development by themselves. To accomplish this, the *Atoh1-HA* coding region was replaced with *GFP*, which would allow the identification of cells expressing transfected miRNA constructs. Furthermore, as the design of p183F-GFP is the same as the p183F-Atoh1-HA (Figure 5A), phenotypic analysis using vectors with or without *Atoh1-HA* is less likely to be confounded by changes in the processing of the RNA transcripts that may affect transcript levels.

To test whether functional GFP protein is expressed from p183F-GFP, HEK293T cells were transfected and observed not only for direct GFP fluorescence but also after enhancing the signal with anti-GFP antibodies. After 24 h, green emissions were detected from the majority of fixed, transfected cells both before (not shown) and after immunofluorescence (Figure 5B).

Figure 5. Plasmid containing *miR-183* family and the *GFP* gene produces functional miRNAs and GFP protein in HEK293T cells. (**A**) Vector design of p183F-GFP; (**B**) Visualization of GFP in cells transfected with p183F-GFP. Scale bar = 100 microns; (**C**) The miR-183 family is expressed from p183F-GFP in mammalian and avian cells. Cells transfected with p183F-GFP showed expression of mature miR-183, -96, and -182. Control (untransfected cells) and pGFP transfected cells exhibit no detectable miRNA. U6 levels serve as the loading control; (**D–F**) Luciferase activity is decreased by expression of miRNAs from p183F-GFP expressing vector; (**D**) Cells co-transfected with p183F-GFP and the psiCHECK-2 reporter containing sites complementary to miR-183 show a significant decrease in luciferase activity compared to cells co-transfected with the pGFP and reporter; (**E**) and (**F**) show results of experiments similar to (**D**) except the reporter contained different complementary binding sites: 96 for (**E**) and 182 for (**F**). Each bar represents mean (±standard error) for each group. Each experiment was replicated at least three times. * $p < 0.05$, ** $p < 0.005$, *** $p < 0.0001$.

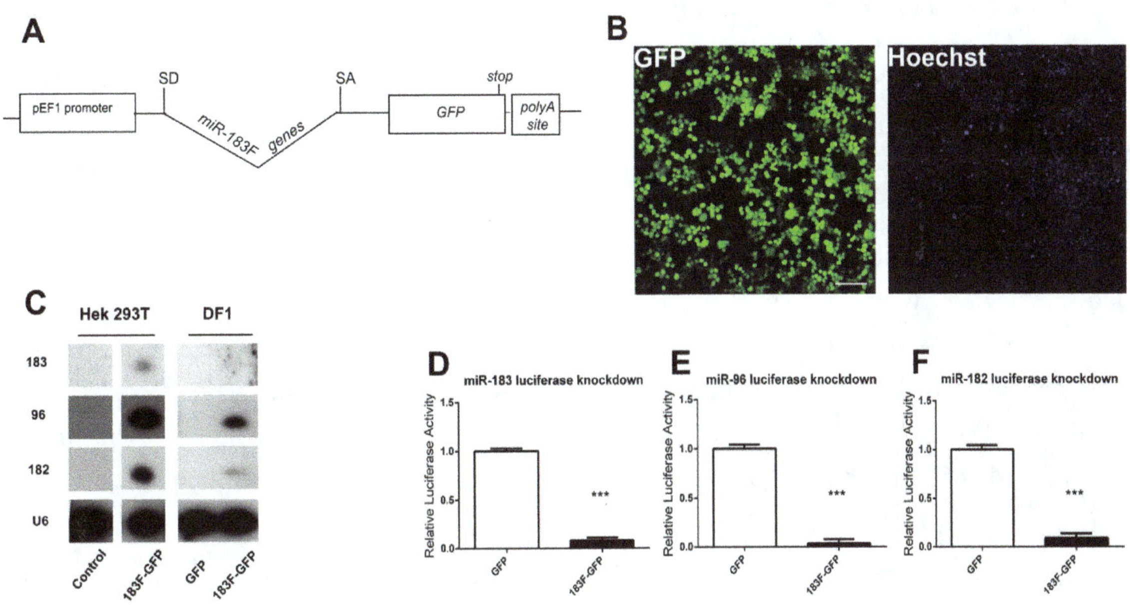

All three mature miRNAs of the miR-183 family could be detected in HEK293T cells transfected with p183F-GFP but not in cells transfected with a GFP vector lacking the miRNA intron, as assessed by Northern blots (Figure 5C). Notably, miRNA expression levels appear to remain consistent regardless of the identity of downstream coding sequence (Figure 5C; HEK293T cells).

To ascertain whether avian cells are able to process and express mammalian miRNAs, small RNAs from DF1 cells (chicken embryo fibroblast cells) transfected with p183F-GFP or pGFP were analyzed by Northern blotting. While both p183F-GFP and pGFP transfected groups expressed GFP fluorescence (data not shown), only those transfected with p183F-GFP showed bands corresponding to miR-182, -96 and -183 (Figure 5C; DF1 cells). The relative levels of the miR-183 family members appeared lower in transfected DF-1 cells than HEK293T cells. This discrepancy could result from the species difference of the transfected cells (chicken *vs.* human, respectively) or the difference in their respective tissue origins (embryonic day 10 fibroblasts *versus* fetal kidney, respectively). Nevertheless, our data demonstrated clearly that the miRNAs from p183F-GFP can be processed and produced in avian cells, allowing the option of using them in avian-specific vectors, like the RCAS retroviral vector [37].

Using the miRNA luciferase reporters discussed above, the function of the three miRNAs expressed from p183F-GFP was tested. Compared to HEK293T cells transfected with pGFP (which lacks the miRNA-producing intron), p183F-GFP transfection showed significant decrease in the expression of all three targets (miR-96: 97% knockdown, $p < 0.0001$; miR-182: 91% knockdown, $p < 0.0001$; miR-183: 92% knockdown, $p < 0.0001$) (Figure 5D–F). These observations demonstrated that miRNAs processed from the intron can successfully knockdown miRNA-specific targets.

2.5. MiRNAs Produced from Expression Vectors Bind with Specificity

To demonstrate the specificity of the knockdown mediated by these intronic miRNAs, we generated another luciferase-based reporter with two sites complementary to miR-9, a miRNA unrelated to the miR-183 family. Two intronic-miR-9 expression vectors with different downstream protein coding sequences (*Atoh1-HA or GFP*) were also generated to ensure this miR-9 reporter functions properly.

Figure 6. The miRNAs produced from both miRNA expression vectors bind to their targets with specificity. (**A**,**B**) Luciferase activity is decreased when vectors containing miR-9 are co-expressed with the miR-9 luciferase reporter but not when co-expressed with the miR-183 family expressing plasmids; (**A**) Luciferase activity of transfected cells containing the miR-9 luciferase reporter with pAtoh1-HA were compared to cells co-transfected with the miR-9 reporter and p183F-Atoh1-HA or p9-Atoh1-HA; (**B**) Experiments similar to those in A except control cells were co-transfected with miR-9 reporter and pGFP, while experimental cells were co-transfected with reporter and p183F-GFP or p9-GFP. * $p < 0.05$, ** $p < 0.005$, *** $p < 0.0001$. Bars represent mean (±standard error) for each group. Each experiment was replicated at least three times.

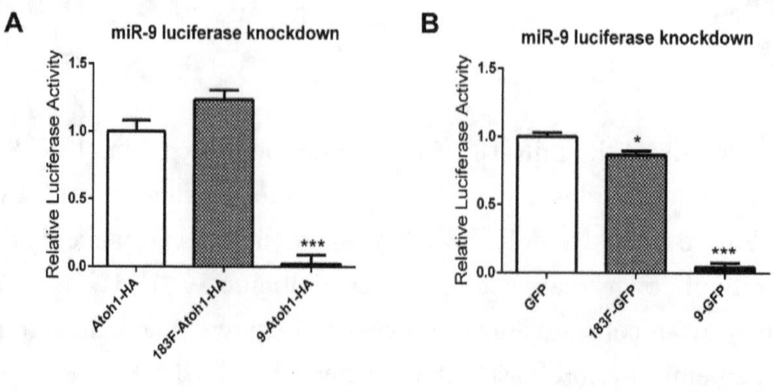

In cells transfected with miR-9 reporter, co-transfection of p9-Atoh1-HA or p9-GFP resulted in greater than 95% decrease in *Renilla* luciferase expression (98% decrease, $p < 0.0001$ for p9-Atoh1-HA; 96% decrease, $p < 0.0001$ for p9-GFP), demonstrating that these miR-9 expression vectors are functional. The comparable knockdown with both suggests that miR-9 vectors are similar to the 183F-expressing plasmid series in being effectively processed from the artificial intron regardless of the identity of downstream coding sequences. Most importantly, co-transfection of p183F-Atoh1-HA or p183F-GFP, while capable of knocking down the expression of 183F-based reporters (see above), showed only negligible effects on the *Renilla* luciferase level from the miR-9 reporter (23% increase, $p = 0.10$ for p183F-Atoh1-HA; 13% decrease, $p = 0.02$ for p183F-GFP) (Figure 6). These data suggest that intronic miRNAs produced from 183F- and miR-9-expressing vectors regulate the expression of their target genes with high specificity.

3. Experimental Section

3.1. Bifunctional Plasmid Construction

The murine *Atoh1* coding sequence was PCR amplified from the pEF1-Atoh-IRES-GFP vector, a gift from John Brigande [33]. To facilitate cloning and protein detection, one primer introduced an EcoRI site, while the other primer added an influenza hemaglutinin (HA) tag (YPYDVPDYA) to the *C*-terminus of Atoh1 coding sequence and a *Not*I site (see Table S1). *Atoh1-HA* was cloned into pEF1X (provided by Cliff Ragsdale) [30] as an *EcoR*I–*Not*I fragment and the entire fusion was verified by sequencing (Purdue Genomics Center).

To construct an artificial miR183-containing intron, a *Sal*I–*Hind*III fragment containing a splice donor, three restriction sites *Xba*I, *BamH*I and *Xho*I, polypyrimidine tract, branch point, and a splice acceptor was generated by PCR and cloned into pME-MCS [38], generating pMCS-SDA. PCR primers were generated based on published sequences [31,32]. Then, PCR amplification was used to extract the primary miRNA DNA of all three members of the *miR-183* family from the mouse genome, and to flank the genomic DNA with *Spe*I and *Sal*I. This sequence was then inserted between the *Xba*I and *Xho*I sites found within the intron contained within pME-MCS-sda to create pME-MCS-sd-miR183F-sa. Then, Kpn1 was used to extract the artificial intron containing about 800 bp of mouse *miR-183* family genomic primary miRNA sequence from pME-MCS-sd-miR183F-sa. The *Kpn*I site was used to insert the intron with the *miR-183* family upstream of the Atoh1-HA fusion protein in pEF1X to generate pEF1X-sd-miR183F-sa-Atoh1-HA.

pEF1X-sd-miR183F-sa-GFP vector was constructed by extracting *GFP* from pAAV2.1-CMV-eGFP3-WPRE, provided by Alberto Auricchio [39], via PCR (see Table S3). Using the *Spe*I and *Not*I sites found on the 5' and 3' ends respectively, *GFP* was inserted downstream of the miRNA-containing artificial intron in pMCS-sd-miR183F-sa to create pMCS-sd-miR183F-sa-GFP. The pEF1X vector was converted to a Gateway Destination vector (Invitrogen, Carlsbad, CA, USA) by inserting cassette B from the Gateway Conversion Kit (Invitrogen, Carlsbad, CA, USA to create pEF1X-cB. A LR recombination reaction between pMCS-miR183F-GFP and pEF1X-cB generated pEF1X-sd-miR183F-sa-GFP.

pEF1X-sd-miR9-sa-Atoh1-HA and pEF1X-sd-miR9-sa-GFP were constructed in a similar manner as the aforementioned vectors except *Mir9-1* genomic sequence was inserted into the artificial intron instead of the *miR-183* family. PCR was used to extract the endogenous mouse *Mir9-1* sequence and flanking regions using primer sequences described in Shibata *et al.* [9].

For each miRNA reporter, primers were designed to contain two sequences that are fully complementary to the mature miRNA of interest (miR-183, -96, -182, or -9; see Table S4). These sites were separated by a 17 nucleotide spacer sequence. For all cases, the forward primer contains a *Xho*I site, while the reverse primer houses a *Not*I site to allow the resulting PCR fragments to be introduced downstream of the *Renilla* luciferase gene located in the psiCHECK-2 vector (Promega, Madison, WI, USA).

3.2. Mutation of Atoh1-HA Fusion Protein

To introduce the N162I substitution in Atoh-1, site-directed mutagenesis was performed with Quikchange 2XL (Agilent Technologies, Santa Clara, CA, USA) according to the manufacturer's instructions. Primers were designed to induce a point mutation to change the amino acid 162 from Asparagine to Isoleucine in the Atoh1-HA fusion protein encoded by pEF1X-sd-miR183F-sa-Atoh1-HA creating pEF1X-sd-183F-sa-Atoh1(N162I)-HA (see Table S2).

3.3. HEK293T Plasmid Transfection

HEK293T were cultured with modified DMEM supplemented with L-glutamine, antibiotics, and 10% calf serum. Using Lipofectamine 2000 (Invitrogen, Carlsbad, CA, USA), cells seeded in 6-well plates were transfected with plasmids of interest. Collection time was assay dependent.

3.4. HEK293T Immunostain and Imaging

Cells transfected with pEF1X-sd-miR183F-sa-Atoh1-HA or pEF1X-sd-miR183F-sa-GFP were fixed 24 h post-transfection with 4% paraformaldehyde. The following primary antibodies (1:1000) were used: for detection of the HA tag, anti-HA.11 mouse IgG_1 monoclonal (Covance, Indianapolis, IN, USA); for detection of GFP, anti-GFP rabbit polyclonal (Molecular Probes, Eugene, OR, USA). Secondary antibodies (1:500) used were Alexa Fluor (Molecular Probes, Eugene, OR, USA) 488 anti-mouse IgG_1 and Alexa Fluor 488 anti-rabbit IgG. Immunostained cells were imaged under theE800 fluorescence microscope (Nikon, Elgin, IL, USA) with the 20× objective.

3.5. Atoh1 and MiRNA Luciferase Assays

The cells, 24 h after transfection, were lysed and luciferase activity was assessed using the dual luciferase assay kit (Promega, Madison, WI, USA) in the Luminoskan Ascent luminometer (Thermo Fisher Scientific, Waltham, MA, USA). For the Atoh1 luciferase assays, the firefly luciferase luminescence readings (at 560nm) were normalized to the *Renilla* luciferase readout (at 480 nm) to account for variation in transfection efficiency. In the case of the miRNA luciferase assays the ratio is inverted: the *Renilla* luciferase readout was normalized to the firefly luciferase readout. These ratios are expressed as relative luciferase activity. Experimental values were referenced to the control values which were arbitrarily set to one. Each treatment condition was conducted at least in duplicate. The experiments were repeated at least three times.

3.6. Northern Blots

HEK 293T/17 cells (abbreviated HEK293T cells; ATCC #CRL-11268) or UMNSAH-DF1 cells (abbreviated DF-1 cells; ATTC #CRL-12203) seeded in 35 mm plates were lysed ~30 h post-transfection and small RNAs were collected according to manufacturer's instructions using the PureLink miRNA Isolation Kit (Invitrogen, Carlsbad, CA, USA). Small RNA (300 ng) was probed for miR-183, -96, or -182 using the High Sensitive miRNA Northern Blot Assay Kit (Signosis, Santa Clara, CA, USA), a chemiluminescence system, according to manufacturer's instructions.

3.7. Statistical Analysis

All results are reported as mean ± standard error. The mean of each group is computed from measurements collected from at least three independent experiments. Statistical significance was determined by using a one-way analysis of variance with block (ANOVA), which was followed by Tukey's or Tukey-Kramer's multiple comparisons test (SAS 9.3, SAS Institute, www.sas.com). *p*-values below 0.05 were considered statistically significant.

4. Conclusions

Previous research has focused primarily on generating vectors to overexpress siRNAs: small non-coding RNAs that are a perfect complement to their targets. Researchers have shown that it is possible to express multiple siRNAs from one vector using a universal hairpin scaffold within or without an intron under the control of a PolII promoter [40–42]. Others have experimented with siRNA expression designs that allow joint delivery of siRNAs and protein coding regions under control of the same promoter without placement of the siRNA within an intron [43,44]. While these vectors do produce both small RNAs and protein, research shows that it is not the most efficient means of simultaneous delivery due to interference between miRNA/siRNA processing and protein production [45]. Thus, researchers sought to design a dual-delivery vector that would more reliably express a protein-coding gene and siRNA by inserting the siRNA within an artificial intron [31,32,45–47]. Capitalizing on the success of these vectors, others created miRNA vectors with a similar design to deliver one or two miRNAs within an endogenous intron upstream of a cell-surface marker [12]. We took that design one step further and created a dual-delivery vector to express a traceable transcription factor and multiple miRNA genes (in this case an entire miRNA family, using its endogenous sequence, contained within a single artificial intron). Luciferase assays showed that the transcription factor (Atoh1) fused to the HA tag is produced from the vector and is transcriptionally active, while immunofluorescence proved the fusion protein can be detected with HA.11 antibody. With this design, not only can the cells receiving the bifunctional vector be monitored, but artificially-expressed Atoh1 can be distinguished from the endogenous Atoh1 via its tag. Northern blots demonstrated that both human and chicken cells can process the single artificial intron containing 800 bp of endogenous sequence to produce each of the 3 mature miRNAs. These miRNAs were also shown to be functional and bind with specificity to their artificial target sequences.

In order to overexpress just the miRNA family and monitor its expression, we created a vector that replaced *Atoh1-HA* with *GFP*. We opted to keep the miRNAs in an intron upstream (rather than downstream) of a coding exon because research has shown that expressing siRNAs within introns upstream of reporters leads to more stable *GFP* mRNA and better GFP expression [45]. Indeed, GFP expression was robust and efficient: every cell present within the transfected wells was GFP-positive. Northern blots also showed that each intronic miRNA is expressed, while luciferase assays illustrated that these miRNA repress their targets and bind with specificity.

The expression construct reported here uses a polII-based promoter to control the expression of an entire miRNA family within the context of an artificial intron and its downstream *GFP* reporter gene. Naturally co-expressed miRNA family members may function optimally if their overexpression is coordinated with one another for healthy cell function. This same design was used to express a functional transcription factor (in the place of *GFP*) and the miR-183 family, while still maintaining the ability to monitor transfected cells by creating the Atoh1-HA fusion protein. Each cassette was transferred into a Gateway-compatible (Invitrogen, Carlsbad, CA, USA) shuttle vector (unpublished) to facilitate their transfer into alternative delivery vectors such as lentivirus or adenovirus. Such vectors will allow delivery of the cassette into tissues that are difficult to access or to transfect, including into the mouse organ of Corti depleted of HCs. Thus, we believe this novel design can be manipulated for multiple overexpression uses to study a variety of different complex cellular systems and possibly for future therapeutic purposes, including hair cell regeneration.

Acknowledgments

This research was funded by NIH (R01DC002756), NIH/NIDCD (1F31DC011687), and NOHR (National Organization for Hearing Research). We would like to thank the following researchers for their reagents: John Brigande (Oregon Health Sciences University), Nissim Ben-Arie (The Hebrew University of Jerusalem), Alberto Auricchio (Telethon Institute of Genetics and Medicine), and Cliff Ragsdale (University of Chicago).

References

1. Ying, S.Y.; Lin, S.L. Intron-mediated RNA interference and microRNA biogenesis. In *Methods of Molecular Biology, siRNA and miRNA Gene Silencing*; Sioud, M., Ed.; Humana Press: New York, NY, USA,2009; Volume 487, pp. 387–413.
2. He, L.; Thomson, J.M.; Hemann, M.T.; Hernando-Monge, E.; Mu, D.; Goodson, S.; Powers, S.; Cordon-Cardo, C.; Lowe, S.W.; Hannon, G.J.; *et al.* A microRNA polycistron as a potential human oncogene. *Nature* **2005**, *435*, 828–833.
3. Kasinski, A.L.; Slack, F.J. Epigenetics and genetics. MicroRNAs en route to the clinic: Progress in validating and targeting microRNAs for cancer therapy. *Nat. Rev. Cancer* **2011**, *11*, 849–864.
4. Lewis, M.A.; Steel, K.P. MicroRNAs in mouse development and disease. *Semin Cell Dev. Biol.* **2010**, *21*, 774–780.
5. Furukawa, N.; Sakurai, F.; Katayama, K.; Seki, N.; Kawabata, K.; Mizuguchi, H. Optimization of a microRNA expression vector for function analysis of microRNA. *J. Control Release* **2011**, *150*, 94–101.
6. Packer, A.N.; Xing, Y.; Harper, S.Q.; Jones, L.; Davidson, B.L. The bifunctional microRNA

miR-9/miR-9* regulates REST and CoREST and is downregulated in Huntington's disease. *J. Neurosci.* **2008**, *28*, 14341–14346.

7. Rissland, O.S.; Hong, S.-J.; Bartel, D.P. MicroRNA destabilization enables dynamic regulation of the miR-16 family in response to cell-cycle changes. *Mol. Cell* **2011**, *43*, 993–1004.
8. Otaegi, G.; Pollock, A.; Sun, T. An optimized sponge for microRNA miR-9 affects spinal motor neuron development *in vivo*. *Front. Neurosci.* **2011**, *5*, 146.
9. Shibata, M.; Kurokawa, D.; Nakao, H.; Ohmura, T.; Aizawa, S. *MicroRNA-9* modulates Cajal-Retzius cell differentiation by suppressing Foxg1 expression in mouse medial pallium. *J. Neurosci.* **2008**, *28*, 10415–10421.
10. Zhao, C.; Sun, G.; Li, S.; Shi, Y. A feedback regulatory loop involving *microRNA-9* and nuclear receptor TLX in neural stem cell fate determination. *Nat.Struct. Mol. Biol.* **2009**, *16*, 365–371.
11. Chiang, H.R.; Schoenfeld, L.W.; Ruby, J.G.; Auyeung, V.C.; Spies, N.; Baek, D.; Johnston, W.K.; Russ, C.; Luo, S.; Babiarz, J.E.; *et al.* Mammalian microRNAs: Experimental evaluation of novel and previously annotated genes. *Gene Dev.* **2010**, *24*, 992–1009.
12. Amendola, M.; Passerini, L.; Pucci, F.; Gentner, B.; Bacchetta, R.; Naldini, L. Regulated and multiple miRNA and siRNA delivery into primary cells by a lentiviral platform. *Mol. Ther.* **2009**, *17*, 1039–1052.
13. Pierce, M.L.; Weston, M.D.; Fritzsch, B.; Gabel, H.W.; Ruvkun, G.; Soukup, G.A. *MicroRNA-183* family conservation and ciliated neurosensory organ expression. *Evo. Dev.* **2008**, *10*, 106–113.
14. Mihelich, B.L.; Khramtsova, E.A.; Arva, N.; Vaishnav, A.; Johnson, D.N.; Giangreco, A.A.; Martens-Uzunova, E.; Bagasra, O.; Kajdacsy-Balla, A.; Nonn, L. MiR-183-96-182 cluster is overexpressed in prostate tissue and regulates zinc homeostasis in prostate cells. *J. Biol. Chem.* **2011**, *286*, 44503–44511.
15. Hannafon, B.N.; Sebastiani, P.; de las Morenas, A.; Lu, J.; Rosenberg, C.L. Expression of microRNA and their gene targets are dysregulated in preinvasive breast cancer. *Breast Cancer Res.* **2011**, *13*, R24.
16. Zhu, W.; Liu, X.; He, J.; Chen, D.; Hunag, Y.; Zhang, Y.K. Overexpression of members of the microRNA-183 family is a risk factor for lung cancer: A case control study. *BMC Cancer* **2011**, *11*, 393.
17. Mencía, A.; Modamio-Høybjør, S.; Redshaw, N.; Morín, M.; Mayo-Merino, F.; Olavarrieta, L.; Aguirre, L.A.; del Castillo, I.; Steel, K.P.; Dalmay, T.; *et al.* Mutations in the seed region of human miR-96 are responsible for nonsyndromic progressive hearing loss. *Nat. Genet.* **2009**, *41*, 609–613.
18. Soldà, G.; Robusto, M.; Primignani, P.; Castorina, P.; Benzoni, E.; Cesarani, A.; Ambrosetti, U.; Asselta, R.; Duga, S. A novel mutation within the MIR96 gene causes non-syndromic inherited hearing loss in an Italian family by altering pre-miRNA processing. *Hum. Mol. Genet.* **2012**, *21*, 577–585.
19. Lewis, M.A.; Quint, E.; Glazier, A.M.; Fuchs, H.; de Angelis, M.H.; Langford, C.; van Dongen, S.; Abreu-Goodger, C.; Piipari, M.; Redshaw, N.; *et al.* An ENU-induced mutation of miR-96 associated with progressive hearing loss in mice. *Nat. Genet.* **2009**, *41*, 614–618.

20. Kuhn, S.; Johnson, S.L.; Furness, D.N.; Chen, J.; Ingham, N.; Hilton, J.M.; Steffes, G.; Lewis, M.A.; Zampini, V.; Hackney, C.M.; *et al.* MiR-96 regulates the progression of differentiation in mammalian cochlear inner and outer hair cells. *Proc. Natl. Acad. Sci. USA* **2011**, *108*, 2355–2360.
21. Xu, S.; Witmer, P.D.; Lumayag, S.; Kovacs, B.; Valle, D. MicroRNA (miRNA) transcriptome of mouse retina and identification of a sensory organ-specific miRNA cluster. *J. Biol. Chem.* **2007**, *282*, 25053–2366.
22. Lumayag, S.; Haldin, C.E.; Corbett, N.J.; Wahlin, K.J.; Cowan, C.; Turturro, S.; Larsen, P.E.; Kovacs, B.; Witmer, P.D.; Valle, D.; *et al.* Inactivation of the *microRNA-183/96/182* cluster results in syndromic retinal degeneration. *Proc. Natl. Acad. Sci. USA* **2013**, *110*, E507–E516.
23. Weston, M.D.; Pierce, M.L.; Jensen-Smith, H.C.; Fritzsch, B.; Rocha-Sanchez, S.; Beisel, K.W.; Soukup, G.A. *MicroRNA-183* family expression in hair cell development and requirement of microRNAs for hair cell maintenance and survival. *Dev. Dynam.* **2011**, *240*, 808–819.
24. Weston, M.D.; Pierce, M.L.; Rocha-Sanchez, S.; Beisel, K.W.; Soukup, G.A. MicroRNA gene expression in the mouse inner ear. *Brain Res.* **2006**, *1111*, 95–104.
25. Sacheli, R.; Nguyen, L.; Borgs, L.; Vandenbosch, R.; Bodson, M.; Lefebvre, P.; Malgrange, B. Expression patterns of miR-96, miR-182 and miR-183 in the development inner ear. *Gene Expr. Patterns* **2009**, *9*, 364–370.
26. Li, H.; Kloosterman, W.; Fekete, D.M. *MicroRNA-183* family members regulate sensorineural fates in the inner ear. *J. Neurosci.* **2010**, *30*, 3254–3263.
27. Izumikawa, M.; Minoda, R.; Kawamoto, K.; Abrashkin, K.A.; Swiderski, D.L.; Dolan, D.F.; Brough, D.E.; Raphael, Y. Auditory hair cell replacement and hearing improvement by *Atoh1* gene therapy in deaf mammals. *Nat. Med.* **2005**, *11*, 271–276.
28. Yang, S.-M.; Chen, W.; Guo, W.-W.; Jia, S.; Sun, J.-H.; Liu, H.-Z.; Young, W.-Y.; He, D.Z.Z. Regeneration of stereocilia of hair cells by forced Atoh1 expression in the adult mammalian cochlea. *PLoS One* **2012**, *7*, e46355.
29. Groves, A.K.; Zhang, K.D.; Fekete, D.M. The genetics of hair cell development and regeneration. *Annu. Rev. Neurosci.* **2013**, *36*, 361–381.
30. Agarwala, S.; Sanders, T.A.; Ragsdale, C.W. Sonic hedgehog control of size and shape in midbrain pattern formation. *Science* **2001**, *291*, 2147–2150.
31. Lin, S.L.; Chang, D.; Wu, D.Y.; Ying, S.Y. A novel RNA splicing-mediated gene silencing mechanism potential for genome evolution. *Biochem. Biophy. Res. Commun.* **2003**, *310*, 754–760.
32. Lin, S.-L.; Ying, S.-Y. New drug design for gene therapy—Taking advantage of introns. *Lett. Drug Des. Discov.* **2004**, *1*, 256–262.
33. Gubbels, S.P.; Woessner, D.W.; Mitchell, J.C.; Ricci, A.J.; Brigande, J.V. Functional auditory hair cells produced in the mammalian cochlea by in utero gene transfer. *Nature* **2008**, *455*, 537–541.
34. Krizhanovsky, V.; Soreq, L.; Kliminski, V.; Ben-arie, N. Math1 target genes are enriched with evolutionarily conserved clustered E-box binding sites. *J. Mol. Neurosci.* **2006**, *28*, 211–229.
35. Jarman, A.P.; Grell, E.H.; Ackerman, L.; Jan, L.J.; Jan, Y.N. Atonal is the proneural gene drosophila photoreceptors. *Nature* **1994**, *369*, 398–400.
36. Mulvaney, J.; Dabdoub, A. Atoh1, an essential transcription factor in neurogenesis and intestinal

and inner ear development: Function, regulation, and context dependency. *JARO* **2012**, *13*, 281–293.

37. Hughes, S.H. The RCAS vector system. *Folia Biol.* **2004**, *50*, 107–119.
38. Kwan, K.M.; Fujimoto, E.; Grabher, C.; Mangum, B.D.; Hardy, M.E.; Campbell, D.S.; Parant, J.M.; Yost, H.J.; Kanki, J.P.; Chien, C.-B. The Tol2kit: A multisite gateway-based construction kit for Tol2 transposon transgenesis constructs. *Dev. Dynam.* **2007**, *236*, 3088–3099.
39. Karali, M.; Manfredi, A.; Puppo, A.; Marrocco, E.; Gargiulo, A.; Allocca, M.; Corte, M.D.; Rossi, S.; Giunti, M.; Bacci, M.L.; *et al.* MicroRNA-restricted transgene expression in the retina. *PLoS One* **2011**, *6*, e22166.
40. Chung, K.-H. Polycistronic RNA polymerase II expression vectors for RNA interference based on BIC/miR-155. *Nucleic Acids Res.* **2006**, *34*, e53.
41. Shin, K.-J.; Wall, E.A.; Zavzavadjian, J.R.; Santat, L.A.; Liu, J.; Hwang, J.-I.; Rebres, R.; Roach, T.; Seaman, W.; Simon, M.I.; *et al.* A single lentiviral vector platform for microRNA-based conditional RNA interference and coordinated transgene expression. *Proc. Natl. Acad. Sci. USA* **2006**, *103*, 13759–13764.
42. Zhu, X.; Santat, L.A.; Chang, M.S.; Liu, J.; Zavzavadjian, J.R.; Wall, E.A.; Kivork, C.; Simon, M.I.; Fraser, I.D. A versatile approach to multiple gene RNA interference using microRNA-based short hairpin RNAs. *BMC Mol. Biol.* **2007**, *8*, 98.
43. Stegmeier, F.; Hu, G.; Rickles, R.J.; Hannon, G.J.; Elledge, S.J. A lentiviral microRNA-based system for single-copy polymerase II-regulated RNA interference in mammalian cells. *Proc. Natl. Acad. Sci. USA* **2005**, *102*, 13212–13217.
44. Wang, X.; Liu, P.; Liu, H.; Yang, W.; Liu, Z.; Zhuo, Z.; Gao, Y. Delivery of interferons and siRNA targeting STAT3 using lentiviral vectors suppresses the growth of murine melanoma. *Cancer Gene Ther.* **2012**, *19*, 822–827.
45. Du, G.; Yonekubo, J.; Zeng, Y.; Osisami, M.; Frohman, M.A. Design of expression vectors for RNA interference based on miRNAs and RNA splicing. *FEBS J.* **2006**, *273*, 5421–5427.
46. Karwacz, K.; Bricogne, C.; MacDonald, D.; Arce, F.; Bennett, C.L.; Collins, M.; Escors, D. PD-L1 co-stimulation contributes to ligand-induced T cell receptor down-modulation on CD8+ T cells. *EMBO Mol. Med.* **2011**, *3*, 581–592.
47. Arce, F.; Breckpot, K.; Stephenson, H.; Karwacz, K.; Ehrenstein, M.R.; Collins, M.; Escors, D. Selective ERK activation differentiates mouse and human tolerogenic dendritic cells, expands antigen-specific regulatory T cells, and suppresses experimental inflammatory arthritis. *Arthritis Rheum.* **2011**, *63*, 84–95.

8

Angiotensin II Regulates microRNA-132/-212 in Hypertensive Rats and Humans

Tilde V. Eskildsen [1,2,†], Pia L. Jeppesen [1,2,†], Mikael Schneider [2], Anne Y. Nossent [1,2,3], Maria B. Sandberg [2], Pernille B. L. Hansen [1], Charlotte H. Jensen [1,2], Maria L. Hansen [2,4], Niels Marcussen [5], Lars M. Rasmussen [2], Peter Bie [1], Ditte C. Andersen [1,2] and Søren P. Sheikh [1,2,*]

1 Department of Cardiovascular and Renal Research, Institute of Molecular Medicine, University of Southern Denmark, DK-5000 Odense, Denmark; E-Mails: teskildsen@health.sdu.dk (T.V.E.); piajeppesens@hotmail.com (P.L.J.); a.y.nossent@lumc.nl (A.Y.N.); plaerkegaard@health.sdu.dk (P.B.L.H.); charken@health.sdu.dk (C.H.J.); pbie@health.sdu.dk (P.B.); dandersen@health.sdu.dk (D.C.A.)

2 Department of Clinical Biochemistry and Pharmacology, Odense University Hospital, Sdr. Boulevard 29, DK-5000 Odense, Denmark; E-Mails: mikaelschneider@gmail.com (M.S.); maria.sandberg@ouh.regionsyddanmark.dk (M.B.S.); mlhansen@health.sdu.dk (M.L.H.); lars.melholt.rasmussen@ouh.regionsyddanmark.dk (L.M.R.)

3 Department of Vascular Surgery, Einthoven Laboratory for Experimental Vascular Medicine, Leiden University Medical Center, 2333 ZA Leiden, The Netherlands

4 Department of Cardiac and Thoracic Surgery, Odense University Hospital, Sdr. Boulevard 29, DK-5000 Odense, Denmark

5 Department of Pathology, University of Southern Denmark, DK-5000 Odense, Denmark; E-Mail: nmarcussen@health.sdu.dk

† These authors contributed equally to this work.

* Author to whom correspondence should be addressed; E-Mail: soeren.sheikh@ouh.regionsyddanmark.dk;

Abstract: MicroRNAs (miRNAs), a group of small non-coding RNAs that fine tune translation of multiple target mRNAs, are emerging as key regulators in cardiovascular development and disease. MiRNAs are involved in cardiac hypertrophy, heart failure and remodeling following cardiac infarction; however, miRNAs involved in hypertension have not been thoroughly investigated. We have recently reported that specific miRNAs play an integral role in Angiotensin II receptor (AT_1R) signaling, especially after activation of the Gαq signaling pathway. Since AT_1R blockers are widely used to treat hypertension, we undertook a detailed analysis of potential miRNAs involved in Angiotensin II (AngII) mediated hypertension in rats and hypertensive patients, using miRNA microarray and qPCR analysis. The miR-132 and miR-212 are highly increased in the heart, aortic wall and kidney of rats with hypertension (159 ± 12 mm Hg) and cardiac hypertrophy following chronic AngII infusion. In addition, activation of the endothelin receptor, another Gαq coupled receptor, also increased miR-132 and miR-212. We sought to extend these observations using human samples by reasoning that AT_1R blockers may decrease miR-132 and miR-212. We analyzed tissue samples of mammary artery obtained from surplus arterial tissue after coronary bypass operations. Indeed, we found a decrease in expression levels of miR-132 and miR-212 in human arteries from bypass-operated patients treated with AT_1R blockers, whereas treatment with β-blockers had no effect. Taken together, these data suggest that miR-132 and miR-212 are involved in AngII induced hypertension, providing a new perspective in hypertensive disease mechanisms.

Keywords: hypertension; Angiotensin II; AT_1R; AT_1 receptor blocker; microRNA

---

1. Introduction

Persistent elevation of systemic blood pressure (hypertension) is one of the most prevalent medical conditions involving the cardiovascular system and affects as many as one billion people worldwide [1]. Hypertension is an undisputed risk factor for cardiovascular diseases, including stroke, cardiac failure and renal diseases [1]. Several mechanisms have been implicated in the pathogenesis of hypertension, including increased activity of the sympathetic nervous system, dysfunction of the vascular endothelium, vascular smooth muscle and cardiac hypertrophy, as well as overactivation of the renin-angiotensin-aldosterone system (RAAS) [2]. Angiotensin II (AngII) controls blood pressure and fluid homeostasis through its receptors, AT_1R and AT_2R, and through stimulation of aldosterone [1]. AngII receptors are expressed in tissues that have an impact on blood pressure control, including heart, kidney and vasculature [3,4]. The classical AngII responses in the cardiovascular and renal systems are mediated mainly by AT_1R signaling [3–5], including heterotrimeric G-protein activation and downstream signaling through the canonical MAP kinases ERK1/2, which, in turn, regulate gene transcription [4]. Accordingly, specific inhibitors of AngII pathways, including AT_1R blockers, dramatically lower blood pressure in hypertensive patients and slow the progression of cardiovascular disease [1,3].

We speculated that altered expression of microRNAs (miRNA) may be part of the pathogenesis behind AngII-related hypertension. MicroRNAs are small non-coding RNAs that regulate gene expression by pairing to and destabilizing the mRNAs of protein coding genes, resulting in decreased mRNA levels [6]. The mammalian miRNAs are highly conserved, and each miRNA is predicted to target mRNAs of hundreds of distinct genes, fine-tuning and optimizing the expression patterns of most protein-coding genes [7]. Theoretically, these miRNAs are ideally suited to co-regulate gene expression events in cellular responses to vasopressors, such as AngII. Most miRNAs are solitary and expressed under the control of their own promoters and regulatory sequences, while others are arranged as clusters and may be co-regulated with additional members of the cluster [7]. For example, miR-132 and miR-212 are clustered closely in the genome and are transcribed together under the regulation of cAMP response element binding protein [8], which is a known AngII regulated gene [9,10]. Several miRNAs are aberrantly expressed in cardiovascular diseases [2,11–13]. miR-21, miR-155 and miR-221/222 have recently been shown to regulate AngII signaling in cardiac fibroblasts [14–16] and in endothelial cells [17], while miR-29 regulates fibrotic pathways [18]. We have recently shown that AngII, via the Gαq pathway, regulates five miRNAs during *in vitro* stimulation of primary cardiac fibroblasts and of HEK293N cells overexpressing the AT_1-receptor [19]. Most of the miRNA studies are based on *in vitro* experiments, and very few studies have examined the relation between AngII mediated hypertension and miRNA regulation *in vivo*.

In this study, we hypothesized that *in vivo* AngII mimics the "five miRNA" expression signature obtained by AT_1R overexpression [19]. We examined the miRNA expression in heart, aorta and kidney from a rat model with a constant intravenous infusion of AngII and expanded these results to data concerning miRNA expression in the mammary artery of patients treated with AT_1R blockers. Our results suggest that miR-132 and miR-212 are involved in AngII-induced Gαq-signaling pathway leading to hypertension. Further understanding of the importance of these miRNAs will come from future miRNA knockdown experiments or knockout in whole animals.

2. Results

2.1. High Blood Pressure, Cardiac Hypertrophy and Fibrosis Are Sustained in the Rat Model

Infusion of AngII for 10 days resulted in a stable and significant elevation in blood pressure to 159 ± 12 mm Hg ($p < 0.001$, $n = 7$) at day 10, as compared to control rats that remained constant at 98 ± 4 mm Hg ($n = 8$) (Figure 1A). Likewise, we found that short time (4 h) AngII infusion resulted in an acute and significant 29 mm Hg increase in blood pressure ($p < 0.001$, $n = 6$) (Figure 1B). AngII hypertensive rats exhibited cardiac hypertrophy, as evidenced by a significant 17% increased left ventricle to body weight ratio ($p < 0.01$, $n = 7$) *versus* control rats ($n = 6$) (Figure 1C). The mass of the left ventricle increased from 614 ± 82 mg ($n = 6$) in control rats to 780 ± 75 mg ($n = 7$) in the hypertensive rats ($p < 0.01$), whereas no increase was observed for the right ventricle or atria weight (Figure 1C). Left ventricular hypertrophy was further validated by a significantly higher expression level of *B-type natriuretic peptide* (*BNP*) ($p < 0.001$) (Figure 1E). Likewise, cardiac fibrosis after infusion of AngII for 10 days was confirmed by increased collagen deposition (Figure 1D) and an increased expression of genes generally associated with fibrosis, including *Fibronectin* ($p < 0.01$) and *Procollagen I* ($p < 0.001$) (Figure 1E). These results thus showed that continuous AngII infusion for 10 days resulted in clear and sustained hypertension, leading to hypertrophic and fibrotic changes of the heart.

Figure 1. Characterization of AngII-induced hypertensive rat. (**A**) Mean daily averages of mean arterial blood pressure from seven rats treated with chronic infusion of 30 ng/kg/min AngII for 10 days (□) and six rats treated with acute infusion of 30 ng/kg/min AngII for 4 h (Δ), compared to eight control rats (O); (**B**) Mean hourly averages of mean arterial blood pressure from acute infusion of rats (Δ) for four h compared to control rats (O). Data are shown as the mean ± SD. Arrows shows the start of AngII infusion (day 0). Statistical significance was tested by two-way ANOVA for either control *versus* AngII for 10 days or control *versus* AngII for 4 h. *** $p < 0.001$. A and B, duplicate figure [20]; (**C**) Weight to body weight ratio (mg/g) of chronic ($n = 7$), acute ($n = 6$) and control ($n = 8$) rat hearts divided into left ventricle, right ventricle and atria. Data is presented as the mean ± SD, and statistical significance was tested by one-way ANOVA using Tukey's multiple comparison test. ** $p < 0.01$; (**D**) Representative sections of left ventricles of AngII affected hearts compared to control hearts, stained with Sirius Red for collagen deposition; (**E**) qRT-PCR for the early marker of hypertrophy, *BNP,* the fibrotic markers, *Fibronectin* and *Procollagen-I,* and the two stably expressed reference genes, *Gapdh and Rpl13a* (M: 0.140 and CV: 0.049). Data is presented as the mean ± SD, and statistical significance was tested by un-paired *t*-test. * $p < 0.05$, ** $p < 0.01$ and *** $p < 0.001$.

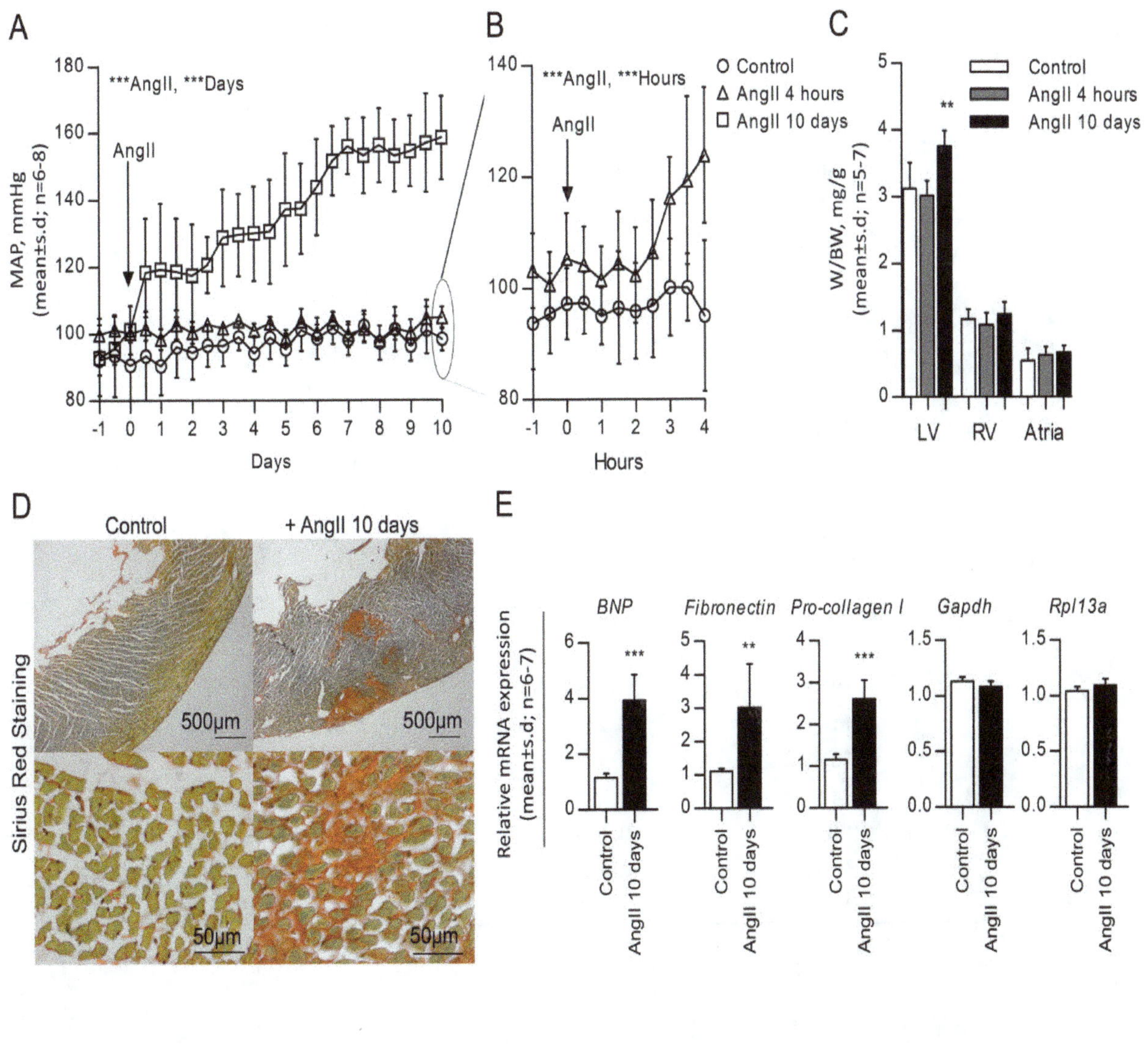

2.2. Chronic AngII-Mediated Hypertension in Rats Increases miR-132/-212 Cluster Expression in Blood Pressure Regulating Organs: Heart, Aorta and Kidney

Table 1. miRNA microarray analysis showing significantly altered miRNA expression in the left ventricles from rats infused with AngII for 10 days compared to controls. Data is presented as $\log_2$ fold expression (n = 6–7) and sorted by p-value.

Microarray: miRNAs regulated by AngII		
Name of miRNA	**Log fold change**	***p*-value**
		$^{***}p < 0.001$
21	0.7016	1.89×10^{-5}
132	0.1261	5.90×10^{-5}
105	0.1883	9.28×10^{-5}
155	0.1425	0.00012
221	0.3414	0.00052
223	0.4459	0.00085
208b	0.5560	0.00089
		$^{**}p < 0.01$
222	0.1681	0.0022
147b	0.1151	0.0032
26b	−0.1153	0.0034
15b	0.2580	0.0057
613	−0.1047	0.0065
31 *	0.1054	0.0075
520b	0.0938	0.0082
30c-1 *	−0.1409	0.0084
18b	0.1334	0.0092
		$^{*}p < 0.05$
301a	0.1537	0.010
143	0.1242	0.011
434-5p	0.1692	0.012
484	0.0855	0.014
155	0.1135	0.014
379	0.0813	0.015
29c	−0.2193	0.017
936	0.1682	0.018
199a-5p	0.1523	0.021
201	−0.1204	0.021
101	−0.1860	0.021
363 *	0.2352	0.021
760	−0.0944	0.022
944	0.1056	0.023
200b *	0.1079	0.024
30b	−0.0888	0.025
322	−0.1890	0.026

Table 1. *Cont.*

Microarray: miRNAs regulated by AngII		
Name of miRNA	**Log fold change**	***p*-value**
337-3p	0.1065	0.026
29c *	−0.1247	0.028
302c *	0.1453	0.030
193a-3p	0.1665	0.030
888	−0.0852	0.033
299-3p	0.2128	0.035
142-3p	0.3121	0.035
517 *	0.0746	0.041
31	0.3286	0.042
194	−0.0860	0.042
701	0.0671	0.042
545	−0.0834	0.043
609	0.0617	0.044
141 *	−0.0928	0.045
211	0.0588	0.047
373	0.0712	0.048
218-1 *	−0.0678	0.049

* means the passenger strand of the miRNA-miRNA* duplex; Passenger miRNA, may also have constitute a bioactive miRNA itself.

Fifty miRNAs were identified as differentially expressed ($p < 0.05$) in hearts of sustained hypertension (10 days of AngII), as compared to controls (Table 1), whereas no miRNAs were differentially expressed in acute hypertensive rats (4 h of AngII). AngII affects the blood pressure by two separate mechanisms: firstly, an acute contractile effect on the arterial walls arising within a few hours, followed by a chronic compensative response arising after a few days. Gene expression is primarily effected by the long-term AngII signaling. However, we sought to investigate if miRNAs could rapidly change, due to the acute effect, and found no regulation of miRNAs in the short period of AngII infusion. Of the 50 miRNAs differentially regulated in chronic hypertension, miR-21 was the most upregulated and was used as a positive control in our analysis. Interestingly, among the many dysregulated miRNAs, the second most significantly regulated miRNA was miR-132. Since the miR-132 gene is clustered with the miR-212 gene and they are likely expressed together [21], we included miR-212 in further analyses. In summary, we found that the expression of miR-21, -132 and -212 were up-regulated by 3.4-, 1.4- and 1.8-fold, respectively, in the left ventricle of AngII-induced hypertensive animals, as compared to controls (Figures 2A and S1). Besides the heart, the arterial wall and kidneys are involved in systemic blood pressure homeostasis, and we, therefore, examined whether the miR-132/-212 levels were affected also in these tissues. miR-132 and -212 were increased 6.4- and 3.2-fold, respectively, in aortas from hypertensive animals, as compared to controls (Figure 2A). Likewise, we observed a significant regulation of miR-132 (1.4-fold) and -212 (1.5-fold) in the kidneys of AngII-infused rats *versus* controls (Figure 2A). Furthermore, miR-132 was found to be significantly upregulated in the plasma of AngII-induced hypertensive animals, whereas no regulation was observed for plasma miR-212 levels compared to the control rats (Figure 2B). Lack of miR-212 upregulation in the plasma of AngII-induced hypertensive rats compared to control rats might be caused by the low concentration and high variability of the miR-212 levels found in the plasma. Of note, the miRNA expression levels were found to significantly correlate with AngII-induced hypertension in the kidneys, a strong tendency of correlation in aorta and a lower tendency for correlation in the heart as tested by linear regression and correlation analysis (Figure 2C). By contrast, no changes in miR-132/-212 levels were found in any of the three tissues in the acutely hypertensive rats (Figure 2A). These data confirmed the microRNA array data obtained for chronic and acute hypertensive rats. Altogether, these results strongly suggest that the miR-132/-212 cluster may be a general and novel mediator of AngII-induced hypertension.

Figure 2. Validation of miRNA regulation in AngII-induced hypertensive rat heart, aorta and kidney. (**A**) qRT-PCR identification of miRNAs in left ventricle of hearts, aortas and kidneys from AngII affected rat hearts. Statistical significance was tested by one way ANOVA. * $p < 0.05$. Values are shown as relative expression with the mean ± SD; $n = 4–7$; (**B**) qRT-PCR identification of miRNAs in plasma of AngII affected rat hearts. Statistical significance was tested by un-paired *t*-test. ** $p < 0.01$. Values are shown as relative expression with the mean ± SD; $n = 4–5$. miRNA expression in the three organs and in plasma was individually normalized to two reference genes stably expressed among the samples. Reference genes used for normalization: Heart, miR-17 and miR-191 (M: 0.356, CV: 0.123); aorta, miR-103 and miR-191 (M: 0.832, CV: 0.290); kidney, miR-17 and miR-191 (M: 0.145, CV: 0.050); and plasma, miR-17 and miR-103 (M: 1.143, CV: 0.390); (**C**) Correlation analysis for miRNA expression levels at day 10 of AngII or isotone glucose infusion. Statistical significance was tested by linear regression (R^2) and correlation analysis (Pearson's *r*). Data are shown as two individual correlations per miRNA is each organ for control (○) and AngII for 10 days (●). ** $p < 0.01$; $n = 7–11$.

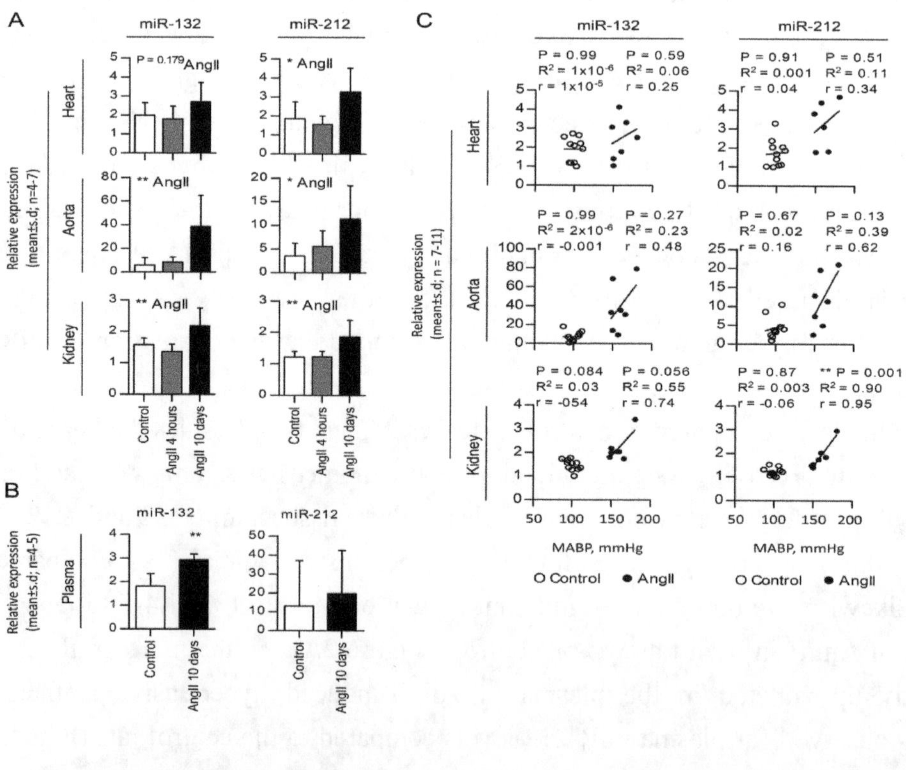

2.3. miR-132 and -212 Regulation in Response to AngII in Mice

Studying mice in this particular field of hypertension research is of interest, because their genome can easily be modified. Infusion of AngII at 15, 30 and 60 ng/kg/min for seven days resulted in a stable and significant elevation in blood pressure to 119 ± 10, 127 ± 5 and 128 ± 10 mm Hg ($p < 0.001$, $n = 3$), respectively, at day seven, as compared to control mice that remained constant at 95 ± 2 mm Hg ($n = 4$) (Figure 3A). AngII-infused mice did not exhibit cardiac hypertrophy when examining the left ventricle to body weight ratio ($n = 3$) *versus* control mice ($n = 4$) (Figure 3B). Likewise, miR-132 and miR-212 were not significantly increased in AngII-infused mice at any of the AngII concentrations when compared to control mice. These results thus showed that continuous AngII infusion for seven days in mice resulted in a modest increase in blood pressure without hypertrophic changes of the heart and no regulation of the miR-132/-212 cluster.

Figure 3. Validation of miRNA regulation in AngII-induced hypertensive mice hearts. (**A**) Mean daily averages of mean arterial blood pressure from three mice treated with chronic infusion of 60, 30 or 15 ng/kg/min AngII for seven days (□, Δ or ◊, respectively), compared to three control mice (O). Data are shown as the mean ± SD. AngII infusion is started at day 0. Statistical significance was tested by two-way ANOVA. *** $p < 0.001$, *$p < 0.05$; (**B**) Left ventricle (LV) weight to body weight ratio (mg/g) of mice infused with 60, 30 or 15 ng/kg/min of AngII for seven days ($n = 3$) and control ($n = 3$) mice. Data is presented as the mean ± SD, and statistical significance was tested by one-way ANOVA using Tukey's multiple comparison test; (**C**) qRT-PCR identification of miRNAs in left ventricle of hearts from AngII affected mice hearts. Statistical significance was tested by one-way ANOVA using Tukey's multiple comparison test. Values are shown as relative expression with the mean ± SD; $n = 3$. Reference genes used for normalization: miR-103 and miR-191 (M: 0.826, CV: 0.288).

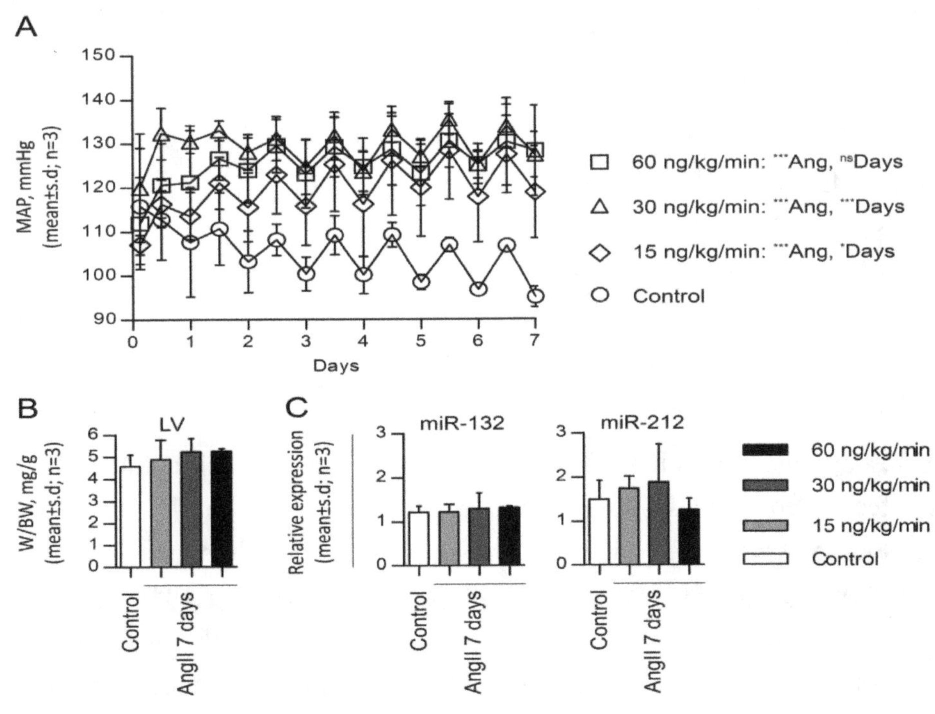

Figure 4. Endothelin 1-induced hypertensive rats. (**A**) Mean daily averages of mean arterial blood pressure from five rats treated with chronic infusion of 5 pmol/kg/min ET-1 for 10 days (□), compared to three control rats (O). Statistical significance was tested by two-way ANOVA. *** $p < 0.001$; (**B**) qRT-PCR for the early hypertrophy marker, *BNP* and the fibrosis markers, *Fibronectin* and *Procollagen-I*. Statistical significance was tested by un-paired *t*-test. Data is shown as the mean ± SD, n = 3–5. Reference genes used for normalization: *GAPDH* and *Rpl13a* (M: 0.153, CV: 0.053); (**C**) qRT-PCR identification of miRNAs in the left ventricle of hearts, aortas and kidneys from ET-1 affected rat hearts. Statistical significance was tested by un-paired *t*-test. * $p < 0.05$. Values are shown as relative expression with the mean ± SD, n = 3–5. miRNA expression was individually normalized to two reference genes stably expressed among the samples. Reference genes used for normalization: heart; miR-17 and miR-191 (M: 0.309, CV: 0.107); aorta, miR-103 and miR-191 (M: 0.667, CV: 0.232); kidney, miR-17 and miR-191 (M: 0.180, CV: 0.062).

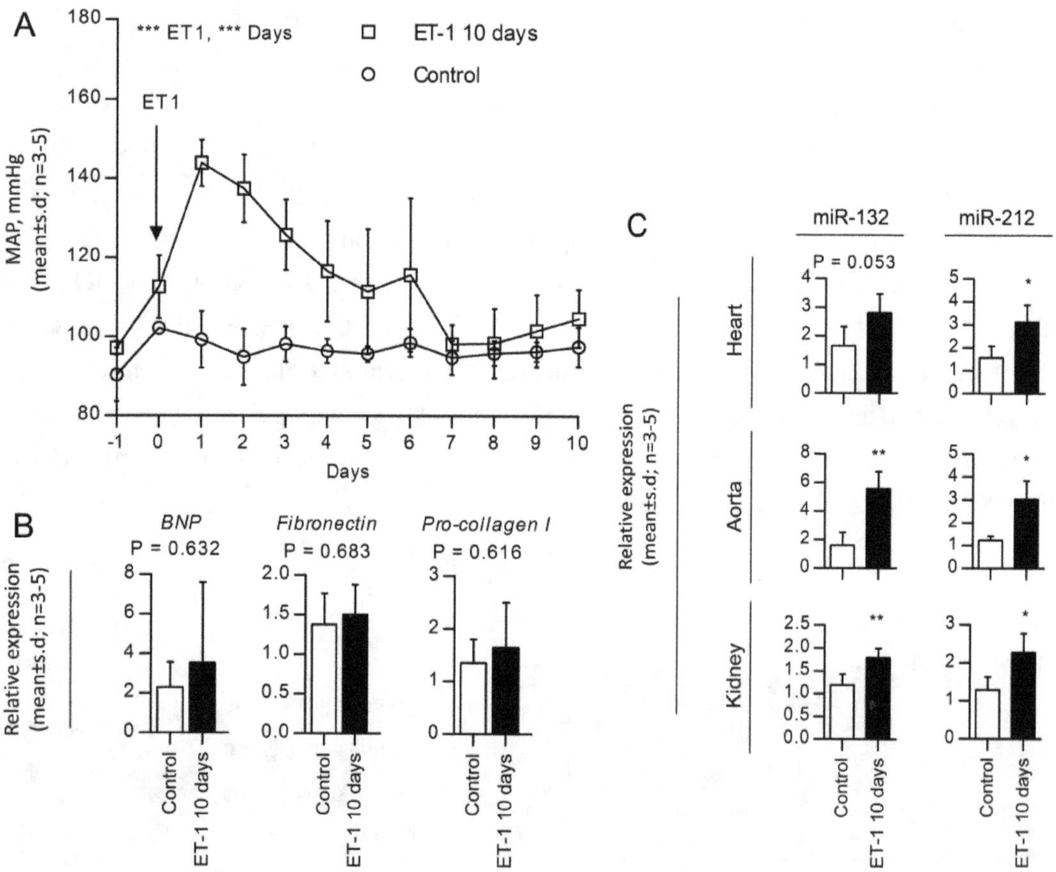

2.4. miR-132 and -212 Regulations in Response to ET-1, Vasopressor-Induced Hypertension

In order to test whether the miR-132/-212 cluster response is specific for AT_1R induction, we examined the effect of continuous infusion of a second vasopressor (ET-1) [22]. In contrast to the sustained high blood pressure in AngII-infused rats, ET-1-mediated hypertension peaked at day one (144 ± 6 mm Hg (n = 5) and, hereafter, gradually declined, reaching 105 ± 7 mm Hg at day 10 (Figure 4A). In line with this, we did not observe any cardiac hypertrophy or fibrosis in ET-1 treated hearts. Thus, we found no change in *BNP*, *Fibronectin* or *Procollagen-I* expression levels (Figure 4B).

However, despite the partially compensated blood pressure and the absence of cardiac hypertrophy and fibrosis, we found significantly higher levels of miR-132 and miR-212 in heart (1.7- and 2.0-fold), aorta (3.5- and 2.5-fold) and kidney (1.5- and 1.8-fold) of the ET-1 treated animals as compared to controls (Figure 4C). In line with the previous findings by Jeppesen *et al.* [19], this suggests that the upregulation of miR-132/-212 is a direct consequence of Gαq vasopressor stimulation pathways and not a result of secondary heart hypertrophy and fibrosis. In summary, these data indicate that upregulation of the miR-132/-212 cluster likely is part of a general response to Gαq-vasopressor stimulation of the ERK1/2 pathway and may be involved in a common AngII- and ET-1-mediated signaling pathway leading to hypertension.

Figure 5. AngII receptor blockers and β-blockers in internal mammary artery (IMA) patients. (**A**,**C**) Study group information on the AngII receptor blocker (ARB) and β-blocker patient groups, respectively. None of the characteristics within the patient groups of ARBs and β-blockers were significantly different. The groups were matched for age, sex, diabetes and treatment using statins. None of the patients were simultaneously treated with angiotensin-converting enzyme (ACE)-inhibitors; (**B**) Scatter plot represents Log_{10} relative miRNA expression with mean bar, in patients treated with ARBs compared to patients treated with non-ARB (n = 16). (**D**) Scatter plot representing Log_{10} relative miRNA expression with mean bar, in patients treated with β-blocker (n = 9) compared to patients treated with non-β-blockers (n = 23). Expression is normalized to two stably expressed reference genes. Reference genes used for normalization: miR-103 and miR-191 (M: 0.804, CV: 0.279).

A

Internal Mammary Artery	Control	AT_1R blocker
n	16	16
Sex, females/males, n	3/13	3/13
Age, years	71.2 (6.2)	71.5 (7.7)
Systolic blood pressure, mmHg	139.9 (9.7)	141.3 (24.5)
Diastolic blood pressure, mmHg	75.5 (10.6)	77.3 (14.4)
β-blockers	11	12
Diuretics	6	9

C

Internal Mammary Artery	Control	β-blocker
n	23	9
Sex, females/males, n	4/19	2/7
Age, years	70.1 (7.3)	73.7 (4.9)
Systolic blood pressure, mmHg	142.9 (18.6)	133.6 (12.1)
Diastolic blood pressure, mmHg	77.8 (12.3)	72.0 (11.8)
ARB	12	4
Diuretics	11	4

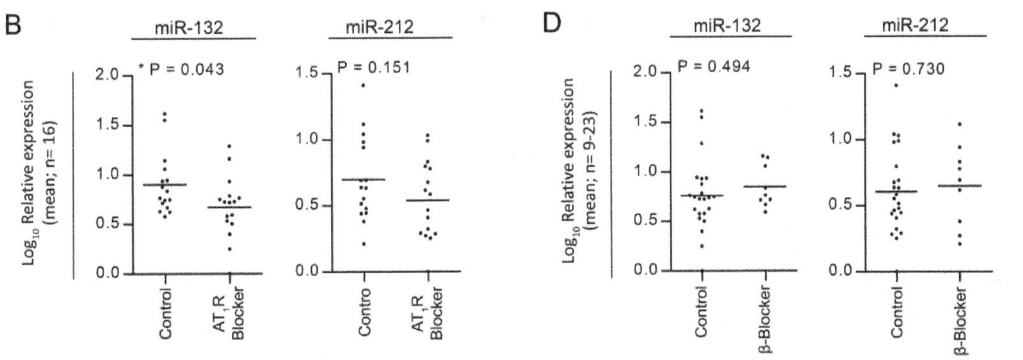

2.5. *Treatment with Angiotensin II Receptor Blocker Attenuates the Expression of the miR-132/-212 Cluster in Human Hypertension*

We measured miR-132/-212 levels in the internal mammary artery (IMA) tissue obtained from patients treated with AngII receptor blocker (ARB) or β-blocker in age and sex matched patients undergoing by-pass surgery (Figure 5A,C). The four patient groups showed similar pre-operative blood pressure (Figure 5A,C). ARB treated patients ($n = 16$) revealed a significant attenuation of miR-132 expression (0.55-fold), as well as a tendency for miR-212 downregulation (0.64-fold), as compared to non-ARB-treated patients (miR-132; 0.93 and miR-212; 1.01) ($n = 16$) (Figure 5B). To further investigate whether this miR-132 and -212 regulation is specific to AngII or related to putative direct influences from blood pressure, we examined patients treated with β-blockers (Figure 5C). Interestingly, we did not find any attenuation or downregulation of miR-132/-212 expression in IMA patients receiving β-blockers, as compared to non-β-blocker-treated patients (Figure 5D), indicating that not all blood pressure-reducing agents can downregulate miR-132/-212 expression, which further supports the notion that AngII mediates a global upregulation of miR-132/-212 in humans. These results suggest that the miR-132/-212 cluster in humans may also be part of the response to Gαq-vasopressors, such as AngII.

3. Discussion

We found increased expression of miR-132 and -212 in the left ventricle, aorta and kidney, as well as in the plasma (Figure 2) after 10 days of sustained AngII-induced hypertension in rats, which is compatible with our pervious published *in vitro* study [19]. Even though miR-132 and miR-212 are expressed from the same precursor, we observed independent regulation in the different tissues in response to the same AngII infusion. This could be due to differences in stability and degradation [21,23]. Moreover, the degree of miR-132/-212 increase shows a tendency to correlate with blood pressure suggesting that these miRNAs could play a novel role in AT1 receptor pharmacology, both *in vitro* and *in vivo*.

Blood pressure has previously been associated with miRNA. Recently, Nossent *et al.* investigated the association between single-nucleotide polymorphisms (SNPs) located in the miRNA binding sites of genes associated with Renin-AngiotensinII-system (RAS), blood pressure and myocardial infarction in a large population study [24]. Several SNPs located in RAS genes were associated with changes in blood pressure and were shown to interfere with miRNA regulation [24].

miR-132/-212 has been described in both the central nervous and cardiovascular systems. In one study, miR-132 was reported to be constitutively expressed and released by pericyte progenitor cells, and transplantation of these cells into mice with myocardial infarction improves cardiac function through proangiogenic activities [25]. Several studies identify miR-132/-212 involvement in the central nervous system, *i.e.*, in neuronal function and plasticity [8,21,26]. In addition, miR-132/-212 has also been found to be involved in neovascularization, inflammation and adipocyte differentiation in the peripheral tissues [21,23,27]. Interestingly, miR-132/-212 was upregulated in the aortas of mice stimulated with AngII from osmotic pumps for 14 days, but this study did not report blood pressure values [28]. We performed similar studies with chronic AngII infusion in mice for seven days, simultaneously measuring blood pressure. In contrast to our rat infusion model, only a modest blood pressure increase was observed in this mice strain, and miR-132/-212 was not significantly upregulated in mouse hearts (Figure 3). Differences in AngII effects between rats and mice have previously been described by Cassis *et al.*, stating that AngII infusion in mice has minimal effects compared to same doses given to rats [29]. Based on these data, we decided that the rat model was more suitable for studying AngII-induced miRNA changes in hypertensive animals. We previously demonstrated that

AT_1R signaling regulates miR-132 and -212 in HEK293N cells and in primary cultures of cardiac fibroblasts through the Gαq dependent pathway [19]. These results were recently confirmed in primary cultures of rat vascular smooth muscle cells [28]. Additionally, we investigated the regulation of the miR-132/-212 cluster in endothelial cell lines and in primary cultures of vascular smooth muscle cells (VSMCs) and leukocytes and found no regulation in either of the cell types (data not shown). Furthermore, by inhibition of the Gαq subunit in cardiac fibroblasts, we demonstrated significant decreases in miR-132 and -212 expressions, pointing to Gαq protein activation as the responsible pathway for AngII-induced miRNA regulation *in vitro* [19].

Using a different vasopressor (ET-1) that binds to the endothelin receptor and also activates Gαq-ERK1/2 signaling [30,31], we examined whether this receptor could regulate the miRNAs. Interestingly, even though the ET-1-induced hypertension had a much shorter duration than the sustained hypertension induced by AngII, both miR-132 and miR-212 were upregulated at a point in time when blood pressure was not (Figure 4). From a mechanistic point of view, these findings indicate that miR-132/-212, also *in vivo*, may be regulated through Gαq-ERK1/2 activation, which is one of the mutual steps in the AngII and ET-1 signaling pathways leading to hypertension [30–32]. Following these observations, we examined whether the increased miRNA expression levels could be attenuated by blocking the AngII signaling in humans (Figure 5). We examined the arterial walls from two different groups of patients treated with either one of several ARBs, including Losartan, Candesartan, Irbesartan and Telmisartan, or no ARB. These ARBs are widely used in hypertension treatment and have similar chemical structures, but different pharmacological properties and efficacy [33]. None of the patients were treated with ACE-inhibitors. Our patient groups were selected based on treatment with ARBs and individual ARB-types grouped and analyzed together in order to allow for statistical testing. Most patients in the ARB group probably suffered from hypertension; however, it has not been possible from the patient files to deduce who strictly fulfills the definitions of hypertension. Likewise, in the non-ARB group, the precise prevalence of hypertension is not known. Another limitation of our human observations is the lack of knowledge about the exact time point of medication, which could not be controlled. Despite these limitations, we observed a significant downregulation of miR-132, as well as a robust attenuation of miR-212 in the ARB-treated patients. We next asked whether treatment with β-blockers, often a first choice antihypertensive drug, would also decrease miR-132 and -212, following the notion that it could be reduction in blood pressure *per se*, which may alter the miRNA levels. However, the levels of miR-132/-212 were not downregulated in patients treated with β-blockers. This observation is compatible with the notion that alterations in miRNA is not only a spurious factor found in rodents or related to experimental systems *in vitro*, but that these molecules also could be suggested to play a role in the arterial wall among patients receiving medication that alters the activity in the AngII system. Our data indicates that AT_1R control of these miRNAs is evolutionary conserved between rat and man. Thus, AT_1R activation in rats increases miR-132 and miR-212, while blocking the AT_1R decreases miRNA levels in humans. Since no anti-miR experiments have been conducted, it has not been possible to deduce the specific cause and effect relationship; however, the regulation of miR-132 and miR-212 is likely biological important, because although rats and humans share biological features in blood pressure control, they have multiple differences in the molecular subtypes of ion channels, receptors and signaling pathways in blood vessel cells. Further studies are necessary to assess the relative biological and pharmacological impact of individual miR-132 and miR-212 levels on systemic blood pressure, in the heart, arterial wall and kidney. We speculate that the AT_1R could perform these effects through Gαq and downstream activation of the ERK1/2 pathway and not through blood pressure or aldosterone. Three lines of evidence support this notion. Firstly, the effect was inhibited by the pharmacological blockade of AngII *in vivo* and Gαq and ERK1/2 *in vitro* [19]. Secondly, another Gαq activator (ET-1) reproduced the effects, while, finally, inhibiting Gs signaling with the β-adrenergic blocker had no effects on the miRNA levels.

4. Experimental Section

4.1. Animal Care

Female Sprague-Dawley rats (8 weeks old, Taconic, Ry, Denmark) were housed in air-conditioned rooms with a 12 h dark-light cycle and fed standard diet (Altromin® Standard 1320, Lage, Germany) with free access to tap water. All rat experiments were approved by the Danish National Animal Experiment Inspectorate (Permission #2009/561/1753).

4.2. Angiotensin II (AngII) Model

Under Hypnorm/midazolam anesthesia, chronic catheters were placed in the left femoral vein and artery and connected to a swivel via a skin button between the scapulae, allowing the rat full mobility [34]. Following a 5–7 day recovery period, the arterial line was connected to a pressure transducer (Föhr Medical Instruments, Hessen, Germany), and data were collected continuously using Lab View software (National Instruments, Austin, TX, USA). The rats were infused with 5% glucose solution at a rate of 10 μL/kg/min with or without AngII (Sigma-Aldrich, St. Louis, MO, USA) for 10 days. Rats were randomly divided into three groups: (1) 10 days' continuous infusion of AngII at 30 ng/kg/min (sustained hypertension); (2) 9 days' and 20 h infusion with 5% glucose, followed by a 4 h period of AngII infusion at 30 ng/kg/min (acute hypertension); and (3) 10 days' infusion of 5% glucose (controls). Rats were sacrificed with an overdose of pentobarbital, then perfused with ice-cold sterile isotonic saline via the heart. The heart, kidney and aorta were dissected under sterile conditions. Hearts were carefully sectioned into atria, right ventricle and left ventricle, whereas aortas were cleaned from fat, as well as connective tissue, and kidneys were decapsulated.

4.3. Endothelin 1 (ET-1) Model

The setup was similar to that described above for the AngII model, but rats were randomly divided into two groups: (1) 10 days' infusion of endothelin-1 (ET-1) (Bachem) at a concentration of 5.0 pmol/kg/min in isotonic saline; and (2) 10 days' infusion with isotonic saline, as previously described by Mortensen L.H. & Fink G.D. 1992 [35].

4.4. Mice AngII Model

The mice experiments were approved by the Danish National Animal Experiment Inspectorate (Permission #2009/561/1749). The setup was similar to that described above for the AngII rat model with minor modifications as described below. Mixed gender C57/BL6 mice were anaesthetized using ketamine/xylazine and catheters placed in the femoral artery and vein for measurements of arterial blood pressure and AngII infusions, respectively. The mice recovered for five days before beginning the continuous measurements of mean arterial blood pressure and heart rate for seven days. Mice were randomly divided into four groups: (1–3) 7 days' continuous infusion of AngII at 15, 30 and 60 ng/kg/min respectively and (4) 7 days' of continuous infusion with 5% glucose (controls). Mice were sacrificed and processed, as described above.

4.5. Patients, Internal Mammary Artery Study

Surplus arterial tissue was obtained from the repair vessel, *i.e.*, the internal mammary artery (IMA), from patients undergoing coronary artery by-pass graft surgery at the Department of Cardiac and Thoracic Surgery, Odense University Hospital, Denmark. Four groups of patients were analyzed, based on intake or no intake of either AngII receptor blocker or β-blocker. Study population characteristics are described in (Figure 5A,C). As previously described [36], the intima/media of the artery was carefully dissected immediately after removal, frozen in liquid nitrogen and stored in the Odense Artery Biobank. The study was approved by the Ethical Review Board of Southern Denmark (No. S-20100044), and informed consent was obtained from each subject.

4.6. Sirius Red Staining

Fibrosis was identified by Sirius Red Staining for collagen deposits. Cryosections were equilibrated at room temperature for 30 min, rinsed in a mixture of NBF (37%) and ethanol (93%) for 45 s, rehydrated (5 min) and subsequently counterstained (15 min) with Weigert's Iron Hematoxylin (Sigma-Aldrich, HT-107/109). Following a wash, sections were further incubated (60 min) with 0.1% Sirius red (Sigma-Aldrich, P6744) in saturated picric acid (Sigma-Aldrich, 365548) and finally washed twice in 99% ethanol. Sections were mounted with Pertex and analyzed by bright-field microscopy using a Leica DML332 equipped with a Leica DFC300F camera.

4.7. Microarray and Data Processing

The miRNA expression profiling was performed as two-color common reference hybridizations on LNA-based arrays (miRCURY LNA™ microRNA Array ready to spot probe set, Exiqon, Denmark) spotted in-house on CodeLink™ HD Activated slides (DHD1-0023, SurModics, Eden Prairie, MN, USA), according to the manufacturer's recommendation. Samples were labeled with Hy3, and the common reference (pool of all samples) was labeled with Hy5, by use of miRCURY LNA microRNA Array Power labeling kit (208032-A, Exiqon) and hybridized for 16 h. Slides were washed (208021, Exiqon), scanned on an Agilent (G2565CA) Microarray scanner and analyzed by the Genepix 6.0 software. Normalization and background correction was performed with "R" software using the "vsn" package (Bioconductor, open source software), and quadruplicate spots were averaged. Differential expression was assayed using the "limma" package (Bioconductor) by fitting the eBayes linear model and contrasting individual treatments with untreated controls. Log_2 fold changes were calculated using the toptable function of the limma package.

4.8. mRNA and miRNA Analysis

Relative qRT-PCR of mRNA and miRNA were performed, as previously described [37,38]. Briefly, total RNA was extracted using the TriReagent protocol (Molecular Research Center, Inc., Cincinnati, Ohio), and RNA purity, integrity and quantity were examined by nanodrop (Nanodrop® Technologies, Wilmington, DE, USA) and Bioanalyzer (Agilent 2100, Santa Clara, California) measurements. Relative quantitative mRNA PCR was performed on reverse transcribed cDNA (High Capacity cDNA RT kit; Applied Biosystems, Foster City, CA, USA). Primer sequences, amplification efficiencies (AE) and standard errors (SE): *BNP* forward: 5'-GATGCAGAAGCTGGAGA-3', reverse: 5'-TCTGCTGGACCCGGAGGGTG-3', AE: 2.046 and SE: 0.011; *Fibronectin* forward: 5'-CAGGGGTCACGTACCTCTTCAAAG-3', reverse: 5'-CGAGGTGGAGTCCAAGTTACCAGA-3', AE: 1.995 and SE: 0.010; *Procollagen-I* forward: 5'-ATCGTGGCTTCTCTGGTCTCCAG-3', reverse: 5'-CAGGGAGACCGTTGAGTCCATCT-3', AE: 2.039, SE: 0.004; *GAPDH* forward: 5'-GTCGGTGTGAACGGATTTGGC-3', reverse: 5'-TGAAGGGGTCGTTGATGGCA-3', AE: 2.079, SE: 0.008; *RPL13A* forward: 5'-GAAAGCGGATGAACACCAACCC-3', reverse: 5'-GGGATCCCATCCAACACCTTGA-3', AE: 1.958, SE: 0.003.

For miRNA qRT-PCR primers specific for rat and human miR-21 (#000397), miR-132 (#000457), miR-212 (#002551), miR-17 (#002308), miR-103 (#000439), miR-191 (#002299) and let-7f (#000382) were purchased from Applied Biosystems. Amplification and detection were performed using 7900HT Fast Real-Time PCR System (Applied Biosystems, Foster City, CA, USA). As recommended by others [39,40] and previously described [37,38], we used the qBase$^+$ software to normalize all qRT-PCR data against multiple stably expressed control genes (information applied in respective figure legends).

4.9. Statistical Analysis

Results are represented as the mean ± SD. All analyses comprised independent experiments and two-way ANOVA, and two-tailed student's *t*-tests were performed, as indicated [GraphPad Prism (5.0 version) software, La Jolla, CA, USA] to test significant levels. In the animal experiments, data on the effect of AngII and ET-1 were obtained from 6–8 and 3–5 animals, respectively. Data not following Gaussian distribution, as tested by D'Agostino's normality test, were subjected to a Log_{10} transformation before statistical analysis and normality evidenced to follow Gaussian distribution prior to significance tests. Differences were considered to be significant at $p < 0.05$.

5. Conclusions

In conclusion, we found that miR-132 and -212 are increased in AngII-induced hypertension *in vivo*, in organs associated with blood pressure control, which partly mimics the "five miRNA" expression signature obtained by AT_1R overexpression [19]. Cleary, the *in vivo* model adds to our understanding, as we can narrow down the miRNA changes after AngII, to miR-132 and -212. Importantly, our results show that functional inhibition of the AT_1R reversed the miR-132 and -212 expression levels, demonstrating that AngII is responsible for the regulation, possibly via the Gαq-dependent pathway.

Acknowledgments

We thank Lene Bundgaard Andersen, Marlene Louise Christensen, Tonja Lyngse Jørgensen and Vivi Monrad for excellent technical assistance. SPS is funded by the John and Birthe Meyer Foundation, LMR is funded by the Novo Nordisk foundation and TVE and PLJ are funded by The Danish Cardiovascular Academy and the Danish heart foundation.

References

1. Coffman, T.M. Under pressure: The search for the essential mechanisms of hypertension. *Nat. Med.* **2011**, *17*, 1402–1409.
2. Batkai, S.; Thum, T. MicroRNAs in hypertension: Mechanisms and therapeutic targets. *Curr. Hypertens. Rep.* **2012**, *14*, 79–87.
3. Crowley, S.D.; Gurley, S.B.; Coffman, T.M. AT(1) receptors and control of blood pressure: The kidney and more. *Trends Cardiovasc. Med.* **2007**, *17*, 30–34.
4. Bader, M.; Ganten, D. Update on tissue renin-angiotensin systems. *J. Mol. Med. (Berl.)* **2008**, *86*, 615–621.
5. Hansen, J.L.; Aplin, M.; Hansen, J.T.; Christensen, G.L.; Bonde, M.M.; Schneider, M.; Haunso, S.; Schiffer, H.H.; Burstein, E.S.; Weiner, D.M.; *et al.* The human angiotensin AT(1) receptor supports G protein-independent extracellular signal-regulated kinase 1/2 activation and cellular proliferation. *Eur. J. Pharmacol.* **2008**, *590*, 255–263.
6. Bartel, D.P. MicroRNAs: Genomics, biogenesis, mechanism, and function. *Cell* **2004**, *116*, 281–297.
7. Ambros, V. The functions of animal microRNAs. *Nature* **2004**, *431*, 350–355.
8. Vo, N.; Klein, M.E.; Varlamova, O.; Keller, D.M.; Yamamoto, T.; Goodman, R.H.; Impey, S. A cAMP-response element binding protein-induced microRNA regulates neuronal morphogenesis. *Proc. Natl. Acad. Sci. USA* **2005**, *102*, 16426–16431.
9. Neyses, L.; Nouskas, J.; Luyken, J.; Fronhoffs, S.; Oberdorf, S.; Pfeifer, U.; Williams, R.S.; Sukhatme, V.P.; Vetter, H. Induction of immediate-early genes by angiotensin II and endothelin-1 in adult rat cardiomyocytes. *J. Hypertens.* **1993**, *11*, 927–934.
10. Christensen, G.L.; Kelstrup, C.D.; Lyngso, C.; Sarwar, U.; Bogebo, R.; Sheikh, S.P.; Gammeltoft, S.; Olsen, J.V.; Hansen, J.L. Quantitative phosphoproteomics dissection of seven-transmembrane receptor signaling using full and biased agonists. *Mol. Cell Proteomics* **2010**, *9*, 1540–1553.
11. Olson, E.N. Gene regulatory networks in the evolution and development of the heart. *Science* **2006**, *313*, 1922–1927.
12. Van Rooij, E.; Sutherland, L.B.; Qi, X.; Richardson, J.A.; Hill, J.; Olson, E.N. Control of stress-dependent cardiac growth and gene expression by a microRNA. *Science* **2007**, *316*, 575–579.
13. Care, A.; Catalucci, D.; Felicetti, F.; Bonci, D.; Addario, A.; Gallo, P.; Bang, M.L.; Segnalini, P.; Gu, Y.; Dalton, N.D.; *et al.* MicroRNA-133 controls cardiac hypertrophy. *Nat. Med.* **2007**, *13*, 613–618.
14. Thum, T.; Gross, C.; Fiedler, J.; Fischer, T.; Kissler, S.; Bussen, M.; Galuppo, P.; Just, S.; Rottbauer, W.; Frantz, S.; *et al.* MicroRNA-21 contributes to myocardial disease by stimulating MAP kinase signalling in fibroblasts. *Nature* **2008**, *456*, 980–984.

15. Van Rooij, E.; Marshall, W.S.; Olson, E.N. Toward microRNA-based therapeutics for heart disease: the sense in antisense. *Circ. Res.* **2008**, *103*, 919–928.
16. Martin, M.M.; Lee, E.J.; Buckenberger, J.A.; Schmittgen, T.D.; Elton, T.S. MicroRNA-155 regulates human angiotensin II type 1 receptor expression in fibroblasts. *J. Biol. Chem.* **2006**, *281*, 18277–18284.
17. Zhu, N.; Zhang, D.; Chen, S.; Liu, X.; Lin, L.; Huang, X.; Guo, Z.; Liu, J.; Wang, Y.; Yuan, W.; *et al.* Endothelial enriched microRNAs regulate angiotensin II-induced endothelial inflammation and migration. *Atherosclerosis* **2011**, *215*, 286–293.
18. Van Rooij, E.; Sutherland, L.B.; Thatcher, J.E.; DiMaio, J.M.; Naseem, R.H.; Marshall, W.S.; Hill, J.A.; Olson, E.N. Dysregulation of microRNAs after myocardial infarction reveals a role of miR-29 in cardiac fibrosis. *Proc. Natl. Acad. Sci. USA* **2008**, *105*, 13027–13032.
19. Jeppesen, P.L.; Christensen, G.L.; Schneider, M.; Nossent, A.Y.; Jensen, H.B.; Andersen, D.C.; Eskildsen, T.; Gammeltoft, S.; Hansen, J.L.; Sheikh, S.P. Angiotensin II type 1 receptor signalling regulates microRNA differentially in cardiac fibroblasts and myocytes. *Br. J. Pharmacol.* **2011**, *164*, 394–404.
20. Nossent, A.Y.; Eskildsen, T.V.; Andersen, L.B.; Bie, P.; Brønnum, H.; Schneider, M.; Andersen, D.C.; Welten, S.M.J.; Jeppesen, P.L.; Hamming, J.F.; *et al.* The 14q32 MicroRNA-487b targets the anti-apoptotic insulin receptor substrate 1 in hypertension induced remodeling of the aorta. *Ann. Surg.* **2013**, in press.
21. Tognini, P.; Pizzorusso, T. MicroRNA212/132 family: Molecular transducer of neuronal function and plasticity. *Int. J. Biochem. Cell Biol.* **2012**, *44*, 6–10.
22. Mortensen, L.H.; Pawloski, C.M.; Kanagy, N.L.; Fink, G.D. Chronic hypertension produced by infusion of endothelin in rats. *Hypertension* **1990**, *15*, 729–733.
23. Anand, S.; Majeti, B.K.; Acevedo, L.M.; Murphy, E.A.; Mukthavaram, R.; Scheppke, L.; Huang, M.; Shields, D.J.; Lindquist, J.N.; Lapinski, P.E.; *et al.* MicroRNA-132-mediated loss of p120RasGAP activates the endothelium to facilitate pathological angiogenesis. *Nat. Med.* **2010**, *16*, 909–914.
24. Nossent, A.Y.; Hansen, J.L.; Doggen, C.; Quax, P.H.; Sheikh, S.P.; Rosendaal, F.R. SNPs in microRNA binding sites in 3'-UTRs of RAAS genes influence arterial blood pressure and risk of myocardial infarction. *Am. J. Hypertens.* **2011**, *24*, 999–1006.
25. Katare, R.; Riu, F.; Mitchell, K.; Gubernator, M.; Campagnolo, P.; Cui, Y.; Fortunato, O.; Avolio, E.; Cesselli, D.; Beltrami, A.P.; *et al.* Transplantation of human pericyte progenitor cells improves the repair of infarcted heart through activation of an angiogenic program involving micro-RNA-132. *Circ. Res.* **2011**, *109*, 894–906.
26. Nudelman, A.S.; DiRocco, D.P.; Lambert, T.J.; Garelick, M.G.; Le, J.; Nathanson, N.M.; Storm, D.R. Neuronal activity rapidly induces transcription of the CREB-regulated microRNA-132, *in vivo*. *Hippocampus* **2010**, *20*, 492–498.
27. Strum, J.C.; Johnson, J.H.; Ward, J.; Xie, H.; Feild, J.; Hester, A.; Alford, A.; Waters, K.M. MicroRNA 132 regulates nutritional stress-induced chemokine production through repression of SirT1. *Mol. Endocrinol.* **2009**, *23*, 1876–1884.
28. Jin, W.; Reddy, M.A.; Chen, Z.; Putta, S.; Lanting, L.; Kato, M.; Park, J.T.; Chandra, M.; Wang, C.; Tangirala, R.; *et al.* Small RNA sequencing reveals microRNAs that modulate angiotensin II effects in vascular smooth muscle cells. *J. Biol. Chem.* **2012**, *287*, 15672–15683.

29. Cassis, L.A.; Huang, J.; Gong, M.C.; Daugherty, A. Role of metabolism and receptor responsiveness in the attenuated responses to Angiotensin II in mice compared to rats. *Regul. Pept.* **2004**, *117*, 107–116.
30. Rautureau, Y.; Schiffrin, E.L. Endothelin in hypertension: An update. *Curr. Opin. Nephrol. Hypertens.* **2012**, *21*, 128–136.
31. Gray, G.A.; Webb, D.J. *Molecular Biology and Pharmacology of the Endothelins*; Springer & R.G. Landes, Co.: New York, NY, USA, 1995; pp. 116–145.
32. Roberts, R.E. The extracellular signal-regulated kinase (ERK) pathway: A potential therapeutic target in hypertension. *J. Exp. Pharmacol.* **2012**, *4*, 77–83.
33. Zaman, M.A.; Oparil, S.; Calhoun, D.A. Drugs targeting the renin-angiotensin-aldosterone system. *Nat. Rev. Drug Discov.* **2002**, *1*, 621–636.
34. Bruner, C.A.; Mangiapane, M.L.; Fink, G.D. Subfornical organ. Does it protect against angiotensin II-induced hypertension in the rat? *Circ. Res.* **1985**, *56*, 462–466.
35. Mortensen, L.H.; Fink, G.D. Captopril prevents chronic hypertension produced by infusion of endothelin-1 in rats. *Hypertension* **1992**, *19*, 676–680.
36. Cangemi, C.; Skov, V.; Poulsen, M.K.; Funder, J.; Twal, W.O.; Gall, M.A.; Hjortdal, V.; Jespersen, M.L.; Kruse, T.A.; Aagard, J.; *et al.* Fibulin-1 is a marker for arterial extracellular matrix alterations in type 2 diabetes. *Clin. Chem.* **2011**, *57*, 1556–1565.
37. Andersen, D.C.; Jensen, C.H.; Schneider, M.; Nossent, A.Y.; Eskildsen, T.; Hansen, J.L.; Teisner, B.; Sheikh, S.P. MicroRNA-15a fine-tunes the level of Delta-like 1 homolog (DLK1) in proliferating 3T3-L1 preadipocytes. *Exp. Cell Res.* **2010**, *316*, 1681–1691.
38. Andersen, D.C.; Andersen, P.; Schneider, M.; Jensen, H.B.; Sheikh, S.P. Murine "cardiospheres" are not a source of stem cells with cardiomyogenic potential. *Stem Cells* **2009**, *27*, 1571–1581.
39. Vandesompele, J.; De Preter, K.; Pattyn, F.; Poppe, B.; Van Roy, N.; de Paepe, A.; Speleman, F. Accurate normalization of real-time quantitative RT-PCR data by geometric averaging of multiple internal control genes. *Genome Biol.* **2002**, *3*, RESEARCH0034.
40. Hellemans, J.; Mortier, G.; de Paepe, A.; Speleman, F.; Vandesompele, J. qBase relative quantification framework and software for management and automated analysis of real-time quantitative PCR data. *Genome Biol.* **2007**, *8*, R19.

9

microRNA-Regulated Protein-Protein Interaction Networks and their Functions in Breast Cancer

Chia-Hsien Lee [1,†], Wen-Hong Kuo [2,†], Chen-Ching Lin [1], Yen-Jen Oyang [1], Hsuan-Cheng Huang [3,*] and Hsueh-Fen Juan [1,4,*]

[1] Graduate Institute of Biomedical Electronics and Bioinformatics, National Taiwan University, Taipei 106, Taiwan; E-Mails: tfglove@gmail.com (C.-H.L.); kdragongo@gmail.com (C.-C.L.); yjoyang@csie.ntu.edu.tw (Y.-J.O.)

[2] Department of Physiology, College of Medicine, National Taiwan University, Taipei 100, Taiwan; E-Mail: brcancer@gmail.com

[3] Institute of Biomedical Informatics and Center for Systems and Synthetic Biology, National Yang-Ming University, Taipei 112, Taiwan

[4] Institute of Molecular and Cellular Biology and Department of Life Science, National Taiwan University, Taipei 106, Taiwan

[†] These authors contributed equally to this work.

[*] Authors to whom correspondence should be addressed; E-Mails: hsuancheng@ym.edu.tw (H.-C.H.); yukijuan@ntu.edu.tw (H.-F.J.);

Abstract: MicroRNAs, which are small endogenous RNA regulators, have been associated with various types of cancer. Breast cancer is a major health threat for women worldwide. Many miRNAs were reported to be associated with the progression and carcinogenesis of breast cancer. In this study, we aimed to discover novel breast cancer-related miRNAs and to elucidate their functions. First, we identified confident miRNA-target pairs by combining data from miRNA target prediction databases and expression profiles of miRNA and mRNA. Then, miRNA-regulated protein interaction networks (PINs) were constructed with confident pairs and known interaction data in the human protein reference database (HPRD). Finally, the functions of miRNA-regulated PINs were elucidated by functional enrichment analysis. From the results, we identified some previously reported breast cancer-related miRNAs and functions of the PINs, e.g., miR-125b, miR-125a, miR-21, and miR-497. Some novel miRNAs without known association to breast cancer were also found, and the putative functions of their PINs were also elucidated. These include miR-139 and miR-383. Furthermore, we validated our results by receiver operating characteristic (ROC) curve analysis using our miRNA expression profile data, gene expression-based outcome for breast cancer online (GOBO) survival analysis, and a literature search. Our results may provide new insights for research in breast cancer-associated miRNAs.

Keywords: miRNA; breast cancer; protein interaction network; functional analysis

---

1. Introduction

Breast cancer is a global health threat for women. According to a 2008 survey [1], breast cancer was the leading cause of cancer deaths in women. Our knowledge of possible risk factors has led to developments in diagnostic methods, drugs, and surgery procedures for treatment [2,3]; however, the details of breast carcinoma progression, and perhaps most importantly, how to cure breast cancer, remain elusive.

Previous research has identified a number of risk factors for breast cancer. Early menarche, late menopause, obesity, late first full pregnancy, and hormone replacement therapy were considered as high risk factors for breast cancer [2]. Breast cancer risk has also been reported to be related to fat intake in diets rich in red meats and high-fat dairy foods [3].

MicroRNAs (miRNAs) [4], are short endogenous non-coding RNAs which are able to regulate gene expression. After miRNA precursors are transcribed from the genome or generated from spliceosomes, they are exported to the cytoplasm and further processed by the Dicer complex [5]. The mature miRNA is then bound to Argonaute protein, forming a miRNA-protein complex known as the RNA-induced silencing complex (RISC) [6], miRNP, or RNAi (RNA interference) enzyme complex [7,8]. The RISC has been reported to down-regulate target genes by translational repression [9] or mRNA cleavage [10].

Like other protein-based regulators, miRNAs have been associated with cancer. Calin *et al.*, reported that miR-15 and miR-16 were deleted in leukemia [11], which was believed to be one of the earliest reports associating miRNAs with cancer [12]. After this report, many miRNAs were found to act as tumor suppressors or oncogenes (also known as oncomirs). For example, miR-21 was identified as an oncomir in hepatocellular cancer [13], breast cancer [14], and kidney cancer [15]. On the other hand, let-7c was found to be a tumor suppressor in prostate cancer [16], and miR-181a was reported as a tumor suppressor in glioma [17]. Further, miR-125b [18], and miR-145 [19] were identified as tumor suppressors in breast cancer, and miR-125a was found to repress tumor growth in breast cancer [20]. Thus, it is highly likely that miRNAs play an important role in breast cancer.

Since miRNA functions by regulating its target genes, we may deduce the effects of miRNAs by analyzing their regulated networks. To use such a method to elucidate miRNA functions, targets of miRNAs should be deduced. Currently, predictions in most target prediction database are based on sequence and statistical methods [21]. For example, in TargetScan, seed base pairing, target site context, conservation of target site and miRNA, and site accessibility are considered in the prediction process [22].

Another method to elucidate miRNA targets is to integrate expression profiles of miRNA and mRNA. In the work of Huang *et al.* [23], a Bayesian-based algorithm, GenMiR++, was developed to predict possible targets of 104 miRNAs in humans. They also verified their results by RT-PCR and microarray experiments. However, the power of other sequence-based target prediction algorithms was not utilized in their work.

It is also possible to combine sequence-based target prediction and expression-based target prediction methods. By integrating expression data into sequence-based predictions, possible false positives can be reduced. Previously, miRNA-mRNA interactions were explored with splitting-averaging Bayesian networks [24]. In that work, expression profiles of miRNA and mRNA from public databases, miRNA target prediction databases, and miRNA sequence information were integrated together to discover miRNA-mRNA interaction networks.

Here, we combined expression profiles of miRNA and mRNA, and three target prediction databases, TargetScan, PicTar and miRanda, to obtain confident miRNA-mRNA relationships and construct miRNA-regulated protein-protein networks for breast cancer. Furthermore, we explored the functions of the miRNAs by inspecting the underlying protein interaction networks (PINs) of the miRNAs with functional enrichment analysis. This method, as described in Figure 1, was used to elucidate the functions of gastric cancer-related miRNAs in our previous work [25]. In that study, a gastric cancer-associated miRNA, miR-148a, was identified and validated as being involved in tumor proliferation, invasion, migration, and the survival rate of the patients. By using a similar method, we aimed to elucidate breast cancer-related miRNA-regulated PINs and their functions.

Figure 1. Analysis flow chart used in this work. After expression profiles and target prediction databases were fetched and preprocessed, they were subjected to the analysis process described here and in the "Experimental Section".

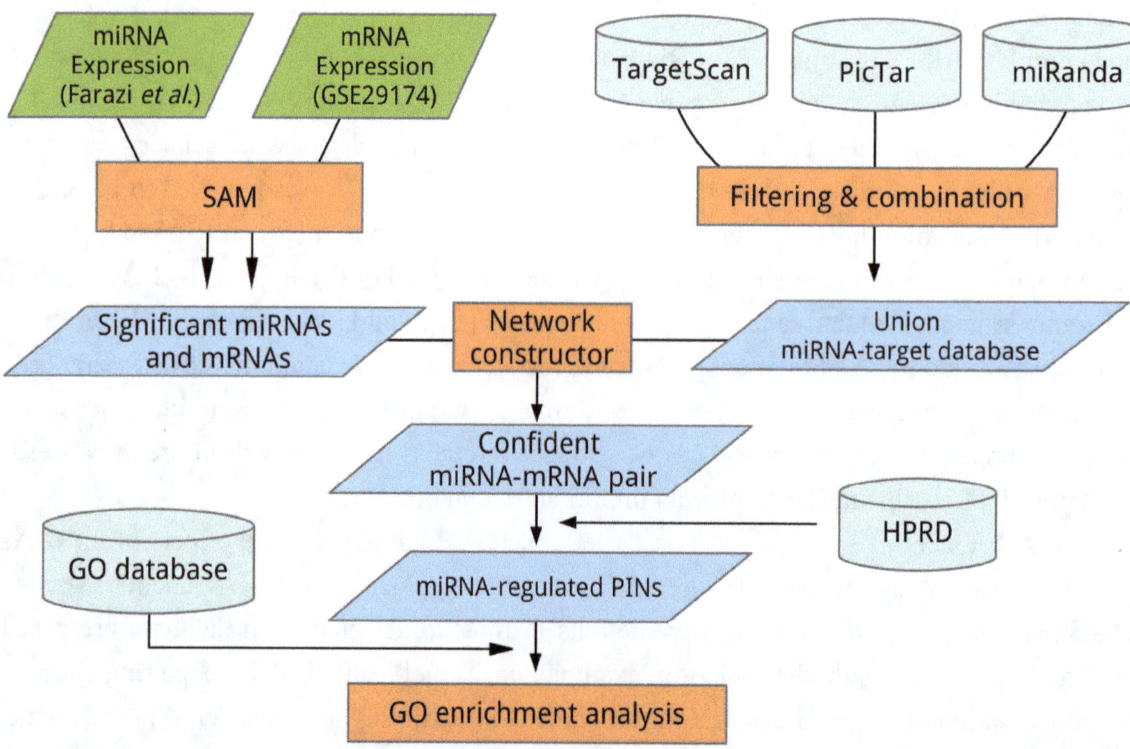

2. Results and Discussion

To construct miRNA-regulated PINs, differentially expressed miRNAs and genes from the dataset from Farazi *et al.* [26] were extracted following proper processing of the expression profiles. From our selected public miRNA dataset, we found 89 down-regulated miRNAs (93 prior to fold-change filtering) and only 1 up-regulated miRNA (Table S1). In gene expression dataset GSE29174, we found a total of 1268 down-regulated genes and 587 up-regulated genes before applying the fold change filter. There were 726 down-regulated genes (Table S2) and 437 up-regulated genes (Table S3) after significantly and differentially expressed genes were filtered by fold change (fold change >2).

From the results of SAM analysis, we identified some well-known breast cancer-related miRNAs (Table S1). For example, miR-214-3p [27] and miR-335-5p [28] have been previously reported to be down-regulated in breast cancer. Let-7c was found to be down-regulated in this work, while let-7a, another member of the let-7 family, was found to be down-regulated in another work [29]. MicroRNAs of the let-7 family were also reportedly down-regulated in several types of cancer [30]. We also found that miR-21-5p, the sole up-regulated miRNA in our list, was also previously found to be up-regulated [14,31]. However, changes in the expression of most of the miRNAs in our down-regulated list have not been reported in the literature. Therefore, we could not rule out the possibility that these miRNAs were novel breast cancer-related miRNAs. There are also some well-known miRNAs not presented in our list (for such a list, one may see [32–34]). The reason that some known miRNAs, for example, miR-19a, miR-155 and miR-205, did not show up in our result might be that we used a very stringent threshold (described in Experimental Section) when selecting differentially expressed miRNAs for PIN construction.

Since the miRNAs of the miRNA-regulated PINs were differentially expressed between normal and tumor tissues, and we identified some cancer-related functions in our functional enrichment analysis, the miRNAs may potentially be useful diagnostic markers for breast cancer. To verify this, we applied ROC curve analysis on the miRNA expression profile that was not used in constructing the miRNA-regulated PINs. Notably, our results (Figures 2 and S1 and Table S4) showed that let-7c (Figure 2), miR-497-5p, miR-125b-5p, and some other miRNAs of miRNA-regulated PINs, performed well when used as breast cancer diagnostic markers.

Figure 2. Receiver operating characteristic (ROC) curve of let-7c from our miRNA array dataset. For ROC curves of other miRNAs, see Figure S1.

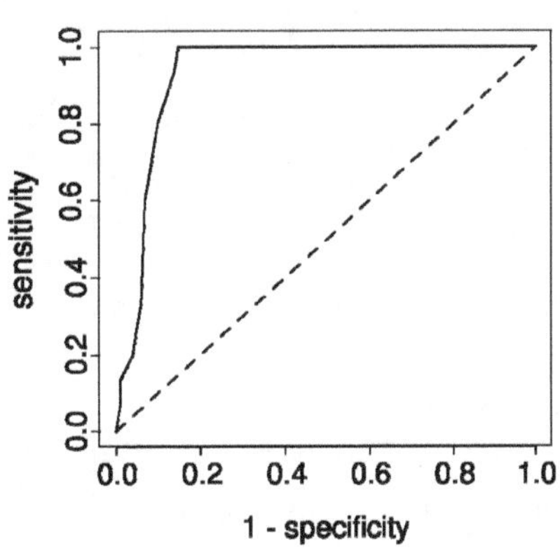

Following elucidation of differentially expressed miRNAs and genes, miRNA-regulated PINs could then be constructed. We identified and constructed partial networks, containing the miRNA and its direct target, with the differentially expressed miRNAs and genes as described in the "Experimental Section". We then extended the network by appending known interactions from the HPRD database. Finally, 18 miRNA-regulated PINs were constructed by the steps described above (Figure 3, Figures S2–S13, and Table S5).

Figure 3. The let-7c-regulated protein interaction network (PIN). This is one of the 18 miRNA-regulated PINs constructed in this work. Figures of all other miRNA-regulated PINs are displayed in Supplementary Figures S2–S13.

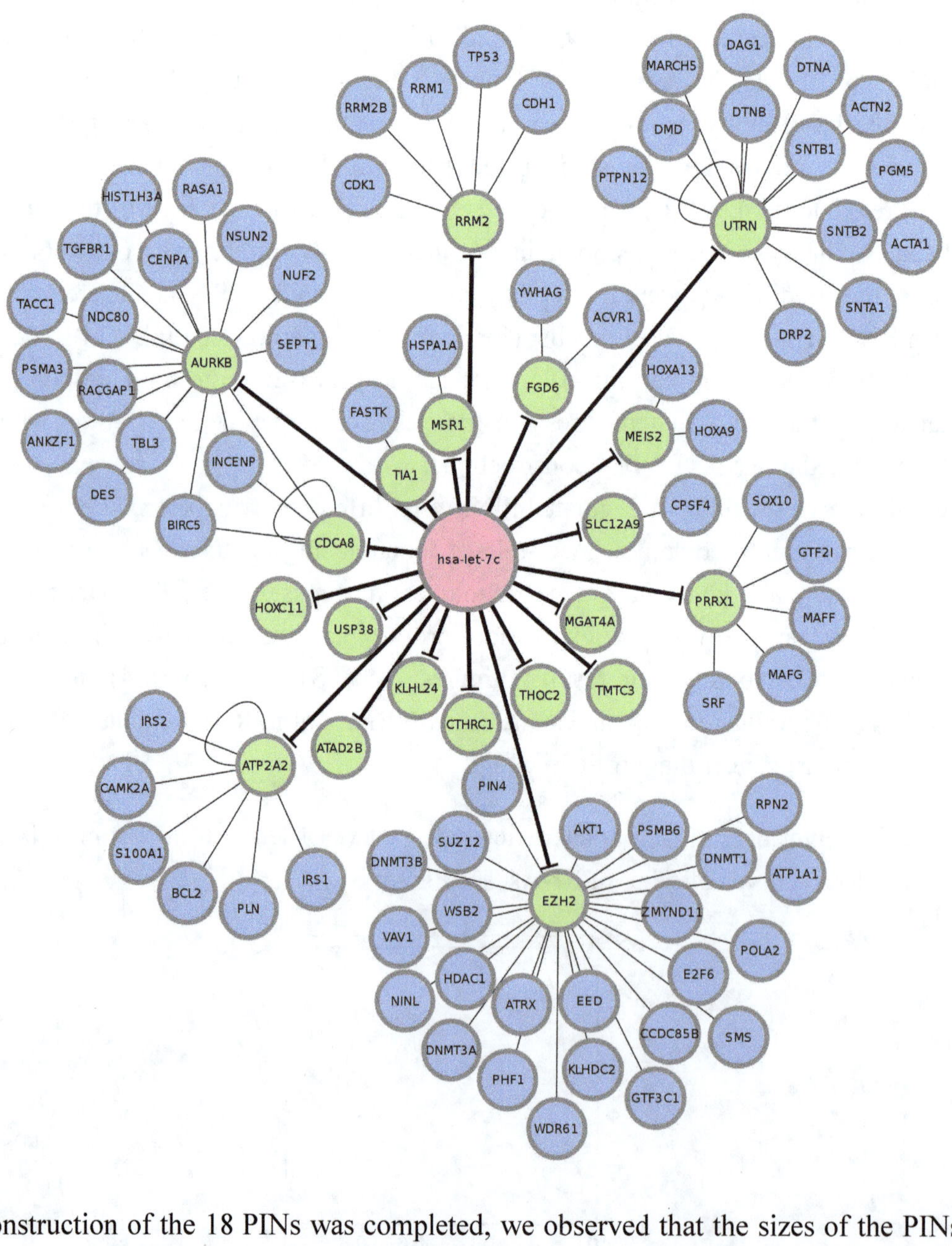

After construction of the 18 PINs was completed, we observed that the sizes of the PINs were not similar: some of the miRNAs seemed to regulate larger sized PINs, while other miRNAs affected only a small number of genes. Small miRNA-regulated PINs may be caused by the strict *q*-value threshold set during SAM analysis, the processing steps performed on the target prediction databases discussed previously, and possibly by lack of protein-protein interaction data for some proteins in the HPRD. Although the HPRD may be considered the most comprehensive source of protein-protein interaction data [35], some proteins may not have been considered and researched by other investigators, and therefore, interaction data for those proteins would not be included in the HPRD. However, it may be true that some of the miRNA-regulated PINs were small in breast cancer, since the construction of the PINs were based on differentially expressed miRNAs and genes between normal tissues and tumor samples, and those miRNAs with small PINs may not be as important as others with larger PINs.

To elucidate the functions of a miRNA with a regulated PIN, GO enrichment analysis was applied to the miRNA-regulated PINs. We did not consider the miRNAs with ≤5 genes in their regulated PINs, and some of the miRNA-regulated PINs had no enriched functions using the defined threshold, FDR < 0.0001. To exclude GO terms that describe a broad range of concepts, we only included high level GO terms, *i.e.*, larger than 5.

The results of the GO enrichment analysis for let-7c related to cancer are listed in Table 1. (Results of all miRNA-regulated PINs were in Table S6). We defined a GO term as cancer-related if a GO term contained "cell proliferation", "cell death", "apoptosis", "signaling", "microtubule", and "actin". We noted that 7 miRNAs had enriched GO terms related to apoptosis, cell death, and cell proliferation, *i.e.*, miR-520d-3p, miR-497-5p, miR-125b-5p, miR-21-5p, miR-31-5p, let-7c, and miR-125-5p. Further, some miRNA-regulated PINs may have functions other than cell survival. For example, the nerve growth factor receptor pathway was enriched in miR-regulated PINs of miR-520d-3p, miR-497-5p, miR-125a-5p, miR-125b-5p, and miR-31-5p, and the epidermal growth factor receptor pathway was enriched in miR-regulated PINs of miR-520d-3p, miR-21-5p, and miR-497-5p. Most of the miRNAs had been previously described and were known to be implicated in breast cancer. Let-7c was not only likely to be down-regulated in breast cancer [29], but was also found to be a tumor suppressor in prostate cancer [16]. Another reported tumor suppressor was miR-125b-5p, which was found to be down-regulated in breast tumor tissue [18], and this finding was consistent with our functional enrichment results. The miR-125a-5p-regulated PIN was found to be able to inhibit apoptosis and regulate epithelial cell proliferation, and has been reported to repress cell growth [36].

Table 1. Selected enriched functions of let-7c. Member genes of the let-7c-regulated network annotated with corresponding enriched functions are listed.

	MIMAT0000064(hsa-let-7c)	
GO term	**Genes**	**Adj. *p*-value**
GO:0043067, Regulation of programmed cell death	RRM2B, ACVR1, TP53, RASA1, TGFBR1, PSMA3, BIRC5, ACTN2, HOXA13, IRS2, FASTK, VAV1, PSMB6, BCL2, CDK1, HDAC1, SOX10, TIA1, AKT1, AURKB	3.43×10^{-8}
GO:0043069, Negative regulation of programmed cell death	RRM2B, ACVR1, TP53, RASA1, TGFBR1, BIRC5, IRS2, BCL2, CDK1, HDAC1, SOX10, AKT1, AURKB	6.58×10^{-6}
GO:0060548, Negative regulation of cell death	RRM2B, ACVR1, TP53, RASA1, TGFBR1, BIRC5, IRS2, BCL2, CDK1, HDAC1, SOX10, AKT1, AURKB	8.92×10^{-6}
GO:0015630, Microtubule cytoskeleton	INCENP, SNTB2, SEPT1, TACC1, BIRC5, RACGAP1, PIN4, CDCA8, CDK1, PHF1, AKT1, AURKB, NINL, CCDC85B	5.15×10^{-5}

We also found that some breast cancer-specific functions were enriched in our results. For example, in the miR-497-5p-regulated PIN, the term "androgen receptor signaling pathway" was enriched. Although it is not clear whether androgens are related to breast cancer, androgen receptors are known to be up-regulated in breast cancer and related to node invasiveness [37].

To further verify the results of the enriched cancer-related functions, we used GOBO for survival analysis. Our hypothesis is that expression of genes annotated with enriched cancer-related terms may be related to survival outcome of patients. With exception to some cell death/proliferation-related terms, it is already known that some pathways or functions are also related to clinical outcomes. For example, cell proliferation-related GO terms have a high probability of affecting survival of cancer tissues, and patient outcome may worsen if cancer tissue survives. In addition, some signaling pathways were known to enhance invasiveness, migration abilities, or were associated with reduced patient survival. For example, the BMP signaling pathway is known to confer various tumor cells with enhanced migration and invasion abilities [38], and nerve growth factor receptor (NGFR) was found to be associated with overall survival of breast cancer [38]. Furthermore, Toll-like receptor 4 has been reported to promote adhesion and invasive migration in breast cancer [39]. Finally, the cytoskeleton plays an important role in regulating cell motility in all cells. Actin filaments are known to participate in the invasive migration of cancer cells [40]. Since some of these functions were present in our enriched terms, we wished to test if the expression of a gene set annotated to the cancer-related enriched terms in the PIN would be related to clinical outcome of patients.

Figure 4. Validation of let-7c result. Gene expression-based outcome for breast cancer online (GOBO) survival analysis of let-7c-regulated PIN members marked with the following functions: (**A**) Microtubule cytoskeleton; (**B**) Negative regulation of programmed cell death; and (**C**) negative regulation of cell death. **Red**: samples with high expression of selected gene set (PIN members); **grey**: samples with low expression of selected gene set (PIN members).

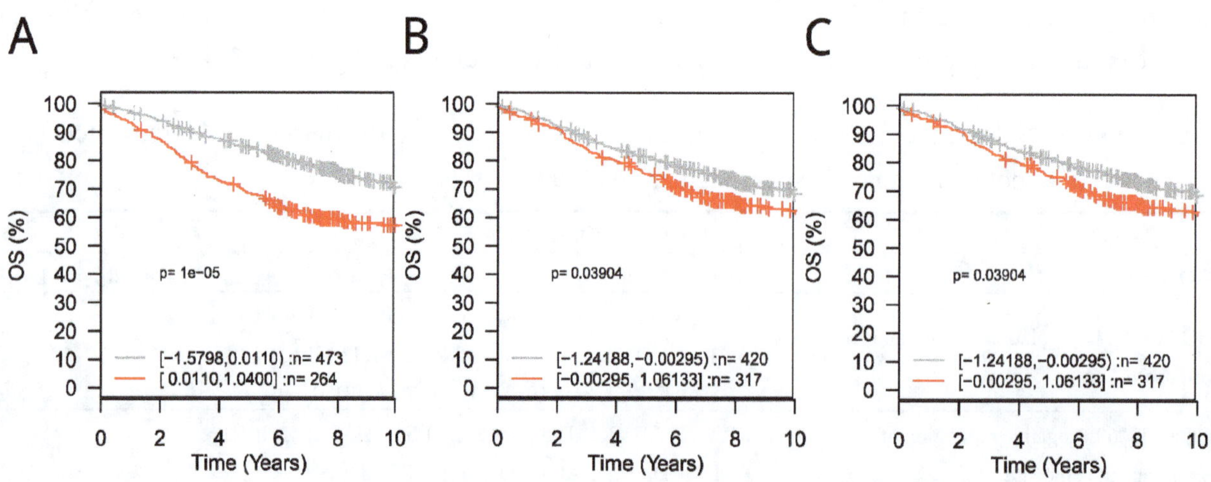

As shown in Figure 4, Figure S14 and Table S7, only some of the enriched terms were significantly associated with clinical outcome. This may be because the changes in these key genes occurred at the protein level, such as in protein expression or even post-translational modification; therefore, mRNA expression-centric tools like GOBO cannot explore association of such genes to clinical outcomes.

Alternatively, it is possible the miRNA did not regulate the whole pathway, or the miRNA did not target the key part of the pathway directly, and thus the clinical outcome of gene sets of the enriched GO terms in this condition cannot be determined. However, some functions associated with clinical outcomes were observed. For example, proteins annotated with the terms "microtubule cytoskeleton", "negative regulation of programmed cell death", and "negative regulation of cell death" in the let-7c-regulated PIN were related to 10 year survival rate of patients, as reported by GOBO (Figure 4). Also, the enriched term "regulation of epithelial cell proliferation" for both miR-125a-5p and miR-125b-5p were found to be associated with the 10-year survival rate of patients. Therefore, these results further supported the GO enrichment analysis discussed previously.

3. Experimental Section

3.1. miRNA Microarray Experiments

We performed a miRNA microarray to obtain the expression profiles for receiver operating characteristic (ROC) curve analysis. This dataset was deposited in Gene Expression Omnibus (GEO, http://www.ncbi.nlm.nih.gov/geo/, Series Accession: GSE45666). In total, there were 15 normal samples and 101 tumor samples in the expression profile. Detailed pathophysiogical characteistics of these samples were in Table S8 and GSE45666. All human tissue samples collected from breast cancer patients were approved and human subject confidentiality was protected by the Institute Review Board of National Taiwan University Hospital (IRB, 20071211R).

Total RNA was extracted from tissues collected from the patients using Trizol® Reagent (Invitrogen, Carlsbad, CA, USA, USA) according to the manufacturer's protocol. Purified RNA was quantified at OD260nm by using a ND-1000 spectrophotometer (Nanodrop Technology, Wilmington, DE, USA) and qualitated by using a Bioanalyzer 2100 with the RNA 6000 Nano LabChip kit (Agilent Technologies, Santa Clara, CA, USA).

After RNA extraction, 100 ng of total RNA was dephosphorylated and labeled with pCp-Cy3 using the Agilent miRNA Complete Labeling and Hyb Kit in conjunction with the microRNA Spike-In kit (Agilent Technologies, Santa Clara, CA, USA). Briefly, 2X hybridization buffer (Agilent Technologies) was added to the labeled mixture to a final volume of 45 μL. The mixture was heated for 5 min at 100 °C and immediately cooled to 0 °C. Each 45 μL sample was hybridized onto an Agilent human miRNA Microarray Release 12.0, 8 × 15 K (Agilent Technologies) at 55 °C for 20 h. After hybridization, slides were washed for 5 min in Gene Expression Wash Buffer 1 at room temperature, then for 5 min in Gene Expression Wash Buffer 2 at 37 °C. Slides were scanned on an Agilent microarray scanner (Agilent Technologies, model G2505C) at 100% and 5% sensitivity settings. Feature Extraction (Agilent Technologies) software version 10.7.3.1 (Agilent Technologies, Santa Clara, CA, USA) was used for image analysis.

3.2. mRNA Expression Profiles

For miRNA-regulated protein-protein interaction network construction, the mRNA expression profile was fetched from GEO (Series Accession: GSE29174). This dataset was produced by Farazi *et al.* [26], which was the only dataset publicly available with large size of tumor samples and reasonably sized normal samples. In total, 161 clinical samples were collected from breast cancer patients by biopsy: 110 invasive ductal carcinoma (IDC), 11 normal, 17 ductal carcinoma *in situ* (DCIS), 1 mucinous A, 8 atypical medullary, 4 apocrine, 8 metaplastic, and 2 adenoid, as classified by Farazi *et al.* The110 IDC samples were classified as the tumor group and the 11 normal samples were classified as the normal group in this study.

3.3. miRNA Expression Profiles

The miRNA expression profile used in this work was fetched from Table S4A of the work of Farazi *et al.* [26]. There were 189 samples in this dataset, with 6 of them from cell lines, and another 183 samples from patient tissues. In the 183 clinical samples collected from breast cancer patients by biopsy in this dataset, there were 128 IDC, 11 normal, 18 DCIS, 1 mucinous A, 8 atypical medullary, 4 apocrine, 9 metaplastic, and 2 adenoid cases, as classified by Farazi *et al.* The 128 IDC samples were classified as the tumor group and the 11 normal samples as were classified as the normal group.

3.4. Data Analysis

The overall workflow design was similar to the previous work of Tseng *et al.* [25] (see Figure 1). However, we applied the workflow on breast cancer expression profiles instead of gastric cancer as in the work of Tseng *et al.* Additionally, we used 3 miRNA target prediction databases, TargetScan (v6.0) [22,41,42], PicTar [43,44], and miRanda (release August 2010) [45] here, while only TargetScan was used in the previous study.

To construct the networks, we first elucidated differentially expressed miRNAs and mRNAs from published datasets. Both of miRNA (generated with miRNA-Seq technique) and mRNA array expression data were produced by Farazi *et al.* [26] from the same batch of clinical tumor samples. To obtain a list of differentially expressed genes and miRNAs between normal and tumor groups, we used significance analysis of microarrays (SAM) [46] implemented in R package samr (version 2.0; Stanford University, Stanford, CA, USA). We set the false discovery rate (FDR) as ≤0.0001%, fold change as ≥2.5 for miRNA, and fold change as ≥1.9 for genes as thresholds to reduce false positives. If a gene/miRNA was expressed higher in the normal group compared to the tumor group, we defined that gene/miRNA as a down-regulated gene/miRNA, and *vice versa.*

Following this, we paired the miRNAs and mRNAs with different expression trends. For example, an up-regulated miRNA would be paired with a down-regulated mRNA. If such pairs could be found in 2 (or more) of the 3 miRNA target prediction databases, they were then added to their corresponding miRNA-regulated network. To further extend the coverage of our network, we incorporated the human protein reference database (HPRD) [47], which contains experimentally verified interaction data, into our miRNA-regulated PINs.

Finally, we used gene ontology (GO) enrichment analysis to explore the function of the miRNA-regulated PINs. Hypergeometric tests were used to determine if a GO term was enriched in a PIN. In this section, we excluded PINs with less than 5 proteins. We also excluded GO terms with levels of less than 5 to avoid non-specific GO terms. Since we tested multiple GO terms on each miRNA-regulated PIN, we adjusted the significance of the test with the FDR method developed by Benjamini *et al.* [48]. We used the adjusted p-value < 0.0001 as our threshold. A GO term would be excluded if its p-value was larger than 0.0001. Therefore, for each network, we ran the GO enrichment analysis, collected the calculated p-values, and adjusted these values using the methods described above.

3.5. ROC and GOBO Survival Analysis

After the PINs were constructed, we attempted to verify our results by literature search, ROC, and GOBO survival analysis [49]. To determine if the miRNAs we found could serve as classification markers for discriminating between normal and tumor samples, we applied ROC analysis on our miRNA array data (described in Section 2.1). ROC analysis is usually used to evaluate the efficiency of a classifier or a biological marker. R package ROCR [50] (version 1.0.4) was used to plot the ROC curve and calculate the area under curve (AUC). The standard error of AUC was then calculated as described in the work of Hanlye and McNeil [51]. The p-value of AUC was thus calculated with standard error obtained in the previous step. To further validate if the PINs we found were related to cancer, we used survival analysis implemented in GOBO [49] (available at http://co.bmc.lu.se/gobo), which provides a large amount of breast cancer gene expression profiles collected from public databases with clinical outcome data. In both ROC analysis and GOBO survival analysis, we considered our results significant when the p-value was smaller than 0.05.

4. Conclusions

Using integrative analysis of miRNA and mRNA expression profiles, we have identified not only breast cancer-related miRNAs and genes, but also putative roles for miRNAs in cancer as elucidated from miRNA-regulated PINs constructed in this work. Here, some previously known functions of miRNAs were again presented in our results, e.g., the relationship between the miRNAs, let-7c, miR-125a-5p, miR-125b-5p, and miR-21-5p, and breast cancer were demonstrated in this research. Furthermore, we have identified additional miRNAs and their related functions that have not been previously reported or discussed, providing valuable resources for further research in breast cancer.

Acknowledgments

This work was supported by the National Science Council of Taiwan (NSC 97-2311-B-002-010-MY3 and NSC 99-2621-B-002-005-MY3), National Taiwan University Cutting-Edge Steering Research Project (NTU-CESRP-102R7602C3) and the National Health Research Institutes, Taiwan (NHRI-EX98-9819PI).

Supplementary Information

Table S1. Significantly differentially expressed miRNAs found in miRNA dataset in Farazi *et al.* [26]. There are 89 down-regulated miRNAs and 1 up-regulated miRNA in this list. *Q*-values reported by SAM were 0 for all miRNAs in this list.

miRBase Accession	miRNA Name	Fold Change	miRBase Accession	miRNA Name	Fold Change
MIMAT0004761	hsa-miR-483-5p	0.01	**MIMAT0000077**	hsa-miR-22-3p	0.21
MIMAT0004552	hsa-miR-139-3p	0.01	**MIMAT0000089**	hsa-miR-31-5p	0.21
MIMAT0000738	hsa-miR-383	0.02	**MIMAT0004612**	hsa-miR-186-3p	0.21
MIMAT0002856	hsa-miR-520d-3p	0.02	**MIMAT0004592**	hsa-miR-125b-1-3p	0.22
MIMAT0002811	hsa-miR-202-3p	0.03	**MIMAT0001639**	hsa-miR-409-3p	0.22
MIMAT0002177	hsa-miR-486-5p	0.04	**MIMAT0015032**	hsa-miR-3158-3p	0.22
MIMAT0022721	hsa-miR-1247-3p	0.05	**MIMAT0004496**	hsa-miR-23a-5p	0.22
MIMAT0002175	hsa-miR-485-5p	0.06	**MIMAT0000690**	hsa-miR-296-5p	0.22
MIMAT0000265	hsa-miR-204-5p	0.07	**MIMAT0000731**	hsa-miR-378a-5p	0.23
MIMAT0000752	hsa-miR-328	0.07	**MIMAT0000448**	hsa-miR-136-5p	0.23
MIMAT0000421	hsa-miR-122-5p	0.07	**MIMAT0004796**	hsa-miR-576-3p	0.23
MIMAT0000447	hsa-miR-134	0.08	**MIMAT0010133**	hsa-miR-2110	0.23
MIMAT0000722	hsa-miR-370	0.09	**MIMAT0004951**	hsa-miR-887	0.23
MIMAT0004513	hsa-miR-101-5p	0.09	**MIMAT0003239**	hsa-miR-574-3p	0.25
MIMAT0000446	hsa-miR-127-3p	0.10	**MIMAT0005901**	hsa-miR-1249	0.25
MIMAT0000097	hsa-miR-99a-5p	0.10	**MIMAT0000510**	hsa-miR-320a	0.26
MIMAT0004566	hsa-miR-218-2-3p	0.10	**MIMAT0002172**	hsa-miR-376b	0.26
MIMAT0000729	hsa-miR-376a-3p	0.11	**MIMAT0000250**	hsa-miR-139-5p	0.27
MIMAT0009197	hsa-miR-205-3p	0.11	**MIMAT0005825**	hsa-miR-1180	0.27
MIMAT0004615	hsa-miR-195-3p	0.11	**MIMAT0000437**	hsa-miR-145-5p	0.28
MIMAT0005899	hsa-miR-1247-5p	0.11	**MIMAT0004601**	hsa-miR-145-3p	0.28
MIMAT0000720	hsa-miR-376c	0.12	**MIMAT0003322**	hsa-miR-652-3p	0.28
MIMAT0000762	hsa-miR-324-3p	0.12	**MIMAT0000756**	hsa-miR-326	0.28
MIMAT0004679	hsa-miR-296-3p	0.12	**MIMAT0000098**	hsa-miR-100-5p	0.29
MIMAT0004614	hsa-miR-193a-5p	0.12	**MIMAT0003296**	hsa-miR-627	0.29
MIMAT0003880	hsa-miR-671-5p	0.12	**MIMAT0002820**	hsa-miR-497-5p	0.31
MIMAT0004795	hsa-miR-574-5p	0.12	**MIMAT0004507**	hsa-miR-92a-1-5p	0.31
MIMAT0004599	hsa-miR-143-5p	0.13	**MIMAT0000271**	hsa-miR-214-3p	0.32
MIMAT0000423	hsa-miR-125b-5p	0.13	**MIMAT0004702**	hsa-miR-339-3p	0.33
MIMAT0004957	hsa-miR-760	0.13	**MIMAT0004611**	hsa-miR-185-3p	0.33
MIMAT0004911	hsa-miR-874	0.14	**MIMAT0000064**	hsa-let-7c	0.34
MIMAT0004603	hsa-miR-125b-2-3p	0.15	**MIMAT0004673**	hsa-miR-29c-5p	0.35
MIMAT0004952	hsa-miR-665	0.15	**MIMAT0000733**	hsa-miR-379-5p	0.35
MIMAT0018205	hsa-miR-3928	0.15	**MIMAT0004594**	hsa-miR-132-5p	0.35
MIMAT0004767	hsa-miR-193b-5p	0.15	**MIMAT0000765**	hsa-miR-335-5p	0.35
MIMAT0002861	hsa-miR-518e-3p	0.15	**MIMAT0002819**	hsa-miR-193b-3p	0.36
MIMAT0004604	hsa-miR-127-5p	0.16	**MIMAT0000088**	hsa-miR-30a-3p	0.36
MIMAT0002807	hsa-miR-491-5p	0.16	**MIMAT0005951**	hsa-miR-1307-3p	0.36
MIMAT0004689	hsa-miR-377-5p	0.16	**MIMAT0004597**	hsa-miR-140-3p	0.37
MIMAT0004762	hsa-miR-486-3p	0.16	**MIMAT0004556**	hsa-miR-10b-3p	0.37
MIMAT0000732	hsa-miR-378a-3p	0.17	**MIMAT0000272**	hsa-miR-215	0.37
MIMAT0017981	hsa-miR-3605-5p	0.18	**MIMAT0004511**	hsa-miR-99a-3p	0.37
MIMAT0004605	hsa-miR-129-2-3p	0.19	**MIMAT0000443**	hsa-miR-125a-5p	0.38
MIMAT0006789	hsa-miR-1468	0.20	**MIMAT0004482**	hsa-let-7b-3p	0.38
MIMAT0000737	hsa-miR-382-5p	0.21	**MIMAT0000076**	hsa-miR-21-5p	6.58

Table S2. Down-regulated genes found in dataset GSE29174. There are 726 down-regulated genes in this list. *Q*-values reported by SAM were 0 for all genes in this list.

NCBI gene ID	Gene Symbol	Fold Change	NCBI gene ID	Gene Symbol	Fold Change
2949	GSTM5	0.06	**619373**	MBOAT4	0.17
10894	LYVE1	0.06	**130399**	ACVR1C	0.17
5950	RBP4	0.07	**1646**	AKR1C2	0.17
762	CA4	0.09	**80763**	C12orf39	0.17
54997	TESC	0.09	**2159**	F10	0.18
3489	IGFBP6	0.09	**84889**	SLC7A3	0.18
3952	LEP	0.09	**1308**	COL17A1	0.18
213	ALB	0.09	**83699**	SH3BGRL2	0.18
3131	HLF	0.10	**84417**	C2orf40	0.18
4023	LPL	0.10	**4081**	MAB21L1	0.18
10633	RASL10A	0.11	**3484**	IGFBP1	0.18
364	AQP7	0.11	**5239**	PGM5	0.19
1908	EDN3	0.11	**4969**	OGN	0.19
1811	SLC26A3	0.11	**2719**	GPC3	0.19
91851	CHRDL1	0.11	**116362**	RBP7	0.19
729359	PLIN4	0.13	**948**	CD36	0.19
1149	CIDEA	0.13	**5764**	PTN	0.19
5959	RDH5	0.13	**3043**	HBB	0.19
5348	FXYD1	0.14	**56920**	SEMA3G	0.20
5346	PLIN1	0.14	**94274**	PPP1R14A	0.20
10249	GLYAT	0.14	**57447**	NDRG2	0.20
158800	RHOXF1	0.14	**84795**	PYROXD2	0.20
221476	PI16	0.14	**84649**	DGAT2	0.20
3040	HBA2	0.14	**2690**	GHR	0.20
6939	TCF15	0.14	**22802**	CLCA4	0.20
79645	EFCAB1	0.14	**5179**	PENK	0.20
80343	SEL1L2	0.14	**6663**	SOX10	0.20
9413	FAM189A2	0.15	**6649**	SOD3	0.21
26289	AK5	0.15	**54922**	RASIP1	0.21
25891	PAMR1	0.15	**8406**	SRPX	0.21
3679	ITGA7	0.15	**1446**	CSN1S1	0.21
1264	CNN1	0.15	**7123**	CLEC3B	0.22
92304	SCGB3A1	0.15	**9647**	PPM1F	0.22
2167	FABP4	0.15	**1842**	ECM2	0.22
23285	KIAA1107	0.15	**3909**	LAMA3	0.22
7145	TNS1	0.16	**8639**	AOC3	0.23
4881	NPR1	0.16	**2934**	GSN	0.23
1028	CDKN1C	0.16	**9370**	ADIPOQ	0.23
1036	CDO1	0.16	**3202**	HOXA5	0.23
130271	PLEKHH2	0.16	**9452**	ITM2A	0.23
8736	MYOM1	0.16	**6290**	SAA3P	0.23
8908	GYG2	0.16	**4604**	MYBPC1	0.23

Table S2. *Cont.*

NCBI gene ID	Gene Symbol	Fold Change	NCBI gene ID	Gene Symbol	Fold Change
79785	RERGL	0.16	**1128**	CHRM1	0.23
221091	LRRN4CL	0.17	**83878**	USHBP1	0.24
3991	LIPE	0.17	**63970**	TP53AIP1	0.24
27175	TUBG2	0.24	**79192**	IRX1	0.28
1346	COX7A1	0.24	**3400**	ID4	0.28
6376	CX3CL1	0.24	**57519**	STARD9	0.29
50486	G0S2	0.24	**57666**	FBRSL1	0.29
6285	S100B	0.24	**3590**	IL11RA	0.29
443	ASPA	0.24	**57664**	PLEKHA4	0.29
947	CD34	0.25	**197257**	LDHD	0.29
84632	AFAP1L2	0.25	**66036**	MTMR9	0.29
3866	KRT15	0.25	**2321**	FLT1	0.29
147463	ANKRD29	0.25	**126**	ADH1C	0.29
2878	GPX3	0.25	**1363**	CPE	0.29
7079	TIMP4	0.25	**56131**	PCDHB4	0.29
54345	SOX18	0.25	**22915**	MMRN1	0.29
51277	DNAJC27	0.25	**7069**	THRSP	0.29
84870	RSPO3	0.25	**57161**	PELI2	0.30
55323	LARP6	0.25	**770**	CA11	0.30
6387	CXCL12	0.25	**53342**	IL17D	0.30
137835	TMEM71	0.25	**79987**	SVEP1	0.30
5212	VIT	0.25	**857**	CAV1	0.30
26577	PCOLCE2	0.25	**222166**	C7orf41	0.30
845	CASQ2	0.25	**27190**	IL17B	0.30
6422	SFRP1	0.25	**116159**	CYYR1	0.30
10351	ABCA8	0.26	**4487**	MSX1	0.30
10840	ALDH1L1	0.26	**9068**	ANGPTL1	0.30
65983	GRAMD3	0.26	**10411**	RAPGEF3	0.30
84327	ZBED3	0.26	**3199**	HOXA2	0.30
57124	CD248	0.26	**2944**	GSTM1	0.30
3235	HOXD9	0.26	**2920**	CXCL2	0.30
2192	FBLN1	0.26	**201134**	CEP112	0.31
91653	BOC	0.26	**220001**	VWCE	0.31
4147	MATN2	0.26	**83888**	FGFBP2	0.31
126669	SHE	0.27	**6366**	CCL21	0.31
2788	GNG7	0.27	**6711**	SPTBN1	0.31
129804	FBLN7	0.27	**85378**	TUBGCP6	0.31
270	AMPD1	0.27	**26040**	SETBP1	0.31
79656	BEND5	0.27	**4692**	NDN	0.31
58503	PROL1	0.27	**25890**	ABI3BP	0.31
3316	HSPB2	0.27	**23531**	MMD	0.31
729440	CCDC61	0.27	**30846**	EHD2	0.31
54438	GFOD1	0.27	**6196**	RPS6KA2	0.31
5243	ABCB1	0.27	**2009**	EML1	0.31

Table S2. *Cont.*

NCBI gene ID	Gene Symbol	Fold Change	NCBI gene ID	Gene Symbol	Fold Change
810	CALML3	0.27	**6289**	SAA2	0.31
6898	TAT	0.27	**345275**	HSD17B13	0.31
5648	MASP1	0.28	**2701**	GJA4	0.32
25999	CLIP3	0.28	**112609**	MRAP2	0.32
125875	CLDND2	0.28	**727**	C5	0.32
7102	TSPAN7	0.28	**477**	ATP1A2	0.32
1879	EBF1	0.28	**9627**	SNCAIP	0.32
23252	OTUD3	0.28	**4435**	CITED1	0.32
5493	PPL	0.28	**10974**	C10orf116	0.32
83987	CCDC8	0.28	**11005**	SPINK5	0.32
9073	CLDN8	0.28	**80325**	ABTB1	0.33
221981	THSD7A	0.28	**221395**	GPR116	0.33
64102	TNMD	0.28	**10014**	HDAC5	0.33
137872	ADHFE1	0.33	**1489**	CTF1	0.37
27151	CPAMD8	0.33	**35**	ACADS	0.37
387923	SERP2	0.33	**3749**	KCNC4	0.37
145581	LRFN5	0.33	**140738**	TMEM37	0.37
6263	RYR3	0.33	**2791**	GNG11	0.37
2354	FOSB	0.33	**23604**	DAPK2	0.37
51302	CYP39A1	0.33	**10217**	CTDSPL	0.37
4128	MAOA	0.34	**23550**	PSD4	0.37
117248	GALNTL2	0.34	**4306**	NR3C2	0.37
10268	RAMP3	0.34	**119587**	CPXM2	0.37
7730	ZNF177	0.34	**7942**	TFEB	0.37
10873	ME3	0.34	**3815**	KIT	0.37
7461	CLIP2	0.34	**1805**	DPT	0.37
7049	TGFBR3	0.34	**23242**	COBL	0.37
79901	CYBRD1	0.34	**4313**	MMP2	0.37
5152	PDE9A	0.34	**4139**	MARK1	0.37
50805	IRX4	0.34	**9104**	RGN	0.37
8644	AKR1C3	0.34	**2329**	FMO4	0.37
5915	RARB	0.34	**25802**	LMOD1	0.38
2770	GNAI1	0.34	**4239**	MFAP4	0.38
54996	2-Mar	0.35	**10392**	NOD1	0.38
79791	FBXO31	0.35	**6794**	STK11	0.38
54776	PPP1R12C	0.35	**85458**	DIXDC1	0.38
9079	LDB2	0.35	**4123**	MAN2C1	0.38
57104	PNPLA2	0.35	**54476**	RNF216	0.38
30008	EFEMP2	0.35	**9920**	KBTBD11	0.38
91461	PKDCC	0.35	**6329**	SCN4A	0.38
23368	PPP1R13B	0.35	**10253**	SPRY2	0.38
23461	ABCA5	0.35	**1910**	EDNRB	0.38
9572	NR1D1	0.35	**9249**	DHRS3	0.38
23338	PHF15	0.35	**22869**	ZNF510	0.38

Table S2. *Cont.*

NCBI gene ID	Gene Symbol	Fold Change	NCBI gene ID	Gene Symbol	Fold Change
114800	CCDC85A	0.35	**3384**	ICAM2	0.38
2550	GABBR1	0.35	**8613**	PPAP2B	0.38
4638	MYLK	0.35	**1950**	EGF	0.38
2327	FMO2	0.35	**55273**	TMEM100	0.38
139411	PTCHD1	0.35	**6297**	SALL2	0.38
10391	CORO2B	0.35	**9365**	KL	0.38
25854	FAM149A	0.35	**8863**	PER3	0.38
55701	ARHGEF40	0.36	**8404**	SPARCL1	0.38
1759	DNM1	0.36	**2202**	EFEMP1	0.38
22849	CPEB3	0.36	**8369**	HIST1H4G	0.38
57716	PRX	0.36	**5187**	PER1	0.39
1628	DBP	0.36	**30815**	ST6GALNAC6	0.39
80031	SEMA6D	0.36	**256364**	EML3	0.39
259217	HSPA12A	0.36	**57381**	RHOJ	0.39
6909	TBX2	0.36	**761**	CA3	0.39
1511	CTSG	0.36	**83989**	FAM172A	0.39
79971	WLS	0.36	**1408**	CRY2	0.39
90865	IL33	0.36	**2281**	FKBP1B	0.39
11343	MGLL	0.36	**51222**	ZNF219	0.39
55800	SCN3B	0.36	**54540**	FAM193B	0.39
1949	EFNB3	0.36	**4053**	LTBP2	0.39
284217	LAMA1	0.36	**55184**	DZANK1	0.39
22927	HABP4	0.37	**5740**	PTGIS	0.39
23645	PPP1R15A	0.39	**84814**	PPAPDC3	0.42
342574	KRT27	0.39	**79365**	BHLHE41	0.42
83543	AIF1L	0.39	**316**	AOX1	0.42
624	BDKRB2	0.39	**23380**	SRGAP2	0.42
347	APOD	0.39	**84033**	OBSCN	0.42
84935	C13orf33	0.39	**90353**	CTU1	0.42
858	CAV2	0.39	**9013**	TAF1C	0.42
5138	PDE2A	0.40	**474344**	GIMAP6	0.42
114928	GPRASP2	0.40	**84883**	AIFM2	0.42
58190	CTDSP1	0.40	**58480**	RHOU	0.42
513	ATP5D	0.40	**65982**	ZSCAN18	0.42
57684	ZBTB26	0.40	**666**	BOK	0.42
7041	TGFB1I1	0.40	**79762**	C1orf115	0.42
5787	PTPRB	0.40	**525**	ATP6V1B1	0.42
7294	TXK	0.40	**4675**	NAP1L3	0.42
56301	SLC7A10	0.40	**3257**	HPS1	0.43
55937	APOM	0.40	**55781**	RIOK2	0.43
6368	CCL23	0.40	**63947**	DMRTC1	0.43
55020	TTC38	0.40	**1969**	EPHA2	0.43
134265	AFAP1L1	0.40	**25927**	CNRIP1	0.43
4485	MST1	0.40	**57685**	CACHD1	0.43

Table S2. *Cont.*

NCBI gene ID	Gene Symbol	Fold Change	NCBI gene ID	Gene Symbol	Fold Change
51559	NT5DC3	0.40	**29997**	GLTSCR2	0.43
7169	TPM2	0.40	**26051**	PPP1R16B	0.43
51705	EMCN	0.40	**83604**	TMEM47	0.43
8938	BAIAP3	0.40	**2308**	FOXO1	0.43
10365	KLF2	0.40	**55225**	RAVER2	0.43
59	ACTA2	0.40	**54839**	LRRC49	0.43
80309	SPHKAP	0.40	**122953**	JDP2	0.43
3779	KCNMB1	0.41	**29775**	CARD10	0.43
10826	C5orf4	0.41	**166**	AES	0.43
219654	ZCCHC24	0.41	**25924**	MYRIP	0.43
92162	TMEM88	0.41	**2852**	GPER	0.43
7450	VWF	0.41	**51421**	AMOTL2	0.43
10266	RAMP2	0.41	**124936**	CYB5D2	0.43
25875	LETMD1	0.41	**1294**	COL7A1	0.43
1938	EEF2	0.41	**127435**	PODN	0.43
121551	BTBD11	0.41	**84952**	CGNL1	0.43
2119	ETV5	0.41	**83483**	PLVAP	0.43
9696	CROCC	0.41	**1958**	EGR1	0.43
1031	CDKN2C	0.41	**230**	ALDOC	0.43
9037	SEMA5A	0.41	**65987**	KCTD14	0.43
3397	ID1	0.41	**4804**	NGFR	0.44
84707	BEX2	0.41	**64852**	TUT1	0.44
57616	TSHZ3	0.41	**84253**	GARNL3	0.44
1471	CST3	0.41	**5866**	RAB3IL1	0.44
55214	LEPREL1	0.41	**10608**	MXD4	0.44
3914	LAMB3	0.41	**4211**	MEIS1	0.44
57478	USP31	0.41	**83547**	RILP	0.44
3783	KCNN4	0.41	**9172**	MYOM2	0.44
8839	WISP2	0.41	**57192**	MCOLN1	0.44
1583	CYP11A1	0.42	**255877**	BCL6B	0.44
10124	ARL4A	0.42	**56904**	SH3GLB2	0.44
738	C11orf2	0.42	**51285**	RASL12	0.44
29800	ZDHHC1	0.42	**3425**	IDUA	0.44
23135	KDM6B	0.44	**402117**	VWC2L	0.46
171024	SYNPO2	0.44	**81490**	PTDSS2	0.46
10350	ABCA9	0.44	**283748**	PLA2G4D	0.46
3691	ITGB4	0.44	**23523**	CABIN1	0.46
2348	FOLR1	0.44	**6146**	RPL22	0.46
11145	PLA2G16	0.44	**85360**	SYDE1	0.46
554	AVPR2	0.45	**60468**	BACH2	0.46
64072	CDH23	0.45	**57451**	ODZ2	0.46
80177	MYCT1	0.45	**4013**	VWA5A	0.46
5957	RCVRN	0.45	**339768**	ESPNL	0.46
408	ARRB1	0.45	**3860**	KRT13	0.46

Table S2. *Cont.*

NCBI gene ID	Gene Symbol	Fold Change	NCBI gene ID	Gene Symbol	Fold Change
144699	FBXL14	0.45	**7094**	TLN1	0.46
83719	YPEL3	0.45	**4232**	MEST	0.46
22841	RAB11FIP2	0.45	**1410**	CRYAB	0.46
283927	NUDT7	0.45	**57452**	GALNTL1	0.47
293	SLC25A6	0.45	**63935**	PCIF1	0.47
90507	SCRN2	0.45	**25873**	RPL36	0.47
37	ACADVL	0.45	**9812**	KIAA0141	0.47
112744	IL17F	0.45	**51665**	ASB1	0.47
6709	SPTAN1	0.45	**64123**	ELTD1	0.47
8086	AAAS	0.45	**6122**	RPL3	0.47
7423	VEGFB	0.45	**222962**	SLC29A4	0.47
64221	ROBO3	0.45	**23102**	TBC1D2B	0.47
7273	TTN	0.45	**3476**	IGBP1	0.47
2657	GDF1	0.45	**93408**	MYL10	0.47
59271	C21orf63	0.45	**5310**	PKD1	0.47
132160	PPM1M	0.45	**4628**	MYH10	0.47
27244	SESN1	0.45	**221935**	SDK1	0.47
51310	SLC22A17	0.45	**23328**	SASH1	0.47
4828	NMB	0.45	**8522**	GAS7	0.47
54360	CYTL1	0.45	**10023**	FRAT1	0.47
203245	NAIF1	0.45	**7301**	TYRO3	0.47
23166	STAB1	0.45	**2767**	GNA11	0.47
2121	EVC	0.45	**9457**	FHL5	0.47
116496	FAM129A	0.45	**4094**	MAF	0.47
23239	PHLPP1	0.45	**65268**	WNK2	0.47
51673	TPPP3	0.45	**54585**	LZTFL1	0.47
64094	SMOC2	0.45	**375449**	MAST4	0.47
6383	SDC2	0.45	**138311**	FAM69B	0.47
2180	ACSL1	0.45	**160622**	GRASP	0.47
23770	FKBP8	0.45	**22837**	COBLL1	0.47
55901	THSD1	0.46	**51435**	SCARA3	0.47
25895	METTL21B	0.46	**217**	ALDH2	0.47
23731	C9orf5	0.46	**6236**	RRAD	0.47
126393	HSPB6	0.46	**8322**	FZD4	0.47
4056	LTC4S	0.46	**653275**	CFC1B	0.47
79825	CCDC48	0.46	**10908**	PNPLA6	0.47
10810	WASF3	0.46	**57526**	PCDH19	0.47
29911	HOOK2	0.46	**8424**	BBOX1	0.47
583	BBS2	0.46	**9905**	SGSM2	0.48
28984	C13orf15	0.46	**10435**	CDC42EP2	0.48
1465	CSRP1	0.46	**23087**	TRIM35	0.48
55258	THNSL2	0.46	**60314**	C12orf10	0.48
161198	CLEC14A	0.46	**1073**	CFL2	0.48
3699	ITIH3	0.48	**5256**	PHKA2	0.49

Table S2. *Cont.*

NCBI gene ID	Gene Symbol	Fold Change	NCBI gene ID	Gene Symbol	Fold Change
92922	CCDC102A	0.48	**6237**	RRAS	0.49
65057	ACD	0.48	**5288**	PIK3C2G	0.49
9095	TBX19	0.48	**10252**	SPRY1	0.49
6441	SFTPD	0.48	**79026**	AHNAK	0.49
22846	VASH1	0.48	**9693**	RAPGEF2	0.49
51066	C3orf32	0.48	**51226**	COPZ2	0.49
23179	RGL1	0.48	**158326**	FREM1	0.49
4664	NAB1	0.48	**1956**	EGFR	0.49
50511	SYCP3	0.48	**5360**	PLTP	0.49
6430	SRSF5	0.48	**290**	ANPEP	0.49
11078	TRIOBP	0.48	**1756**	DMD	0.49
78991	PCYOX1L	0.48	**5118**	PCOLCE	0.49
6623	SNCG	0.48	**56654**	NPDC1	0.49
23384	SPECC1L	0.48	**9254**	CACNA2D2	0.49
53826	FXYD6	0.48	**55536**	CDCA7L	0.49
9397	NMT2	0.48	**124975**	GGT6	0.49
6041	RNASEL	0.48	**1906**	EDN1	0.49
113510	HELQ	0.48	**81029**	WNT5B	0.49
64788	LMF1	0.48	**2646**	GCKR	0.49
2217	FCGRT	0.48	**9811**	CTIF	0.50
79720	VPS37B	0.48	**145376**	PPP1R36	0.50
6764	ST5	0.48	**222865**	TMEM130	0.50
252969	NEIL2	0.48	**92999**	ZBTB47	0.50
8987	STBD1	0.48	**168002**	DACT2	0.50
41	ACCN2	0.48	**6829**	SUPT5H	0.50
7905	REEP5	0.48	**9992**	KCNE2	0.50
5919	RARRES2	0.48	**58509**	C19orf29	0.50
10544	PROCR	0.48	**79706**	PRKRIP1	0.50
6876	TAGLN	0.48	**1153**	CIRBP	0.50
8436	SDPR	0.49	**9639**	ARHGEF10	0.50
23500	DAAM2	0.49	**4054**	LTBP3	0.50
130132	RFTN2	0.49	**1120**	CHKB	0.50
80310	PDGFD	0.49	**286046**	XKR6	0.50
4215	MAP3K3	0.49	**9590**	AKAP12	0.50
282775	OR5J2	0.49	**64115**	C10orf54	0.50
51161	C3orf18	0.49	**2067**	ERCC1	0.50
29098	RANGRF	0.49	**7507**	XPA	0.50
53336	CPXCR1	0.49	**22897**	CEP164	0.50
9081	PRY	0.49	**652**	BMP4	0.50
9459	ARHGEF6	0.49	**55702**	CCDC94	0.50
2995	GYPC	0.49	**57613**	KIAA1467	0.50
23057	NMNAT2	0.49	**28514**	DLL1	0.50
4669	NAGLU	0.49	**169270**	ZNF596	0.50
6452	SH3BP2	0.49	**83982**	IFI27L2	0.50

Table S2. *Cont.*

NCBI gene ID	Gene Symbol	Fold Change	NCBI gene ID	Gene Symbol	Fold Change
51458	RHCG	0.49	**2247**	FGF2	0.50
1112	FOXN3	0.49	**26248**	OR2K2	0.50
29954	POMT2	0.49	**84303**	CHCHD6	0.50
9612	NCOR2	0.49	**3615**	IMPDH2	0.50
3198	HOXA1	0.49	**1813**	DRD2	0.50
5311	PKD2	0.49	**80148**	PQLC1	0.50
2946	GSTM2	0.49	**390081**	OR52E4	0.50
2109	ETFB	0.49	**352954**	GATS	0.50
56062	KLHL4	0.49	**90871**	C9orf123	0.50
6915	TBXA2R	0.50	**50945**	TBX22	0.52
64288	ZNF323	0.50	**5204**	PFDN5	0.52
5195	PEX14	0.50	**5338**	PLD2	0.52
84557	MAP1LC3A	0.50	**94**	ACVRL1	0.52
6164	RPL34	0.50	**54039**	PCBP3	0.52
8835	SOCS2	0.50	**7691**	ZNF132	0.52
2735	GLI1	0.50	**338**	APOB	0.52
26022	TMEM98	0.50	**84658**	EMR3	0.52
3908	LAMA2	0.50	**283232**	TMEM80	0.52
1825	DSC3	0.50	**5430**	POLR2A	0.52
5730	PTGDS	0.50	**54623**	PAF1	0.52
162515	SLC16A11	0.51	**11070**	TMEM115	0.52
274	BIN1	0.51	**10395**	DLC1	0.52
79654	HECTD3	0.51	**57140**	RNPEPL1	0.52
22863	ATG14	0.51	**79781**	IQCA1	0.52
25949	SYF2	0.51	**1838**	DTNB	0.52
84872	ZC3H10	0.51	**51386**	EIF3L	0.52
23187	PHLDB1	0.51	**56919**	DHX33	0.52
5434	POLR2E	0.51	**57542**	KLHDC5	0.52
6181	RPLP2	0.51	**3628**	INPP1	0.52
6141	RPL18	0.51	**4520**	MTF1	0.52
84747	UNC119B	0.51	**8547**	FCN3	0.52
23399	CTDNEP1	0.51	**60401**	EDA2R	0.52
599	BCL2L2	0.51	**8082**	SSPN	0.52
197258	FUK	0.51	**80755**	AARSD1	0.52
5207	PFKFB1	0.51	**710**	SERPING1	0.52
8131	NPRL3	0.51	**56246**	MRAP	0.52
25839	COG4	0.51	**10555**	AGPAT2	0.52
10816	SPINT3	0.51	**949**	SCARB1	0.52
60485	SAV1	0.51	**23743**	BHMT2	0.52
5681	PSKH1	0.51	**3910**	LAMA4	0.52
80318	GKAP1	0.51	**60370**	AVPI1	0.52
57088	PLSCR4	0.51	**5021**	OXTR	0.52
93129	ORAI3	0.51	**55997**	CFC1	0.52
5829	PXN	0.51	**23144**	ZC3H3	0.52

Table S2. *Cont.*

NCBI gene ID	Gene Symbol	Fold Change	NCBI gene ID	Gene Symbol	Fold Change
56776	FMN2	0.51	**150709**	ANKAR	0.52
85456	TNKS1BP1	0.51	**6591**	SNAI2	0.52
283	ANG	0.51	**10129**	FRY	0.52
7035	TFPI	0.51	**5166**	PDK4	0.52
51232	CRIM1	0.51	**146433**	IL34	0.52
112616	CMTM7	0.51	**118812**	MORN4	0.53
22981	NINL	0.51	**10516**	FBLN5	0.53
8727	CTNNAL1	0.51	**9463**	PICK1	0.53
9902	MRC2	0.51	**127495**	LRRC39	0.53
10900	RUNDC3A	0.51	**7753**	ZNF202	0.53
51299	NRN1	0.51	**79827**	CLMP	0.53
79632	FAM184A	0.52	**203260**	CCDC107	0.53
80820	EEPD1	0.52	**83657**	DYNLRB2	0.53

Table S1. Up-regulated genes found in dataset GSE29174. There are 437 up-regulated genes in this list. *Q*-values reported by SAM were 0 for all genes in this list.

NCBI gene ID	Gene Symbol	Fold Change	NCBI gene ID	Gene Symbol	Fold Change
1300	COL10A1	42.74	**54443**	ANLN	5.79
3007	HIST1H1D	29.72	**6710**	SPTB	5.71
8366	HIST1H4B	25.58	**7272**	TTK	5.64
6286	S100P	25.19	**10635**	RAD51AP1	5.49
1301	COL11A1	24.72	**4069**	LYZ	5.37
3627	CXCL10	17.83	**55183**	RIF1	5.34
4283	CXCL9	15.88	**891**	CCNB1	5.34
1387	CREBBP	12.83	**91543**	RSAD2	5.31
27299	ADAMDEC1	12.78	**81610**	FAM83D	5.24
54986	ULK4	12.46	**64581**	CLEC7A	5.10
55771	PRR11	12.02	**10051**	SMC4	5.02
54790	TET2	11.25	**4085**	MAD2L1	4.96
6241	RRM2	10.60	**55872**	PBK	4.83
3433	IFIT2	10.49	**991**	CDC20	4.82
6999	TDO2	9.73	**9221**	NOLC1	4.74
1656	DDX6	9.72	**2124**	EVI2B	4.66
55088	C10orf118	9.37	**375248**	ANKRD36	4.66
9648	GCC2	9.24	**1164**	CKS2	4.64
6696	SPP1	8.92	**1230**	CCR1	4.62
2803	GOLGA4	8.57	**890**	CCNA2	4.56
83540	NUF2	7.73	**127933**	UHMK1	4.49
10112	KIF20A	7.66	**10274**	STAG1	4.45
9833	MELK	7.59	**597**	BCL2A1	4.43
55165	CEP55	7.50	**55355**	HJURP	4.41
10142	AKAP9	7.44	**54210**	TREM1	4.36
9447	AIM2	7.42	**253558**	LCLAT1	4.26

Table S3. *Cont.*

NCBI gene ID	Gene Symbol	Fold Change	NCBI gene ID	Gene Symbol	Fold Change
2706	GJB2	7.33	**1033**	CDKN3	4.24
6498	SKIL	7.13	**79801**	SHCBP1	4.23
219285	SAMD9L	7.06	**126731**	C1orf96	4.21
10261	IGSF6	7.01	**6772**	STAT1	4.20
2335	FN1	6.95	**55729**	ATF7IP	4.14
699	BUB1	6.75	**6713**	SQLE	4.14
1058	CENPA	6.75	**157570**	ESCO2	4.10
332	BIRC5	6.73	**79871**	RPAP2	4.09
51203	NUSAP1	6.59	**9493**	KIF23	4.09
259266	ASPM	6.54	**4751**	NEK2	4.05
1063	CENPF	6.49	**10631**	POSTN	4.03
165918	RNF168	6.44	**23515**	MORC3	4.02
9232	PTTG1	6.34	**7153**	TOP2A	4.02
5996	RGS1	6.07	**10403**	NDC80	4.00
29089	UBE2T	5.96	**10915**	TCERG1	3.99
22974	TPX2	5.94	**57650**	KIAA1524	3.99
4321	MMP12	5.91	**23049**	SMG1	3.93
983	CDK1	5.89	**80231**	CXorf21	3.87
85444	LRRCC1	5.87	**5111**	PCNA	3.86
29121	CLEC2D	3.83	**79682**	MLF1IP	3.11
4090	SMAD5	3.80	**29123**	ANKRD11	3.09
2123	EVI2A	3.80	**5429**	POLH	3.09
57695	USP37	3.79	**701**	BUB1B	3.07
133418	EMB	3.76	**200030**	NBPF11	3.06
4131	MAP1B	3.76	**55677**	IWS1	3.06
9787	DLGAP5	3.75	**160418**	TMTC3	3.04
9768	KIAA0101	3.74	**9147**	NEMF	3.04
54625	PARP14	3.73	**11320**	MGAT4A	3.04
2215	FCGR3B	3.71	**5238**	PGM3	3.03
9134	CCNE2	3.70	**2820**	GPD2	3.02
3117	HLA-DQA1	3.68	**388886**	FAM211B	3.01
10380	BPNT1	3.67	**7852**	CXCR4	3.00
79056	PRRG4	3.63	**57082**	CASC5	2.99
10673	TNFSF13B	3.63	**22926**	ATF6	2.98
8467	SMARCA5	3.61	**7594**	ZNF43	2.98
115908	CTHRC1	3.61	**968**	CD68	2.97
3428	IFI16	3.61	**7171**	TPM4	2.96
1520	CTSS	3.61	**11004**	KIF2C	2.96
10797	MTHFD2	3.57	**10808**	HSPH1	2.95
55681	SCYL2	3.57	**84909**	C9orf3	2.94
9749	PHACTR2	3.57	**1894**	ECT2	2.93
94240	EPSTI1	3.56	**1629**	DBT	2.92
64151	NCAPG	3.51	**116969**	ART5	2.90
25879	DCAF13	3.51	**3227**	HOXC11	2.88

Table S3. *Cont.*

NCBI gene ID	Gene Symbol	Fold Change	NCBI gene ID	Gene Symbol	Fold Change
116064	LRRC58	3.47	**3149**	HMGB3	2.87
29899	GPSM2	3.47	**10437**	IFI30	2.87
135114	HINT3	3.45	**57489**	ODF2L	2.87
27333	GOLIM4	3.43	**2151**	F2RL2	2.86
55839	CENPN	3.43	**23215**	PRRC2C	2.85
23213	SULF1	3.41	**128710**	C20orf94	2.85
81671	VMP1	3.39	**23594**	ORC6	2.84
9889	ZBED4	3.36	**5205**	ATP8B1	2.83
3092	HIP1	3.34	**51430**	C1orf9	2.80
51512	GTSE1	3.34	**57405**	SPC25	2.80
92797	HELB	3.34	**112401**	BIRC8	2.80
51426	POLK	3.30	**3606**	IL18	2.80
5611	DNAJC3	3.30	**115362**	GBP5	2.80
6596	HLTF	3.28	**50515**	CHST11	2.79
9910	RABGAP1L	3.25	**83461**	CDCA3	2.79
528	ATP6V1C1	3.23	**10744**	PTTG2	2.78
3833	KIFC1	3.23	**51765**	MST4	2.77
197131	UBR1	3.20	**10926**	DBF4	2.76
29923	HILPDA	3.20	**27125**	AFF4	2.75
28998	MRPL13	3.19	**10615**	SPAG5	2.75
58527	C6orf115	3.19	**55143**	CDCA8	2.74
79000	C1orf135	3.19	**51602**	NOP58	2.74
9857	CEP350	3.18	**51478**	HSD17B7	2.73
84296	GINS4	3.18	**2209**	FCGR1A	2.73
81034	SLC25A32	3.15	**9958**	USP15	2.72
55723	ASF1B	3.14	**5469**	MED1	2.72
7110	TMF1	3.14	**8813**	DPM1	2.70
84081	NSRP1	3.14	**6731**	SRP72	2.70
23075	SWAP70	3.12	**9991**	PTBP3	2.70
6726	SRP9	2.69	**79866**	BORA	2.41
55215	FANCI	2.68	**7072**	TIA1	2.40
57590	WDFY1	2.67	**55632**	G2E3	2.40
55142	HAUS2	2.66	**2213**	FCGR2B	2.40
23047	PDS5B	2.66	**3987**	LIMS1	2.39
5373	PMM2	2.66	**829**	CAPZA1	2.39
11065	UBE2C	2.66	**26973**	CHORDC1	2.38
23085	ERC1	2.66	**435**	ASL	2.38
389197	C4orf50	2.65	**29979**	UBQLN1	2.38
11260	XPOT	2.65	**8548**	BLZF1	2.37
29980	DONSON	2.65	**9694**	TTC35	2.37
64399	HHIP	2.64	**55055**	ZWILCH	2.36
6453	ITSN1	2.63	**4481**	MSR1	2.36
29108	PYCARD	2.63	**10213**	PSMD14	2.35
9877	ZC3H11A	2.62	**9966**	TNFSF15	2.35

Table S3. *Cont.*

NCBI gene ID	Gene Symbol	Fold Change	NCBI gene ID	Gene Symbol	Fold Change
81624	DIAPH3	2.62	**51582**	AZIN1	2.35
79723	SUV39H2	2.61	**54843**	SYTL2	2.34
55789	DEPDC1B	2.61	**9039**	UBA3	2.33
10097	ACTR2	2.59	**933**	CD22	2.33
23036	ZNF292	2.58	**5685**	PSMA4	2.33
22936	ELL2	2.57	**9885**	OSBPL2	2.33
8477	GPR65	2.57	**9262**	STK17B	2.33
23397	NCAPH	2.57	**56942**	C16orf61	2.32
3015	H2AFZ	2.54	**10767**	HBS1L	2.32
55749	CCAR1	2.53	**87178**	PNPT1	2.32
25937	WWTR1	2.52	**6303**	SAT1	2.32
360023	ZBTB41	2.51	**7316**	UBC	2.32
5080	PAX6	2.51	**4205**	MEF2A	2.32
4193	MDM2	2.51	**85465**	EPT1	2.31
24137	KIF4A	2.51	**84640**	USP38	2.31
9212	AURKB	2.51	**5810**	RAD1	2.30
168850	ZNF800	2.50	**64397**	ZFP106	2.29
55109	AGGF1	2.49	**5706**	PSMC6	2.29
23185	LARP4B	2.49	**22948**	CCT5	2.29
51571	FAM49B	2.49	**10672**	GNA13	2.29
51077	FCF1	2.49	**339344**	MYPOP	2.28
23167	EFR3A	2.49	**7292**	TNFSF4	2.28
23468	CBX5	2.48	**57103**	C12orf5	2.28
5396	PRRX1	2.48	**388403**	YPEL2	2.28
10096	ACTR3	2.47	**54876**	DCAF16	2.27
10308	ZNF267	2.47	**113235**	SLC46A1	2.27
6782	HSPA13	2.47	**11177**	BAZ1A	2.27
3832	KIF11	2.47	**339175**	METTL2A	2.26
917	CD3G	2.47	**26586**	CKAP2	2.26
80821	DDHD1	2.46	**55785**	FGD6	2.26
52	ACP1	2.46	**24145**	PANX1	2.25
4179	CD46	2.46	**253461**	ZBTB38	2.25
10499	NCOA2	2.44	**23232**	TBC1D12	2.25
60558	GUF1	2.44	**995**	CDC25C	2.25
55676	SLC30A6	2.43	**55974**	SLC50A1	2.25
6646	SOAT1	2.43	**472**	ATM	2.25
5440	POLR2K	2.43	**23008**	KLHDC10	2.24
84955	NUDCD1	2.42	**10024**	TROAP	2.24
54739	XAF1	2.42	**9521**	EEF1E1	2.24
84295	PHF6	2.23	**7402**	UTRN	2.09
7295	TXN	2.23	**55589**	BMP2K	2.08
2710	GK	2.23	**158747**	MOSPD2	2.08
10905	MAN1A2	2.22	**56886**	UGGT1	2.07
6780	STAU1	2.22	**203100**	HTRA4	2.07

Table S3. *Cont.*

NCBI gene ID	Gene Symbol	Fold Change	NCBI gene ID	Gene Symbol	Fold Change
10282	BET1	2.22	**55279**	ZNF654	2.07
134430	WDR36	2.21	**54499**	TMCO1	2.07
4299	AFF1	2.21	**81930**	KIF18A	2.07
6747	SSR3	2.21	**142686**	ASB14	2.06
7334	UBE2N	2.21	**55209**	SETD5	2.06
5965	RECQL	2.21	**9736**	USP34	2.04
4605	MYBL2	2.2	**116285**	ACSM1	2.04
6093	ROCK1	2.19	**2201**	FBN2	2.04
161725	OTUD7A	2.19	**963**	CD53	2.04
23518	R3HDM1	2.18	**55159**	RFWD3	2.03
2239	GPC4	2.18	**9871**	SEC24D	2.03
28977	MRPL42	2.18	**9887**	SMG7	2.02
64859	OBFC2A	2.18	**23376**	UFL1	2.02
3845	KRAS	2.18	**79646**	PANK3	2.01
51388	NIP7	2.18	**50613**	UBQLN3	2.00
7586	ZKSCAN1	2.18	**201595**	STT3B	2.00
10762	NUP50	2.17	**59345**	GNB4	1.99
7328	UBE2H	2.17	**5876**	RABGGTB	1.99
10730	YME1L1	2.17	**79820**	CATSPERB	1.99
23093	TTLL5	2.17	**6637**	SNRPG	1.99
6790	AURKA	2.17	**51330**	TNFRSF12A	1.99
22889	KIAA0907	2.17	**9928**	KIF14	1.99
10875	FGL2	2.17	**286097**	EFHA2	1.98
23161	SNX13	2.17	**9131**	AIFM1	1.98
9169	SCAF11	2.16	**488**	ATP2A2	1.98
1788	DNMT3A	2.15	**23042**	PDXDC1	1.98
9088	PKMYT1	2.15	**7114**	TMSB4X	1.98
23033	DOPEY1	2.13	**9123**	SLC16A3	1.98
89882	TPD52L3	2.13	**54454**	ATAD2B	1.97
6556	SLC11A1	2.13	**23143**	LRCH1	1.97
64216	TFB2M	2.13	**4212**	MEIS2	1.97
3071	NCKAP1L	2.13	**1457**	CSNK2A1	1.97
51068	NMD3	2.13	**80012**	PHC3	1.97
509	ATP5C1	2.13	**128497**	SPATA25	1.96
953	ENTPD1	2.13	**186**	AGTR2	1.96
51105	PHF20L1	2.13	**53981**	CPSF2	1.96
5062	PAK2	2.13	**56996**	SLC12A9	1.96
9205	ZMYM5	2.12	**1584**	CYP11B1	1.96
55157	DARS2	2.12	**133619**	PRRC1	1.96
8520	HAT1	2.11	**4288**	MKI67	1.96
79739	TTLL7	2.11	**9014**	TAF1B	1.96
9495	AKAP5	2.10	**55858**	TMEM165	1.96
3181	HNRNPA2B1	2.10	**2212**	FCGR2A	1.96

Table S3. *Cont.*

NCBI gene ID	Gene Symbol	Fold Change	NCBI gene ID	Gene Symbol	Fold Change
389898	UBE2NL	2.10	**10075**	HUWE1	1.96
29850	TRPM5	2.10	**220988**	HNRNPA3	1.96
3070	HELLS	2.10	**80146**	UXS1	1.95
331	XIAP	2.09	**122011**	CSNK1A1L	1.95
55751	TMEM184C	2.09	**150468**	CKAP2L	1.95
2146	EZH2	2.09	**84624**	FNDC1	1.95
26057	ANKRD17	1.95	**7332**	UBE2L3	1.92
128061	C1orf131	1.95	**3336**	HSPE1	1.92
64090	GAL3ST2	1.94	**54800**	KLHL24	1.92
130507	UBR3	1.93	**2290**	FOXG1	1.91
2298	FOXD4	1.93	**50848**	F11R	1.91
123169	LEO1	1.93	**10627**	MYL12A	1.91
57187	THOC2	1.93	**5074**	PAWR	1.91
148789	B3GALNT2	1.93	**6476**	SI	1.91
58508	MLL3	1.92	**1009**	CDH11	1.90
5701	PSMC2	1.92	**29066**	ZC3H7A	1.90
148066	ZNRF4	1.92	**51319**	RSRC1	1.90
6670	SP3	1.92			

Table S4. Result of ROC curve analysis on our miRNA array data. ROC analysis was done to validate the diagnostic value of the miRNA in the miRNA-regulated PINs. AUC: area under (ROC) curve; * *p*-value < 0.05; *** *p*-value < 0.001.

miRBase Accession	miRNA name	AUC	*p*-value
MIMAT0002856	hsa-miR-520d-3p	0.49	0.549112
MIMAT0000265	hsa-miR-204-5p	0.98	6.47×10^{-10} ***
MIMAT0000272	hsa-miR-215	0.21	0.999782
MIMAT0000271	hsa-miR-214-3p	0.68	0.010387 *
MIMAT0002820	hsa-miR-497-5p	0.99	2.75×10^{-10} ***
MIMAT0000076	hsa-miR-21-5p	0.78	0.000184 ***
MIMAT0000738	hsa-miR-383	0.60	0.106284
MIMAT0000423	hsa-miR-125b-5p	0.99	2.48×10^{-10} ***
MIMAT0000064	hsa-let-7c	0.93	3.79×10^{-8} ***
MIMAT0000089	hsa-miR-31-5p	0.80	8.63×10^{-5} ***
MIMAT0000077	hsa-miR-22-3p	0.27	0.99749
MIMAT0000098	hsa-miR-100-5p	0.98	5.55×10^{-10} ***
MIMAT0000097	hsa-miR-99a-5p	0.99	2.55×10^{-10} ***
MIMAT0000443	hsa-miR-125a-5p	0.31	0.990694
MIMAT0002819	hsa-miR-193b-3p	0.41	0.86128
MIMAT0000250	hsa-miR-139-5p	0.99	2.42×10^{-10} ***
MIMAT0000437	hsa-miR-145-5p	0.96	3.14×10^{-9} ***
MIMAT0000421	hsa-miR-122-5p	0.48	0.597483

Table S5. Summary of constructed miRNA-regulated networks. L0 gene: genes connected directly to the miRNA (*i.e.*, direct target of miRNA); L1 gene: genes not connected directly to the miRNA.

miRBase Accession	miR name	Total gene count	L0 count	L1 count
MIMAT0002819	hsa-miR-193b-3p	16	1	15
MIMAT0000250	hsa-miR-139-5p	28	10	18
MIMAT0000437	hsa-miR-145-5p	86	22	64
MIMAT0000423	hsa-miR-125b-5p	211	16	195
MIMAT0000443	hsa-miR-125a-5p	206	14	192
MIMAT0000097	hsa-miR-99a-5p	14	1	13
MIMAT0000265	hsa-miR-204-5p	64	18	46
MIMAT0000076	hsa-miR-21-5p	91	16	75
MIMAT0000064	hsa-let-7c	96	20	76
MIMAT0000421	hsa-miR-122-5p	5	3	2
MIMAT0000098	hsa-miR-100-5p	14	1	13
MIMAT0000272	hsa-miR-215	3	3	0
MIMAT0000271	hsa-miR-214-3p	14	8	6
MIMAT0000738	hsa-miR-383	34	3	31
MIMAT0002856	hsa-miR-520d-3p	146	23	123
MIMAT0000077	hsa-miR-22-3p	46	11	35
MIMAT0002820	hsa-miR-497-5p	267	32	235
MIMAT0000089	hsa-miR-31-5p	34	3	31

Table S6. Specific enriched GO terms of each miRNA-regulated PINs. Genes annotated with the specific GO term in the PIN were also listed in this table. Adj. *p*-value: multiple-test adjusted p-value calculated by the method described in the work of Benjamini and Yekutieli [48].

MIMAT0002856(hsa-miR-520d-3p)		
GO term	**Genes**	**Adj. *p*-value**
GO:0007169, Transmembrane receptor protein tyrosine kinase signaling pathway	SH3KBP1, HDAC2, RET, ABI1, LYN, GRB2, SORBS1, CLTC, CLTA, CDC42, CASP9, RAF1, SRC, AP2A1, AP2B1, MAPK3, ARHGEF7, PRKCA, RPS6, PRKAR2B, MAPK1, ARHGEF6, CDK1, SH3GL2, EIF4G1, HDAC1, ECT2, MKNK1, CASP3, PRKACA, ADRB2, PRKAR2A, EIF4B, SHC1, RAC1	2.77×10^{-29}
GO:0048011, Nerve growth factor receptor signaling pathway	HDAC2, GRB2, CLTC, CLTA, CASP9, RAF1, SRC, AP2A1, AP2B1, MAPK3, ARHGEF7, PRKCA, PRKAR2B, MAPK1, ARHGEF6, CDK1, SH3GL2, HDAC1, ECT2, CASP3, PRKACA, PRKAR2A, SHC1, RAC1	5.09×10^{-24}
GO:0007173, Epidermal growth factor receptor signaling pathway	SH3KBP1, GRB2, CLTC, CLTA, CDC42, CASP9, RAF1, SRC, AP2A1, AP2B1, MAPK3, ARHGEF7, PRKCA, PRKAR2B, MAPK1, CDK1, SH3GL2, PRKACA, PRKAR2A, SHC1	2.96×10^{-22}
GO:0043067, Regulation of programmed cell death	HDAC2, STK17B, ESR1, ABL1, LYN, TP53, GABRB3, PAK2, LCK, CASP9, RAF1, PLK1, ARHGEF7, PRKCA, RPS6, SH3RF1, MAPK1, IFT57, ARHGAP10, ARHGEF6, CDK1, APAF1, HDAC1, ECT2, CASP3, SOX10, EP300, ARAF, TFAP2A, ADRB2, HCK, KLHL20, CASP8, HIP1, RAC1	4.76×10^{-19}
GO:0042058, Regulation of epidermal growth factor receptor signaling pathway	SH3KBP1, ESR1, GRB2, CLTC, CLTA, CDC42, AP2A1, AP2B1, ARHGEF7, SH3GL2, SHC1	2.36×10^{-12}
GO:0008543, Fibroblast growth factor receptor signaling pathway	GRB2, CASP9, RAF1, SRC, MAPK3, PRKCA, PRKAR2B, MAPK1, CDK1, MKNK1, PRKACA, PRKAR2A, SHC1	2.39×10^{-12}
GO:0043068, Positive regulation of programmed cell death	STK17B, ABL1, LYN, TP53, LCK, CASP9, ARHGEF7, PRKCA, RPS6, SH3RF1, MAPK1, ARHGEF6, APAF1, ECT2, CASP3, EP300, TFAP2A, ADRB2, CASP8, HIP1, RAC1	3.09×10^{-12}
GO:0010942, Positive regulation of cell death	STK17B, ABL1, LYN, TP53, LCK, CASP9, ARHGEF7, PRKCA, RPS6, SH3RF1, MAPK1, ARHGEF6, APAF1, ECT2, CASP3, EP300, TFAP2A, ADRB2, CASP8, HIP1, RAC1	4.49×10^{-12}
GO:0006917, Induction of apoptosis	STK17B, ABL1, TP53, LCK, CASP9, ARHGEF7, PRKCA, SH3RF1, MAPK1, ARHGEF6, APAF1, ECT2, CASP3, EP300, CASP8, HIP1, RAC1	6.91×10^{-11}
GO:0012502, Induction of programmed cell death	STK17B, ABL1, TP53, LCK, CASP9, ARHGEF7, PRKCA, SH3RF1, MAPK1, ARHGEF6, APAF1, ECT2, CASP3, EP300, CASP8, HIP1, RAC1	7.42×10^{-11}

Table S6. *Cont.*

MIMAT0002856(hsa-miR-520d-3p)		
GO term	**Genes**	**Adj. *p*-value**
GO:0042059, Negative regulation of epidermal growth factor receptor signaling pathway	SH3KBP1, GRB2, CLTC, CLTA, CDC42, AP2A1, AP2B1, ARHGEF7, SH3GL2	8.63×10^{-11}
GO:0015630, Microtubule cytoskeleton	STMN1, SORBS1, SMAD4, CLTC, CDC42, LCK, RACGAP1, PLK1, PRKAR2B, YES1, MAPK1, IFT57, CDK1, ECT2, PRKACA, RB1, EP300, CCNB1, CHAF1B, TFAP2A, CASP8, PRKAR2A	4.75×10^{-10}
GO:0060548, Negative regulation of cell death	HDAC2, ESR1, TP53, SMAD4, RAF1, PLK1, PRKCA, RPS6, SH3RF1, CDK1, HDAC1, CASP3, SOX10, ARAF, TFAP2A, HCK, KLHL20	6.27×10^{-8}
GO:0008286, Insulin receptor signaling pathway	GRB2, SORBS1, RAF1, MAPK3, RPS6, MAPK1, CDK1, EIF4G1, EIF4B, SHC1	2.15×10^{-7}
GO:0043069, Negative regulation of programmed cell death	HDAC2, ESR1, TP53, RAF1, PLK1, PRKCA, RPS6, SH3RF1, CDK1, HDAC1, CASP3, SOX10, ARAF, TFAP2A, HCK, KLHL20	3.13×10^{-7}
GO:0008284, Positive regulation of cell proliferation	HDAC2, ESR1, LYN, CDC42, E2F1, PRKCA, MAPK1, CDK1, RHOG, HDAC1, NCK1, SOX10, CCNB1, ADRB2, HCK, SHC1	8.34×10^{-7}
GO:0051988, Regulation of attachment of spindle microtubules to kinetochore	CDC42, RACGAP1, ECT2, CCNB1	3.27×10^{-5}
GO:0008629, Induction of apoptosis by intracellular signals	ABL1, TP53, CASP9, APAF1, CASP3, EP300, CASP8	5.83×10^{-5}

MIMAT0002820(hsa-miR-497-5p)		
GO term	**Genes**	**Adj. *p*-value**
GO:0043067, Regulation of programmed cell death	ESR1, MEN1, ABL1, HIPK3, PPARGC1A, SIAH1, SH3RF1, PAK2, LCK, MED1, PPARG, CBX4, ARHGEF7, YWHAB, RXRA, ACVR1, MAPK1, CASP3, CASP6, AR, PTPRF, MDM2, BRCA1, MLH1, RAB27A, PIAS4, FAF1, RAC1, VHL, SKI, NR4A1, LYN, TP53, PSMC2, GATA1, GATA6, GATA3, RAF1, CDKN1B, PLK1, PSMD11, HOXA13, RPS6, ESR2, ARHGAP10, ARHGEF6, SMAD3, SKIL, RYR2, PSEN1, HCK, TRAF2	2.67×10^{-25}
GO:0043068, Positive regulation of programmed cell death	MEN1, ABL1, SIAH1, SH3RF1, LCK, PPARG, ARHGEF7, YWHAB, RXRA, MAPK1, CASP3, CASP6, PTPRF, BRCA1, MLH1, RAB27A, PIAS4, FAF1, RAC1, NR4A1, LYN, TP53, GATA6, CDKN1B, HOXA13, RPS6, ESR2, ARHGEF6, SMAD3, RYR2, PSEN1, TRAF2	1.74×10^{-17}
GO:0010942, Positive regulation of cell death	MEN1, ABL1, SIAH1, SH3RF1, LCK, PPARG, ARHGEF7, YWHAB, RXRA, MAPK1, CASP3, CASP6, PTPRF, BRCA1, MLH1, RAB27A, PIAS4, FAF1, RAC1, NR4A1, LYN, TP53, GATA6, CDKN1B, HOXA13, RPS6, ESR2, ARHGEF6, SMAD3, RYR2, PSEN1, TRAF2	3.08×10^{-17}

Table S6. *Cont.*

MIMAT0002820(hsa-miR-497-5p)		
GO term	**Genes**	**Adj. *p*-value**
GO:0008285, Negative regulation of cell proliferation	MEN1, MED1, PPARG, RXRA, CASP3, AR, PTPRF, VDR, VHL, SKI, LYN, TP53, TOB1, GATA1, GATA3, RAF1, HNF4A, CDKN1B, BRD7, MED25, ESR2, ABI1, SMAD1, SMAD2, SMAD3, SMAD4, SOX7	3.85×10^{-14}
GO:0015629, Actin cytoskeleton	ABL1, SORBS1, FLNA, SEPT7, ANLN, MACF1, HAP1, SH3PXD2A, IQGAP2, BRCA1, ACTC1, ACTA1, MYL2, MYLK, SORBS2, ARPC4, ARPC5, ACTR2, ACTR3, ARPC1B, WASF1, WASF2, HCK	2.52×10^{-13}
GO:0006917, Induction of apoptosis	ABL1, SH3RF1, LCK, PPARG, ARHGEF7, YWHAB, MAPK1, CASP3, CASP6, BRCA1, MLH1, RAB27A, RAC1, NR4A1, TP53, CDKN1B, ARHGEF6, SMAD3, RYR2, PSEN1, TRAF2	9.69×10^{-11}
GO:0012502, Induction of programmed cell death	ABL1, SH3RF1, LCK, PPARG, ARHGEF7, YWHAB, MAPK1, CASP3, CASP6, BRCA1, MLH1, RAB27A, RAC1, NR4A1, TP53, CDKN1B, ARHGEF6, SMAD3, RYR2, PSEN1, TRAF2	1.06×10^{-10}
GO:0007178, Transmembrane receptor protein serine/threonine kinase signaling pathway	ACVR1, SMURF2, SKI, GDF6, BMP6, ZNF8, GATA4, HNF4A, SMAD1, SMAD2, SMAD3, SMAD4, SMAD5, RYR2	1.22×10^{-10}
GO:0007169, Transmembrane receptor protein tyrosine kinase signaling pathway	SORBS1, CDC42, SRC, MAPK3, ARHGEF7, YWHAB, MAPK1, SH3GL2, CASP3, MDM2, EIF4G1, RAC1, SH3KBP1, NR4A1, LYN, GRB2, RAF1, CDKN1B, RPS6, ABI1, ARHGEF6, MKNK1, PSEN1, EIF4B	1.67×10^{-10}
GO:0090092, Regulation of transmembrane receptor protein serine/threonine kinase signaling pathway	MEN1, ACVR1, SMURF2, SKI, GDF6, TP53, BMP6, GATA4, GATA6, HOXA13, SMAD2, SMAD3, SMAD4, SKIL	7.51×10^{-10}
GO:0030509, BMP signaling pathway	ACVR1, SMURF2, SKI, GDF6, BMP6, ZNF8, SMAD1, SMAD4, SMAD5, RYR2	3.25×10^{-9}
GO:0060548, Negative regulation of cell death	ESR1, HIPK3, PPARGC1A, SH3RF1, MED1, CBX4, ACVR1, CASP3, AR, MDM2, VHL, TP53, GATA1, GATA6, GATA3, RAF1, CDKN1B, PLK1, RPS6, SMAD3, SMAD4, PSEN1, HCK	7.88×10^{-9}
GO:0007173, Epidermal growth factor receptor signaling pathway	CDC42, SRC, MAPK3, ARHGEF7, YWHAB, MAPK1, SH3GL2, MDM2, SH3KBP1, NR4A1, GRB2, RAF1, CDKN1B	9.22×10^{-9}
GO:0030521, Androgen receptor signaling pathway	PPARGC1A, MED14, MED1, AR, BRCA1, MED12, PIAS1, RAN, NR1I3	1.42×10^{-8}
GO:0043069, Negative regulation of programmed cell death	ESR1, HIPK3, PPARGC1A, SH3RF1, MED1, CBX4, ACVR1, CASP3, AR, MDM2, VHL, TP53, GATA1, GATA6, GATA3, RAF1, CDKN1B, PLK1, RPS6, SMAD3, PSEN1, HCK	2.44×10^{-8}
GO:0048011, Nerve growth factor receptor signaling pathway	SRC, MAPK3, ARHGEF7, YWHAB, MAPK1, SH3GL2, CASP3, MDM2, RAC1, NR4A1, GRB2, RAF1, CDKN1B, ARHGEF6, PSEN1	2.64×10^{-8}
GO:0032956, Regulation of actin cytoskeleton organization	ABL1, LRP1, ARPC4, ARPC5, ACTR3, ARPC1B, SMAD3, NCK1, SORBS3, HCK, LIMK1	6.42×10^{-6}

Table S6. *Cont.*

MIMAT0002820(hsa-miR-497-5p)		
GO term	**Genes**	**Adj. *p*-value**
GO:0008543, Fibroblast growth factor receptor signaling pathway	SRC, MAPK3, YWHAB, MAPK1, MDM2, NR4A1, GRB2, RAF1, CDKN1B, MKNK1	9.06×10^{-6}
GO:0042059, Negative regulation of epidermal growth factor receptor signaling pathway	CDC42, ARHGEF7, SH3GL2, PTPRF, SH3KBP1, GRB2, PSEN1	1.11×10^{-5}
GO:0042058, Regulation of epidermal growth factor receptor signaling pathway	ESR1, CDC42, ARHGEF7, SH3GL2, PTPRF, SH3KBP1, GRB2, PSEN1	1.17×10^{-5}
GO:0007179, Transforming growth factor beta receptor signaling pathway	ACVR1, SMURF2, SKI, SMAD1, SMAD2, SMAD3, SMAD4, SMAD5	1.67×10^{-5}
GO:0015630, Microtubule cytoskeleton	STMN1, RIF1, SORBS1, CDC42, LCK, RACGAP1, YES1, YWHAB, MAPK1, SEPT7, KIF23, CDC16, MACF1, BRCA1, FEZ1, NCOR1, PLK1, CHD3, SMAD4, CEP350, CDC27, PSEN1	2.14×10^{-5}
GO:0017015, Regulation of transforming growth factor beta receptor signaling pathway	MEN1, SMURF2, SKI, TP53, SMAD2, SMAD3, SMAD4, SKIL	2.30×10^{-5}
GO:0070302, Regulation of stress-activated protein kinase signaling cascade	MEN1, ZEB2, HIPK3, SH3RF1, CDC42, MAPK3, MAPK1, LYN, NCOR1, TRAF2	2.74×10^{-5}
GO:0001959, Regulation of cytokine-mediated signaling pathway	HSP90AB1, MED1, PPARG, PTPRF, NR1H2, PIAS1, IL36RN, HIPK1	6.35×10^{-5}
GO:0008284, Positive regulation of cell proliferation	ESR1, CDC42, MED1, RARA, MAPK1, AR, MDM2, NR4A1, LYN, FZR1, BMP6, GATA1, GATA4, GATA6, CDKN1B, NCK1, HCLS1, HCK	7.29×10^{-5}

Table S6. *Cont.*

MIMAT0000423(hsa-miR-125b-5p)		
GO term	**Genes**	**Adj. *p*-value**
GO:0043067, Regulation of programmed cell death	HMGA2, PML, PRNP, FGF2, XRCC4, BRCA1, IGFBP3, HDAC3, CTNNB1, CD5, CDK1, NKX2-5, MEF2C, PRKCI, CASP2, PSMA4, PSMA3, CFDP1, CAV1, FAF1, YWHAB, HIF1A, RELA, TCF7L2, TNFSF12, PSEN2, TP53, TOP2A, TNFRSF4, BID, MYC, JUN, OGT, CDKN1A, HOXA13, RNF7, PPP2R4, HDAC2, HDAC1, SNCA, PTEN, NFKBIA, IFI16, NOL3, TRAF2, HSP90B1	4.56×10^{-24}
GO:0060548, Negative regulation of cell death	HMGA2, PRNP, FGF2, XRCC4, HDAC3, CTNNB1, CDK1, NKX2-5, MEF2C, PRKCI, CFDP1, HIF1A, RELA, TCF7L2, PSEN2, TP53, TNFRSF4, MYC, JUN, CDKN1A, RNF7, HDAC2, HDAC1, SNCA, PTEN, MGMT, NFKBIA, NOL3, HSP90B1	8.04×10^{-17}
GO:0008284, Positive regulation of cell proliferation	HMGA2, FGF2, XRCC4, CDC25B, CTNNB1, EGR1, AGGF1, CDK1, NKX2-5, MEF2C, PRKCI, IRS1, HIF1A, RELA, HCLS1, TNFSF12, ARNT, PTPRC, TNFSF4, TNFRSF4, MYC, JUN, FGF1, CDKN1A, HDAC2, HDAC1, NOLC1, PTEN	2.20×10^{-15}
GO:0043069, Negative regulation of programmed cell death	HMGA2, PRNP, XRCC4, HDAC3, CTNNB1, CDK1, NKX2-5, MEF2C, PRKCI, CFDP1, HIF1A, RELA, TCF7L2, PSEN2, TP53, TNFRSF4, MYC, JUN, CDKN1A, RNF7, HDAC2, HDAC1, SNCA, PTEN, NFKBIA, NOL3, HSP90B1	3.62×10^{-15}
GO:0043068, Positive regulation of programmed cell death	HMGA2, PML, BRCA1, IGFBP3, CTNNB1, CD5, MEF2C, PRKCI, CASP2, CAV1, FAF1, YWHAB, TNFSF12, PSEN2, TP53, TOP2A, BID, JUN, OGT, CDKN1A, HOXA13, RNF7, PPP2R4, PTEN, IFI16, TRAF2	4.51×10^{-14}
GO:0010942, Positive regulation of cell death	HMGA2, PML, BRCA1, IGFBP3, CTNNB1, CD5, MEF2C, PRKCI, CASP2, CAV1, FAF1, YWHAB, TNFSF12, PSEN2, TP53, TOP2A, BID, JUN, OGT, CDKN1A, HOXA13, RNF7, PPP2R4, PTEN, IFI16, TRAF2	7.15×10^{-14}
GO:0006916, Anti-apoptosis	PRNP, HDAC3, CDK1, NKX2-5, MEF2C, PRKCI, CFDP1, RELA, TCF7L2, PSEN2, RNF7, HDAC1, SNCA, NFKBIA, NOL3, HSP90B1	1.25×10^{-10}
GO:0008285, Negative regulation of cell proliferation	SERPINF1, SRF, PML, PRNP, FGF2, CSNK2B, IGFBP3, CTNNB1, CAV1, HMGA1, VDR, CDH5, HSF1, COL18A1, TP53, MYC, JUN, CDKN1A, PAK1, PTEN	2.08×10^{-9}
GO:0015630, Microtubule cytoskeleton	STMN1, KIF1C, RANGAP1, CDC25B, BRCA1, HDAC3, CTNNB1, PIN4, HSPH1, RANBP9, CDK1, SPTAN1, YWHAQ, DVL1, FKBP4, YWHAB, CCDC85B, MAPT, PSEN2, TOP2A, SPIB, MYC, OGT, APEX1, PAFAH1B1	2.24×10^{-9}
GO:0048011, Nerve growth factor receptor signaling pathway	HDAC3, CDK1, MEF2C, PRKCI, CASP2, IRS1, YWHAB, RELA, PSEN2, HDAC2, HDAC1, PTEN, ATF1, NFKBIA	1.99×10^{-8}
GO:0050678, Regulation of epithelial cell proliferation	SERPINF1, PGR, FGF2, CTNNB1, AGGF1, CAV1, HIF1A, TNFSF12, ARNT, MYC, JUN, FGF1, PTEN	3.41×10^{-8}
GO:0007169, Transmembrane receptor protein tyrosine kinase signaling pathway	FGF2, HDAC3, CDK1, MEF2C, PRKCI, CASP2, IRS1, FIBP, PTPN1, YWHAB, RELA, PSEN2, FGF1, HDAC2, HDAC1, PTEN, ATF1, NFKBIA, EIF4EBP1	6.39×10^{-8}

Table S6. *Cont.*

MIMAT0000423(hsa-miR-125b-5p)		
GO term	**Genes**	**Adj. *p*-value**
GO:0006917, Induction of apoptosis	PML, BRCA1, CD5, CASP2, CAV1, YWHAB, TNFSF12, PSEN2, TP53, BID, OGT, CDKN1A, RNF7, PTEN, IFI16, TRAF2	1.68×10^{-7}
GO:0012502, Induction of programmed cell death	PML, BRCA1, CD5, CASP2, CAV1, YWHAB, TNFSF12, PSEN2, TP53, BID, OGT, CDKN1A, RNF7, PTEN, IFI16, TRAF2	1.81×10^{-7}
GO:0035666, TRIF-dependent toll-like receptor signaling pathway	ATF2, CDK1, MEF2C, FOS, RELA, JUN, ATF1, NFKBIA	2.40×10^{-6}
GO:0034138, Toll-like receptor 3 signaling pathway	ATF2, CDK1, MEF2C, FOS, RELA, JUN, ATF1, NFKBIA	2.71×10^{-6}
GO:0051693, Actin filament capping	SPTB, SPTBN1, SPTAN1, SPTA1, ADD1, EPB49	3.43×10^{-6}
GO:0002756, MyD88-independent toll-like receptor signaling pathway	ATF2, CDK1, MEF2C, FOS, RELA, JUN, ATF1, NFKBIA	3.82×10^{-6}
GO:0034134, Toll-like receptor 2 signaling pathway	ATF2, CDK1, MEF2C, FOS, RELA, JUN, ATF1, NFKBIA	5.33×10^{-6}
GO:0034130, Toll-like receptor 1 signaling pathway	ATF2, CDK1, MEF2C, FOS, RELA, JUN, ATF1, NFKBIA	5.33×10^{-6}
GO:0030835, Negative regulation of actin filament depolymerization	SPTB, SPTBN1, SPTAN1, SPTA1, ADD1, EPB49	5.58×10^{-6}
GO:0015629, Actin cytoskeleton	WAS, CDH1, BRCA1, SPTB, SPTBN1, SPTAN1, CTDP1, STX1A, SPTA1, PAK1, SNCA, ADD1, EPB41, EPB49	5.58×10^{-6}
GO:0002755, MyD88-dependent toll-like receptor signaling pathway	ATF2, CDK1, MEF2C, FOS, RELA, JUN, ATF1, NFKBIA	7.66×10^{-6}
GO:0034142, Toll-like receptor 4 signaling pathway	ATF2, CDK1, MEF2C, FOS, RELA, JUN, ATF1, NFKBIA	1.14×10^{-5}
GO:0050679, Positive regulation of epithelial cell proliferation	FGF2, CTNNB1, AGGF1, HIF1A, TNFSF12, ARNT, MYC, JUN, FGF1	1.14×10^{-5}

Table S6. *Cont.*

GO term	Genes	Adj. *p*-value
MIMAT0000076(hsa-miR-21-5p)		
GO term	**Genes**	**Adj. *p*-value**
GO:0030834, Regulation of actin filament depolymerization	SPTB, SPTBN1, SPTAN1, SPTA1, ADD1, EPB49	1.27×10^{-5}
GO:0030837, Negative regulation of actin filament polymerization	SPTB, SPTBN1, SPTAN1, SPTA1, ADD1, EPB49	2.71×10^{-5}
GO:0002224, Toll-like receptor signaling pathway	ATF2, CDK1, MEF2C, FOS, RELA, JUN, ATF1, NFKBIA	2.96×10^{-5}
GO:0008629, Induction of apoptosis by intracellular signals	PML, BRCA1, YWHAB, TP53, BID, CDKN1A, RNF7, IFI16	3.52×10^{-5}
GO:0043067, Regulation of programmed cell death	SPRY2, TP53, ADAMTSL4, ETS1, TDGF1, RAF1, HOXA5, HOXA13, MSX1, MSX2, NKX2-5, CBL, INHBB, COL4A3, ACVR1C, TRAF2	1.92×10^{-5}
GO:0007173, Epidermal growth factor receptor signaling pathway	SPRY2, SPRY1, GRB2, PTPN11, TDGF1, RAF1, CBL	9.44×10^{-5}
MIMAT0000250(hsa-miR-139-5p)		
GO term	**Genes**	**Adj. *p*-value**
GO:0035583, Negative regulation of transforming growth factor beta receptor signaling pathway by extracellular sequestering of TGFbeta	LTBP1, FBN1, FBN2	2.61×10^{-5}
MIMAT0000089(hsa-miR-31-5p)		
GO term	**Genes**	**Adj. *p*-value**
GO:0007187, G-protein signaling, coupled to cyclic nucleotide second messenger	GNA12, GNA13, DRD5, MTNR1A, S1PR3, TSHR, S1PR4	8.74×10^{-7}
GO:0019935, Cyclic-nucleotide-mediated signaling	GNA12, GNA13, DRD5, MTNR1A, S1PR3, TSHR, S1PR4	1.60×10^{-6}
GO:0007188, G-protein signaling, coupled to cAMP nucleotide second messenger	GNA12, GNA13, DRD5, S1PR3, TSHR, S1PR4	6.82×10^{-6}

Table S6. *Cont.*

MIMAT0000089(hsa-miR-31-5p)		
GO term	**Genes**	**Adj. *p*-value**
GO:0048011, Nerve growth factor receptor signaling pathway	ARHGEF1, PRKCD, ARHGEF12, PRKACA, PRKCE, MCF2, ARHGEF11	6.93×10^{-6}
GO:0019933, CAMP-mediated signaling	GNA12, GNA13, DRD5, S1PR3, TSHR, S1PR4	1.05×10^{-5}
GO:0043067, Regulation of programmed cell death	PTK2B, PRKCD, TGFBR1, ARHGEF12, CTNNB1, PRKCE, TIA1, MCF2, FASTK, ARHGEF11, F2R	1.25×10^{-5}
GO:0003376, Sphingosine-1-phosphate signaling pathway	S1PR3, S1PR2, S1PR4	3.22×10^{-5}
GO:0007169, Transmembrane receptor protein tyrosine kinase signaling pathway	PTK2B, ARHGEF1, PRKCD, ARHGEF12, PRKACA, PRKCE, MCF2, ARHGEF11	5.20×10^{-5}
GO:0043068, Positive regulation of programmed cell death	TGFBR1, ARHGEF12, CTNNB1, PRKCE, TIA1, MCF2, FASTK, ARHGEF11	7.67×10^{-5}
GO:0010942, Positive regulation of cell death	TGFBR1, ARHGEF12, CTNNB1, PRKCE, TIA1, MCF2, FASTK, ARHGEF11	8.78×10^{-5}
GO:0006917, Induction of apoptosis	TGFBR1, ARHGEF12, PRKCE, TIA1, MCF2, FASTK, ARHGEF11	9.32×10^{-5}
GO:0012502, Induction of programmed cell death	TGFBR1, ARHGEF12, PRKCE, TIA1, MCF2, FASTK, ARHGEF11	9.51×10^{-5}
MIMAT0000437(hsa-miR-145-5p)		
GO term	**Genes**	**Adj. *p*-value**
GO:0030509, BMP signaling pathway	BMP6, ZNF8, ACVR1, SMAD1, SMAD4, RYR2, SMAD5, SMURF2, GDF6	1.37×10^{-11}
GO:0007178, Transmembrane receptor protein serine/threonine kinase signaling pathway	BMP6, ZNF8, ACVR1, SMAD1, SMAD4, RYR2, SMAD5, SMURF2, GDF6	2.56×10^{-8}
GO:0090092, Regulation of transmembrane receptor protein serine/threonine kinase signaling pathway	MEN1, TP53, BMP6, HOXA13, ACVR1, SMAD4, SULF1, SMURF2, GDF6	8.22×10^{-8}

Table S6. *Cont.*

MIMAT0000064(hsa-let-7c)		
GO term	**Genes**	**Adj. *p*-value**
GO:0043067, Regulation of programmed cell death	RRM2B, ACVR1, TP53, RASA1, TGFBR1, PSMA3, BIRC5, ACTN2, HOXA13, IRS2, FASTK, VAV1, PSMB6, BCL2, CDK1, HDAC1, SOX10, TIA1, AKT1, AURKB	3.43×10^{-8}
GO:0043069, Negative regulation of programmed cell death	RRM2B, ACVR1, TP53, RASA1, TGFBR1, BIRC5, IRS2, BCL2, CDK1, HDAC1, SOX10, AKT1, AURKB	6.58×10^{-6}
GO:0060548, Negative regulation of cell death	RRM2B, ACVR1, TP53, RASA1, TGFBR1, BIRC5, IRS2, BCL2, CDK1, HDAC1, SOX10, AKT1, AURKB	8.92×10^{-6}
GO:0015630, Microtubule cytoskeleton	INCENP, SNTB2, SEPT1, TACC1, BIRC5, RACGAP1, PIN4, CDCA8, CDK1, PHF1, AKT1, AURKB, NINL, CCDC85B	5.15×10^{-5}
GO:0043067, Regulation of programmed cell death	HMGA2, PML, PRNP, FGF2, XRCC4, BRCA1, IGFBP3, HDAC3, CTNNB1, CD5, CDK1, NKX2-5, MEF2C, PRKCI, CASP2, PSMA4, PSMA3, CFDP1, CAV1, FAF1, YWHAB, HIF1A, RELA, TCF7L2, TNFSF12, PSEN2, TP53, TOP2A, TNFRSF4, BID, MYC, JUN, OGT, CDKN1A, RNF7, PPP2R4, HDAC2, HDAC1, SNCA, PTEN, NFKBIA, IFI16, NOL3, TRAF2, HSP90B1	1.62×10^{-23}
GO:0060548, Negative regulation of cell death	HMGA2, PRNP, FGF2, XRCC4, HDAC3, CTNNB1, CDK1, NKX2-5, MEF2C, PRKCI, CFDP1, HIF1A, RELA, TCF7L2, PSEN2, TP53, TNFRSF4, MYC, JUN, CDKN1A, RNF7, HDAC2, HDAC1, SNCA, PTEN, MGMT, NFKBIA, NOL3, HSP90B1	4.53×10^{-17}
GO:0008284, Positive regulation of cell proliferation	HMGA2, FGF2, XRCC4, CDC25B, CTNNB1, EGR1, AGGF1, CDK1, NKX2-5, MEF2C, PRKCI, IRS1, HIF1A, RELA, HCLS1, TNFSF12, ARNT, PTPRC, TNFSF4, TNFRSF4, MYC, JUN, FGF1, CDKN1A, HDAC2, HDAC1, NOLC1, PTEN	1.23×10^{-15}

MIMAT0000443(hsa-miR-125a-5p)		
GO term	**Genes**	**Adj. *p*-value**
GO:0043069, Negative regulation of programmed cell death	HMGA2, PRNP, XRCC4, HDAC3, CTNNB1, CDK1, NKX2-5, MEF2C, PRKCI, CFDP1, HIF1A, RELA, TCF7L2, PSEN2, TP53, TNFRSF4, MYC, JUN, CDKN1A, RNF7, HDAC2, HDAC1, SNCA, PTEN, NFKBIA, NOL3, HSP90B1	2.02×10^{-15}
GO:0043068, Positive regulation of programmed cell death	HMGA2, PML, BRCA1, IGFBP3, CTNNB1, CD5, MEF2C, PRKCI, CASP2, CAV1, FAF1, YWHAB, TNFSF12, PSEN2, TP53, TOP2A, BID, JUN, OGT, CDKN1A, RNF7, PPP2R4, PTEN, IFI16, TRAF2	2.80×10^{-13}
GO:0010942, Positive regulation of cell death	HMGA2, PML, BRCA1, IGFBP3, CTNNB1, CD5, MEF2C, PRKCI, CASP2, CAV1, FAF1, YWHAB, TNFSF12, PSEN2, TP53, TOP2A, BID, JUN, OGT, CDKN1A, RNF7, PPP2R4, PTEN, IFI16, TRAF2	4.35×10^{-13}
GO:0006916, Anti-apoptosis	PRNP, HDAC3, CDK1, NKX2-5, MEF2C, PRKCI, CFDP1, RELA, TCF7L2, PSEN2, RNF7, HDAC1, SNCA, NFKBIA, NOL3, HSP90B1	9.04×10^{-11}
GO:0008285, Negative regulation of cell proliferation	SERPINF1, SRF, PML, PRNP, FGF2, CSNK2B, IGFBP3, CTNNB1, CAV1, HMGA1, VDR, CDH5, HSF1, COL18A1, TP53, MYC, JUN, CDKN1A, PAK1, PTEN	1.32×10^{-9}

Table S6. *Cont.*

MIMAT0000443(hsa-miR-125a-5p)		
GO term	**Genes**	**Adj. *p*-value**
GO:0015630, Microtubule cytoskeleton	STMN1, KIF1C, RANGAP1, CDC25B, BRCA1, HDAC3, CTNNB1, PIN4, HSPH1, RANBP9, CDK1, SPTAN1, YWHAQ, DVL1, FKBP4, YWHAB, CCDC85B, MAPT, PSEN2, TOP2A, SPIB, MYC, OGT, APEX1, PAFAH1B1	1.32×10^{-9}
GO:0048011, Nerve growth factor receptor signaling pathway	HDAC3, CDK1, MEF2C, PRKCI, CASP2, IRS1, YWHAB, RELA, PSEN2, HDAC2, HDAC1, PTEN, ATF1, NFKBIA	1.50×10^{-8}
GO:0050678, Regulation of epithelial cell proliferation	SERPINF1, PGR, FGF2, CTNNB1, AGGF1, CAV1, HIF1A, TNFSF12, ARNT, MYC, JUN, FGF1, PTEN	2.66×10^{-8}
GO:0007169, Transmembrane receptor protein tyrosine kinase signaling pathway	FGF2, HDAC3, CDK1, MEF2C, PRKCI, CASP2, IRS1, FIBP, PTPN1, YWHAB, RELA, PSEN2, FGF1, HDAC2, HDAC1, PTEN, ATF1, NFKBIA, EIF4EBP1	4.37×10^{-8}
GO:0006917, Induction of apoptosis	PML, BRCA1, CD5, CASP2, CAV1, YWHAB, TNFSF12, PSEN2, TP53, BID, OGT, CDKN1A, RNF7, PTEN, IFI16, TRAF2	1.22×10^{-7}
GO:0012502, Induction of programmed cell death	PML, BRCA1, CD5, CASP2, CAV1, YWHAB, TNFSF12, PSEN2, TP53, BID, OGT, CDKN1A, RNF7, PTEN, IFI16, TRAF2	1.31×10^{-7}
GO:0035666, TRIF-dependent toll-like receptor signaling pathway	ATF2, CDK1, MEF2C, FOS, RELA, JUN, ATF1, NFKBIA	2.01×10^{-6}
GO:0034138, Toll-like receptor 3 signaling pathway	ATF2, CDK1, MEF2C, FOS, RELA, JUN, ATF1, NFKBIA	2.27×10^{-6}
GO:0051693, Actin filament capping	SPTB, SPTBN1, SPTAN1, SPTA1, ADD1, EPB49	2.97×10^{-6}
GO:0002756, MyD88-independent toll-like receptor signaling pathway	ATF2, CDK1, MEF2C, FOS, RELA, JUN, ATF1, NFKBIA	3.20×10^{-6}
GO:0015629, Actin cytoskeleton	WAS, CDH1, BRCA1, SPTB, SPTBN1, SPTAN1, CTDP1, STX1A, SPTA1, PAK1, SNCA, ADD1, EPB41, EPB49	4.25×10^{-6}
GO:0034134, Toll-like receptor 2 signaling pathway	ATF2, CDK1, MEF2C, FOS, RELA, JUN, ATF1, NFKBIA	4.43×10^{-6}
GO:0034130, Toll-like receptor 1 signaling pathway	ATF2, CDK1, MEF2C, FOS, RELA, JUN, ATF1, NFKBIA	4.43×10^{-6}
GO:0030835, Negative regulation of actin filament depolymerization	SPTB, SPTBN1, SPTAN1, SPTA1, ADD1, EPB49	4.72×10^{-6}

Table S6. *Cont.*

GO term	Genes	Adj. *p*-value
MIMAT0000443(hsa-miR-125a-5p)		
GO term	**Genes**	**Adj. *p*-value**
GO:0002755, MyD88-dependent toll-like receptor signaling pathway	ATF2, CDK1, MEF2C, FOS, RELA, JUN, ATF1, NFKBIA	6.41×10^{-6}
GO:0050679, Positive regulation of epithelial cell proliferation	FGF2, CTNNB1, AGGF1, HIF1A, TNFSF12, ARNT, MYC, JUN, FGF1	9.31×10^{-6}
GO:0034142, Toll-like receptor 4 signaling pathway	ATF2, CDK1, MEF2C, FOS, RELA, JUN, ATF1, NFKBIA	9.31×10^{-6}
GO:0030834, Regulation of actin filament depolymerization	SPTB, SPTBN1, SPTAN1, SPTA1, ADD1, EPB49	1.09×10^{-5}
GO:0030837, Negative regulation of actin filament polymerization	SPTB, SPTBN1, SPTAN1, SPTA1, ADD1, EPB49	2.37×10^{-5}
GO:0002224, Toll-like receptor signaling pathway	ATF2, CDK1, MEF2C, FOS, RELA, JUN, ATF1, NFKBIA	2.48×10^{-5}
GO:0008629, Induction of apoptosis by intracellular signals	PML, BRCA1, YWHAB, TP53, BID, CDKN1A, RNF7, IFI16	2.92×10^{-5}
GO:0050851, Antigen receptor-mediated signaling pathway	WAS, MEF2C, RELA, PSEN2, PTPRC, PAK1, PTEN, NFKBIA	8.55×10^{-5}
GO:0002221, Pattern recognition receptor signaling pathway	ATF2, CDK1, MEF2C, FOS, RELA, JUN, ATF1, NFKBIA	9.15×10^{-5}
GO:0001936, Regulation of endothelial cell proliferation	FGF2, AGGF1, CAV1, HIF1A, TNFSF12, ARNT, JUN	9.47×10^{-5}
MIMAT0000077(hsa-miR-22-3p)		
GO term	**Genes**	**Adj. *p*-value**
GO:0035583, Negative regulation of transforming growth factor beta receptor signaling pathway by extracellular sequestering of TGFbeta	FBN2, FBN1, LTBP1	1.45×10^{-5}
MIMAT0000265(hsa-miR-204-5p)		
GO term	**Genes**	**Adj. *p*-value**
GO:0035583, Negative regulation of transforming growth factor beta receptor signaling pathway by extracellular sequestering of TGFbeta	FBN1, FBN2, LTBP1	6.31×10^{-5}

Table S7. GOBO survival analysis results. Genes which was annotated with the specified GO term of proteins in the PIN would be used as input gene set for GOBO analysis. * $p < 0.05$, ** $p < 0.01$; *** $p < 0.001$.

miRNA	GO term	*p*-value
MIMAT0000064(hsa-let-7c)	**GO:0015630, Microtubule cytoskeleton**	**9.97×10^{-6} *****
	GO:0043067, Regulation of programmed cell death	0.268067
	GO:0043069, Negative regulation of programmed cell death	**0.0390439 ***
	GO:0060548, Negative regulation of cell death	**0.0390439 ***
MIMAT0000076(hsa-miR-21-5p)	GO:0007173, Epidermal growth factor receptor signaling pathway	0.721139
	GO:0043067, Regulation of programmed cell death	0.266312
MIMAT0000077(hsa-miR-22-3p)	GO:0035583, Negative regulation of transforming growth factor beta receptor signaling pathway by extracellular sequestering of TGFbeta	0.940727
MIMAT0000089(hsa-miR-31-5p)	GO:0003376, Sphingosine-1-phosphate signaling pathway	0.062202
	GO:0006917, Induction of apoptosis	**0.048584 ***
	GO:0007169, Transmembrane receptor protein tyrosine kinase signaling pathway	0.050408
	GO:0007187, G-protein signaling, coupled to cyclic nucleotide second messenger	0.289466
	GO:0007188, G-protein signaling, coupled to cAMP nucleotide second messenger	0.687572
	GO:0010942, Positive regulation of cell death	0.356228
	GO:0012502, Induction of programmed cell death	**0.048584 ***
	GO:0019933, CAMP-mediated signaling	0.687572
	GO:0019935, Cyclic-nucleotide-mediated signaling	0.289466
	GO:0043067, Regulation of programmed cell death	0.694486
	GO:0043068, Positive regulation of programmed cell death	0.356228
	GO:0048011, Nerve growth factor receptor signaling pathway	0.154543
MIMAT0000250(hsa-miR-139-5p)	GO:0035583, Negative regulation of transforming growth factor beta receptor signaling pathway by extracellular sequestering of TGFbeta	0.940727
MIMAT0000265(hsa-miR-204-5p)	GO:0035583, Negative regulation of transforming growth factor beta receptor signaling pathway by extracellular sequestering of TGFbeta	0.940727
MIMAT0000423(hsa-miR-125b-5p)	GO:0002224, Toll-like receptor signaling pathway	0.380928
	GO:0002755, MyD88-dependent toll-like receptor signaling pathway	0.380928
	GO:0002756, MyD88-independent toll-like receptor signaling pathway	0.380928
	GO:0006916, Anti-apoptosis	0.0593
	GO:0006917, Induction of apoptosis	0.618064
	GO:0007169, Transmembrane receptor protein tyrosine kinase signaling pathway	0.776269
	GO:0008284, Positive regulation of cell proliferation	0.882324
	GO:0008285, Negative regulation of cell proliferation	0.883393
	GO:0008629, Induction of apoptosis by intracellular signals	0.073118
	GO:0010942, Positive regulation of cell death	0.972892
	GO:0012502, Induction of programmed cell death	0.618064
	GO:0015629, Actin cytoskeleton	0.596528

Table S7. *Cont.*

miRNA	GO term	*p*-value
MIMAT0000423(hsa-miR-125b-5p)	**GO:0015630, Microtubule cytoskeleton**	**0.028245 ***
	GO:0030834, Regulation of actin filament depolymerization	0.654383
	GO:0030835, Negative regulation of actin filament depolymerization	0.654383
	GO:0030837, Negative regulation of actin filament polymerization	0.654383
	GO:0034130, Toll-like receptor 1 signaling pathway	0.380928
	GO:0034134, Toll-like receptor 2 signaling pathway	0.380928
	GO:0034138, Toll-like receptor 3 signaling pathway	0.380928
	GO:0034142, Toll-like receptor 4 signaling pathway	0.380928
	GO:0035666, TRIF-dependent toll-like receptor signaling pathway	0.380928
	GO:0043067, Regulation of programmed cell death	0.643418
	GO:0043068, Positive regulation of programmed cell death	0.972892
	GO:0043069, Negative regulation of programmed cell death	0.492576
	GO:0048011, Nerve growth factor receptor signaling pathway	0.171634
	GO:0050678, Regulation of epithelial cell proliferation	**0.002205 ****
	GO:0050679, Positive regulation of epithelial cell proliferation	0.205483
	GO:0051693, Actin filament capping	0.654383
	GO:0060548, Negative regulation of cell death	0.413665
MIMAT0000437(hsa-miR-145-5p)	GO:0007178, Transmembrane receptor protein serine/threonine kinase signaling pathway	0.196953
	GO:0030509, BMP signaling pathway	0.196953
	GO:0090092, Regulation of transmembrane receptor protein serine/threonine kinase signaling pathway	0.843529
MIMAT0000443(hsa-miR-125a-5p)	GO:0001936, Regulation of endothelial cell proliferation	0.115146
	GO:0002221, Pattern recognition receptor signaling pathway	0.380928
	GO:0002224, Toll-like receptor signaling pathway	0.380928
	GO:0002755, MyD88-dependent toll-like receptor signaling pathway	0.380928
	GO:0002756, MyD88-independent toll-like receptor signaling pathway	0.380928
	GO:0006916, Anti-apoptosis	0.0593
	GO:0006917, Induction of apoptosis	0.618064
	GO:0007169, Transmembrane receptor protein tyrosine kinase signaling pathway	0.776269
	GO:0008284, Positive regulation of cell proliferation	0.882324
	GO:0008285, Negative regulation of cell proliferation	0.883393
	GO:0008629, Induction of apoptosis by intracellular signals	0.073118
	GO:0010942, Positive regulation of cell death	0.972892
	GO:0012502, Induction of programmed cell death	0.618064

Table S7. *Cont.*

miRNA	GO term	*p*-value
MIMAT0000443(hsa-miR-125a-5p)	GO:0015629, Actin cytoskeleton	0.596528
	GO:0015630, Microtubule cytoskeleton	**0.028245** *
	GO:0030834, Regulation of actin filament depolymerization	0.654383
	GO:0030835, Negative regulation of actin filament depolymerization	0.654383
	GO:0030837, Negative regulation of actin filament polymerization	0.654383
	GO:0034130, Toll-like receptor 1 signaling pathway	0.380928
	GO:0034134, Toll-like receptor 2 signaling pathway	0.380928
	GO:0034138, Toll-like receptor 3 signaling pathway	0.380928
	GO:0034142, Toll-like receptor 4 signaling pathway	0.380928
	GO:0035666, TRIF-dependent toll-like receptor signaling pathway	0.380928
	GO:0043067, Regulation of programmed cell death	0.643418
	GO:0043068, Positive regulation of programmed cell death	0.972892
	GO:0043069, Negative regulation of programmed cell death	0.492576
	GO:0048011, Nerve growth factor receptor signaling pathway	0.171634
	GO:0050678, Regulation of epithelial cell proliferation	**0.002205** **
	GO:0050679, Positive regulation of epithelial cell proliferation	0.205483
	GO:0050851, Antigen receptor-mediated signaling pathway	0.103325
	GO:0051693, Actin filament capping	0.654383
	GO:0060548, Negative regulation of cell death	0.413665
MIMAT0002820(hsa-miR-497-5p)	GO:0001959, Regulation of cytokine-mediated signaling pathway	0.06699
	GO:0006917, Induction of apoptosis	0.142401
	GO:0007169, Transmembrane receptor protein tyrosine kinase signaling pathway	0.837635
	GO:0007173, Epidermal growth factor receptor signaling pathway	0.387447
	GO:0007178, Transmembrane receptor protein serine/threonine kinase signaling pathway	0.240167
	GO:0007179, Transforming growth factor beta receptor signaling pathway	**0.017876** *
	GO:0008284, Positive regulation of cell proliferation	0.430255
	GO:0008285, Negative regulation of cell proliferation	0.149994
	GO:0008543, Fibroblast growth factor receptor signaling pathway	0.521978
	GO:0010942, Positive regulation of cell death	0.237692
	GO:0012502, Induction of programmed cell death	0.142401
	GO:0015629, Actin cytoskeleton	0.228804
	GO:0015630, Microtubule cytoskeleton	0.18331
	GO:0017015, Regulation of transforming growth factor beta receptor signaling pathway	0.128773

Table S7. *Cont.*

miRNA	GO term	*p*-value
MIMAT0002820(hsa-miR-497-5p)	GO:0030509, BMP signaling pathway	0.39837
	GO:0030521, Androgen receptor signaling pathway	0.383811
	GO:0032956, Regulation of actin cytoskeleton organization	0.91762
	GO:0042058, Regulation of epidermal growth factor receptor signaling pathway	0.934045
	GO:0042059, Negative regulation of epidermal growth factor receptor signaling pathway	0.789492
	GO:0043067, Regulation of programmed cell death	0.111544
	GO:0043068, Positive regulation of programmed cell death	0.237692
	GO:0043069, Negative regulation of programmed cell death	0.856892
	GO:0048011, Nerve growth factor receptor signaling pathway	0.471986
	GO:0060548, Negative regulation of cell death	0.667437
	GO:0070302, Regulation of stress-activated protein kinase signaling cascade	0.561032
	GO:0090092, Regulation of transmembrane receptor protein serine/threonine kinase signaling pathway	0.182314
MIMAT0002856(hsa-miR-520d-3p)	GO:0006917, Induction of apoptosis	0.489781
	GO:0007169, Transmembrane receptor protein tyrosine kinase signaling pathway	0.171689
	GO:0007173, Epidermal growth factor receptor signaling pathway	0.449696
	GO:0008284, Positive regulation of cell proliferation	0.05916
	GO:0008286, Insulin receptor signaling pathway	0.237933
	GO:0008543, Fibroblast growth factor receptor signaling pathway	0.159318
	GO:0008629, Induction of apoptosis by intracellular signals	0.502822
	GO:0010942, Positive regulation of cell death	0.076906
	GO:0012502, Induction of programmed cell death	0.489781
	GO:0015630, Microtubule cytoskeleton	**0.000292** ***
	GO:0042058, Regulation of epidermal growth factor receptor signaling pathway	0.762633
	GO:0042059, Negative regulation of epidermal growth factor receptor signaling pathway	0.826854
	GO:0043067, Regulation of programmed cell death	0.740092
	GO:0043068, Positive regulation of programmed cell death	0.076906
	GO:0043069, Negative regulation of programmed cell death	0.067499
	GO:0048011, Nerve growth factor receptor signaling pathway	**0.014926** *
	GO:0051988, Regulation of attachment of spindle microtubules to kinetochore	**7.48×10^{-6}** ***
	GO:0060548, Negative regulation of cell death	0.213035

Table S8. Pathophysiogical characteristics of miRNA array data used in ROC curve analysis.

Sample Name	ER	PR	HER	TNM	Stage	Grade
S621T1	0	0	0	pT1N0M0	I	2
S434T1	1	0	1	T2N1M0	IIB	3
S403T1	0	1	0	T2N1M0	IIB	2
S459T1	1	0	0	T4N0M1	IV	1
S455N1	1	0	0	T3N3M1	IV	3
S545T1	0	0	0	pT2N0M0	IIA	3
S173N1	0	0	1	T2N3M0	IIIC	3
S363T1	1	1	0	T2N1M0	IIB	1
S909T1	1	1	1	pT3N3aM0	IIIC	(Unknown)
S645T1	0	1	1	pT1bN0M0	I	3
S898T1	1	0	0	pT2N0(i-)M0	IIA	(Unknown)
S201T1	1	0	0	T1N1M0	IIA	2
S631T1	1	1	1	T2N3aM1	IV	2
S303T1	0	0	1	T2N0M0	IIA	2
S502T1	0	0	0	pT3N0M0	IIB	3
S498N1	1	1	1	pT1cN1aM0	IIA	2
S536T1	1	0	1	T1cN1miM0	IIA	2
S660T1	0	0	1	T1N0M0	I	3
S358N1	0	0	1	T2pN2M0	IIIA	2
S665T1	0	0	0	T2N3M0	IIIC	3
S475T1	0	0	1	pT1cN0M0	I	3
S423T1	1	1	0	T2N3M0	IIIC	1
S507T1	0	0	0	pT1cN1aM0	IIA	3
S891T1	1	0	0	pT2NxM0	IIA	2
S422T1	1	1	0	T2N0M0	IIA	1
S961T1	0	0	1	T2N2aM0	IIIA	2
S622T1	0	0	0	pT2N0M0	IIA	2
S454T1	0	0	0	T1cN0M0	I	2
S433T1	1	0	0	T2N0M0	IIA	1
S673T1	0	0	0	pT2N0M0	IIA	2
S450T1	0	0	1	T2N1M0	IIB	3
S430T1	1	1	0	T2N3M0	IIIC	2
S437T1	1	0	0	T3N1M0	IIIA	3
S574T1	0	0	0	T2N0M0	IIA	2
S401T1	1	1	0	T1N0M0	I	1
S427N1	0	0	0	T3N1M0	IIIA	2
S894T1	0	0	0	pT1cN0M0	I	2
S929T1	1	0	0	pT2N0M0	IIA	2
S173T1	0	0	1	T2N3M0	IIIC	3
S622N1	0	0	0	pT2N0M0	IIA	2
S562T1	1	1	1	pT1aN0M0	I	3
S602T1	0	0	0	pT1cN0M0	I	3
S490T1	1	1	1	pT2N0M0	IIA	2
S677T1	0	0	0	pT2N1M0	IIB	3
S881T1	0	0	0	pT2N1aM0	IIB	3
S619T1	0	0	0	pT2N0M0	IIA	3
S446T1	0	0	0	T2N2M0	IIIA	3
S446T2	0	0	0	T2N2M0	IIIA	3
S453T1	1	1	1	T2N1M0	IIB	2
S562N1	1	1	1	pT1aN0M0	I	3
S557T1	0	0	0	pT2N0M0	IIA	3

Table S8. *Cont.*

Sample Name	ER	PR	HER	TNM	Stage	Grade
S594T1	1	1	1	pT2N3aM0	IIIC	3
S582T1	0	0	0	pT2N0M0	IIA	3
S358T1	0	0	1	T2pN2M0	IIIA	2
S368T1	0	0	1	T3N3M0	IIIC	2
S175T1	1	1	0	T3N3Mx	IIIC	3
S357T1	0	0	0	T2pN1M0	IIIC	3
S653T1	0	0	0	pT3N3aM0	IIIC	2
S722T1	1	1	1	pT1N0M0	I	3
S593T1	0	0	0	pT1N0M0	I	3
S543T1	1	1	1	pT2N1M0	IIB	2
S498T1	1	1	1	pT1cN1aM0	IIA	2
S389T1	1	0	0	T2N1M0	IIB	3
S614T1	0	1	1	TxN0M1	IIIA	(Unknown)
S536N1	1	0	1	T1cN1miM0	IIA	2
S462T1	1	1	0	T3N3M0	IIIC	2
S477T1	0	0	0	pT1cN0M0	I	3
S917T1	0	0	0	T2N0M0	IIA	(Unknown)
S213T1	0	0	1	T1cN1M0	IIA	3
S628T1	0	0	0	T1N0M0	I	2
S291T1	0	0	0	T3pN2M0	IIIA	3
S593N1	0	0	0	pT1N0M0	I	3
S418T1	1	1	0	T2N3M0	IIIC	2
S420T1	1	1	0	T1N0M0	I	2
S363N1	1	1	0	T2N1M0	IIB	1
S629T1	1	0	1	pT1N0M0	I	2
S586T1	1	1	1	pT2N3aM0	IIIC	2
S415T1	0	1	0	T2N2M0	IIIA	2
S439T1	1	1	0	T2N2M0	IIIA	2
S941N1	0	0	0	pT1cN1aM0	IIA	3
S893T1	0	0	0	pT2N1micM0	IIB	3
S380T1	1	0	0	T2N1M0	IIB	2
S906N1	1	1	1	pT3N3aM0	IIIC	2
S918T1	0	0	0	pT2N0M0	IIA	3
S400T1	0	0	1	T2N0M0	IIA	2
S328T1	0	0	0	T1cpN0M0	I	3
S367T1	1	0	0	T1N1M0	IIIA	1
S420N1	1	1	0	T1N0M0	I	2
S922T1	0	0	0	pT1cN0(i-)M0	I	3
S896T1	0	0	1	pT2N1aM0	IIB	2
S410T1	0	1	1	T1N0M0	I	3
S572T1	0	0	1	T4N1aM0	IIIC	3
S448T1	0	0	0	T1N0M0	I	3
S207T1	0	0	1	T2N0M0	IIA	3
S604T1	0	0	1	pT2N1M0	IIB	3
S379T1	0	0	1	T2N2M0	IIIA	2
S906T1	1	1	1	pT3N3aM0	IIIC	2
S941T2	0	0	0	pT1cN1aM0	IIA	3
S941T1	0	0	0	pT1cN1aM0	IIA	3
S375T1	1	1	0	T1N0M0	I	2
S427T1	0	0	0	T3N1M0	IIIA	2
S417T1	0	0	0	pT2N0M0	IIA	3

Table S8. *Cont.*

Sample Name	ER	PR	HER	TNM	Stage
S180T1	1	0	0	T4NxM0	IIIC
S455T1	1	0	0	T3N3M1	IV
S887T1	0	0	0	pT2N0M0	IIA
S445T1	1	0	1	T2N1M0	IIB
S469T1	1	1	0	T2N0M0	IIA
S469T2	1	1	0	T2N0M0	IIA
S483T1	1	1	0	T2N3M0	IIIC
S444T1	1	1	0	T1NxM0	(Unknown)
S909N1	1	1	1	pT3N3aM0	IIIC
S464T1	1	1	0	T1N0M0	I
S894N1	0	0	0	pT1cN0M0	I
S698T1	1	1	1	pT1cN0M0	I
S452T1	0	0	0	T1N0M0	I
S474T1	(Unknown)	(Unknown)	(Unknown)	(Unknown)	(Unknown)

References

1. Jemal, A.; Center, M.M.; DeSantis, C.; Ward, E.M. Global patterns of cancer incidence and mortality rates and trends. *Cancer Epidemiol. Biomark. Prev.* **2010**, *19*, 1893–1907.
2. Veronesi, U.; Boyle, P.; Goldhirsch, A.; Orecchia, R.; Viale, G. Breast cancer. *Lancet* **2005**, *365*, 1727–1741.
3. Cho, E.; Spiegelman, D.; Hunter, D.J.; Chen, W.Y.; Stampfer, M.J.; Colditz, G.A.; Willett, W.C. Premenopausal fat intake and risk of breast cancer. *J. Natl. Cancer Inst.* **2003**, *95*, 1079–1085.
4. Huttenhofer, A.; Schattner, P.; Polacek, N. Non-coding RNAs: Hope or hype? *Trends Genet.* **2005**, *21*, 289–297.
5. Kim, V.N.; Han, J.; Siomi, M.C. Biogenesis of small RNAs in animals. *Nat. Rev. Mol. Cell Biol.* **2009**, *10*, 126–139.
6. Kwak, P.B.; Iwasaki, S.; Tomari, Y. The microRNA pathway and cancer. *Cancer Sci.* **2010**, *101*, 2309–2315.
7. Hutvagner, G.; Zamore, P.D. A microRNA in a multiple-turnover RNAi enzyme complex. *Science* **2002**, *297*, 2056–2060.
8. Mourelatos, Z.; Dostie, J.; Paushkin, S.; Sharma, A.; Charroux, B.; Abel, L.; Rappsilber, J.; Mann, M.; Dreyfuss, G. miRNPs: A novel class of ribonucleoproteins containing numerous microRNAs. *Genes Dev.* **2002**, *16*, 720–728.
9. Nelson, P.T.; Hatzigeorgiou, A.G.; Mourelatos, Z. miRNP:mRNA association in polyribosomes in a human neuronal cell line. *RNA* **2004**, *10*, 387–394.
10. Meister, G.; Landthaler, M.; Patkaniowska, A.; Dorsett, Y.; Teng, G.; Tuschl, T. Human Argonaute2 mediates RNA cleavage targeted by miRNAs and siRNAs. *Mol. Cell* **2004**, *15*, 185–197.
11. Calin, G.A.; Dumitru, C.D.; Shimizu, M.; Bichi, R.; Zupo, S.; Noch, E.; Aldler, H.; Rattan, S.; Keating, M.; Rai, K.; *et al*. Frequent deletions and down-regulation of micro- RNA genes miR15 and miR16 at 13q14 in chronic lymphocytic leukemia. *Proc. Natl. Acad. Sci. USA* **2002**, *99*, 15524–15529.

12. Zhang, B.; Pan, X.; Cobb, G.P.; Anderson, T.A. microRNAs as oncogenes and tumor suppressors. *Dev. Biol.* **2007**, *302*, 1–12.
13. Meng, F.; Henson, R.; Wehbe-Janek, H.; Ghoshal, K.; Jacob, S.T.; Patel, T. MicroRNA-21 regulates expression of the PTEN tumor suppressor gene in human hepatocellular cancer. *Gastroenterology* **2007**, *133*, 647–658.
14. Rask, L.; Balslev, E.; Jorgensen, S.; Eriksen, J.; Flyger, H.; Moller, S.; Hogdall, E.; Litman, T.; Nielsen, B.S. High expression of miR-21 in tumor stroma correlates with increased cancer cell proliferation in human breast cancer. *APMIS* **2011**, *119*, 663–673.
15. Zaman, M.S.; Shahryari, V.; Deng, G.; Thamminana, S.; Saini, S.; Majid, S.; Chang, I.; Hirata, H.; Ueno, K.; Yamamura, S.; *et al.* Up-regulation of microRNA-21 correlates with lower kidney cancer survival. *PLoS One* **2012,** *7*, e31060.
16. Nadiminty, N.; Tummala, R.; Lou, W.; Zhu, Y.; Shi, X.B.; Zou, J.X.; Chen, H.; Zhang, J.; Chen, X.; Luo, J.; *et al.* MicroRNA let-7c is downregulated in prostate cancer and suppresses prostate cancer growth. *PLoS One* **2012,** *7*, e32832.
17. Wang, X.F.; Shi, Z.M.; Wang, X.R.; Cao, L.; Wang, Y.Y.; Zhang, J.X.; Yin, Y.; Luo, H.; Kang, C.S.; Liu, N.; *et al.* MiR-181d acts as a tumor suppressor in glioma by targeting K-ras and Bcl-2. *J. Cancer Res. Clin. Oncol.* **2012,** *138*, 573–584.
18. Zhang, Y.; Yan, L.X.; Wu, Q.N.; Du, Z.M.; Chen, J.; Liao, D.Z.; Huang, M.Y.; Hou, J.H.; Wu, Q.L.; Zeng, M.S.; *et al.* miR-125b is methylated and functions as a tumor suppressor by regulating the ETS1 proto-oncogene in human invasive breast cancer. *Cancer Res.* **2011**, *71*, 3552–3562.
19. Sachdeva, M.; Mo, Y.Y. MicroRNA-145 suppresses cell invasion and metastasis by directly targeting mucin 1. *Cancer Res.* **2010**, *70*, 378–387.
20. Guo, X.; Wu, Y.; Hartley, R.S. MicroRNA-125a represses cell growth by targeting HuR in breast cancer. *RNA Biol.* **2009**, *6*, 575–583.
21. Jin, H.; Tuo, W.; Lian, H.; Liu, Q.; Zhu, X.Q.; Gao, H. Strategies to identify microRNA targets: New advances. *N. Biotechnol.* **2010**, *27*, 734–738.
22. Friedman, R.C.; Farh, K.K.; Burge, C.B.; Bartel, D.P. Most mammalian mRNAs are conserved targets of microRNAs. *Genome Res.* **2009**, *19*, 92–105.
23. Huang, J.C.; Babak, T.; Corson, T.W.; Chua, G.; Khan, S.; Gallie, B.L.; Hughes, T.R.; Blencowe, B.J.; Frey, B.J.; Morris, Q.D. Using expression profiling data to identify human microRNA targets. *Nat. Methods* **2007**, *4*, 1045–1049.
24. Liu, B.; Li, J.; Tsykin, A.; Liu, L.; Gaur, A.B.; Goodall, G.J. Exploring complex miRNA-mRNA interactions with Bayesian networks by splitting-averaging strategy. *BMC Bioinforma.* **2009**, *10*, 408.
25. Tseng, C.W.; Lin, C.C.; Chen, C.N.; Huang, H.C.; Juan, H.F. Integrative network analysis reveals active microRNAs and their functions in gastric cancer. *BMC Syst. Biol.* **2011**, *5*, 99.
26. Farazi, T.A.; Horlings, H.M.; Ten Hoeve, J.J.; Mihailovic, A.; Halfwerk, H.; Morozov, P.; Brown, M.; Hafner, M.; Reyal, F.; van Kouwenhove, M.; *et al.* MicroRNA sequence and expression analysis in breast tumors by deep sequencing. *Cancer Res.* **2011**, *71*, 4443–4453.

27. Derfoul, A.; Juan, A.H.; Difilippantonio, M.J.; Palanisamy, N.; Ried, T.; Sartorelli, V. Decreased microRNA-214 levels in breast cancer cells coincides with increased cell proliferation, invasion and accumulation of the Polycomb Ezh2 methyltransferase. *Carcinogenesis* **2011**, *32*, 1607–1614.
28. Heyn, H.; Engelmann, M.; Schreek, S.; Ahrens, P.; Lehmann, U.; Kreipe, H.; Schlegelberger, B.; Beger, C. MicroRNA miR-335 is crucial for the BRCA1 regulatory cascade in breast cancer development. *Int. J. Cancer* **2011**, *129*, 2797–2806.
29. Sempere, L.F.; Christensen, M.; Silahtaroglu, A.; Bak, M.; Heath, C.V.; Schwartz, G.; Wells, W.; Kauppinen, S.; Cole, C.N. Altered MicroRNA expression confined to specific epithelial cell subpopulations in breast cancer. *Cancer Res.* **2007**, *67*, 11612–11620.
30. Zhu, X.M.; Wu, L.J.; Xu, J.; Yang, R.; Wu, F.S. Let-7c microRNA expression and clinical significance in hepatocellular carcinoma. *J. Int. Med. Res.* **2011**, *39*, 2323–2329.
31. Iyevleva, A.G.; Kuligina, E.; Mitiushkina, N.V.; Togo, A.V.; Miki, Y.; Imyanitov, E.N. High level of miR-21, miR-10b, and miR-31 expression in bilateral *vs.* unilateral breast carcinomas. *Breast Cancer Res. Treat.* **2012**, *131*, 1049–1059.
32. Rutnam, Z.J.; Yang, B.B. The involvement of microRNAs in malignant transformation. *Histol. Histopathol.* **2012**, *27*, 1263–1270.
33. Liu, H. MicroRNAs in breast cancer initiation and progression. *Cell Mol. Life Sci.* **2012**, *69*, 3587–3599.
34. Zhang, Z.J.; Ma, S.L. miRNAs in breast cancer tumorigenesis (Review). *Oncol. Rep.* **2012**, *27*, 903–910.
35. Mathivanan, S.; Periaswamy, B.; Gandhi, T.K.; Kandasamy, K.; Suresh, S.; Mohmood, R.; Ramachandra, Y.L.; Pandey, A. An evaluation of human protein-protein interaction data in the public domain. *BMC Bioinforma.* **2006**, *7*, S19.
36. Halbert, C.H.; Kumanyika, S.; Bowman, M.; Bellamy, S.L.; Briggs, V.; Brown, S.; Bryant, B.; Delmoor, E.; Johnson, J.C.; Purnell, J.; *et al*. Participation rates and representativeness of African Americans recruited to a health promotion program. *Health Educ. Res.* **2010**, *25*, 6–13.
37. Nicolas Diaz-Chico, B.; German Rodriguez, F.; Gonzalez, A.; Ramirez, R.; Bilbao, C.; Cabrera de Leon, A.; Aguirre Jaime, A.; Chirino, R.; Navarro, D.; Diaz-Chico, J.C. Androgens and androgen receptors in breast cancer. *J. Steroid Biochem. Mol. Biol.* **2007**, *105*, 1–15.
38. Sethi, N.; Kang, Y. Dysregulation of developmental pathways in bone metastasis. *Bone* **2011**, *48*, 16–22.
39. Liao, S.J.; Zhou, Y.H.; Yuan, Y.; Li, D.; Wu, F.H.; Wang, Q.; Zhu, J.H.; Yan, B.; Wei, J.J.; Zhang, G.M.; *et al*. Triggering of Toll-like receptor 4 on metastatic breast cancer cells promotes alphavbeta3-mediated adhesion and invasive migration. *Breast Cancer Res. Treat.* **2011**, *133*, 853–863.
40. Jiang, P.; Enomoto, A.; Takahashi, M. Cell biology of the movement of breast cancer cells: Intracellular signalling and the actin cytoskeleton. *Cancer Lett.* **2009**, *284*, 122–130.
41. Garcia, D.M.; Baek, D.; Shin, C.; Bell, G.W.; Grimson, A.; Bartel, D.P. Weak seed-pairing stability and high target-site abundance decrease the proficiency of lsy-6 and other microRNAs. *Nat. Struct. Mol. Biol.* **2011**, *18*, 1139–1146.
42. Lewis, B.P.; Burge, C.B.; Bartel, D.P. Conserved seed pairing, often flanked by adenosines, indicates that thousands of human genes are microRNA targets. *Cell* **2005**, *120*, 15–20.

43. Lall, S.; Grun, D.; Krek, A.; Chen, K.; Wang, Y.L.; Dewey, C.N.; Sood, P.; Colombo, T.; Bray, N.; Macmenamin, P.; *et al.* A genome-wide map of conserved microRNA targets in *C. elegans*. *Curr. Biol.* **2006**, *16*, 460–471.
44. Krek, A.; Grun, D.; Poy, M.N.; Wolf, R.; Rosenberg, L.; Epstein, E.J.; MacMenamin, P.; da Piedade, I.; Gunsalus, K.C.; Stoffel, M.; *et al.* Combinatorial microRNA target predictions. *Nat. Genet.* **2005**, *37*, 495–500.
45. Betel, D.; Wilson, M.; Gabow, A.; Marks, D.S.; Sander, C. The microRNA.org resource: Targets and expression. *Nucleic Acids Res.* **2008**, *36*, D149–D153.
46. Tusher, V.G.; Tibshirani, R.; Chu, G. Significance analysis of microarrays applied to the ionizing radiation response. *Proc. Natl. Acad. Sci. USA* **2001**, *98*, 5116–5121.
47. Keshava Prasad, T.S.; Goel, R.; Kandasamy, K.; Keerthikumar, S.; Kumar, S.; Mathivanan, S.; Telikicherla, D.; Raju, R.; Shafreen, B.; Venugopal, A.; *et al.* Human Protein Reference Database—2009 update. *Nucleic Acids Res.* **2009**, *37*, D767–D772.
48. Benjamini, Y.; Yekutieli, D. The control of the false discovery rate in multiple testing under dependency. *Ann. Stat.* **2001**, *29*, 1165–1188.
49. Ringner, M.; Fredlund, E.; Hakkinen, J.; Borg, A.; Staaf, J. GOBO: Gene expression-based outcome for breast cancer online. *PLoS One* **2011**, *6*, e17911.
50. Sing, T.; Sander, O.; Beerenwinkel, N.; Lengauer, T. ROCR: Visualizing classifier performance in R. *Bioinformatics* **2005**, *21*, 3940–3941.
51. Hanley, J.A.; McNeil, B.J. The meaning and use of the area under a receiver operating characteristic (ROC) curve. *Radiology* **1982**, *143*, 29–36.

10

Non-Coding RNAs: Multi-Tasking Molecules in the Cell

Anita Quintal Gomes [1,2], **Sofia Nolasco** [1,3,4] **and Helena Soares** [1,3,5,*]

[1] Health Technology College of Lisbon—Polytechnic Institute of Lisbon, 1990-096 Lisbon, Portugal; E-Mails: anita.gomes@estesl.ipl.pt (A.Q.G.); sofianolasco@fmv.utl.pt (S.N.)

[2] Institute of Molecular Medicine, Faculty of Medicine, University of Lisbon, 1649-028 Lisbon, Portugal

[3] Gulbenkian Science Institute, 2780-256 Oeiras, Portugal

[4] Interdisciplinary Centre of Research in Animal Health (CIISA), Faculty of Veterinary Medicine, 1300-666 Lisbon, Portugal

[5] Center for Chemistry and Biochemistry, Department of Chemistry and Biochemistry, Faculty of Sciences, University of Lisbon, 1749-016 Lisbon, Portugal

* Author to whom correspondence should be addressed; E-Mail: mhsoares@fc.ul.pt;

Abstract: In the last years it has become increasingly clear that the mammalian transcriptome is highly complex and includes a large number of small non-coding RNAs (sncRNAs) and long noncoding RNAs (lncRNAs). Here we review the biogenesis pathways of the three classes of sncRNAs, namely short interfering RNAs (siRNAs), microRNAs (miRNAs) and PIWI-interacting RNAs (piRNAs). These ncRNAs have been extensively studied and are involved in pathways leading to specific gene silencing and the protection of genomes against virus and transposons, for example. Also, lncRNAs have emerged as pivotal molecules for the transcriptional and post-transcriptional regulation of gene expression which is supported by their tissue-specific expression patterns, subcellular distribution, and developmental regulation. Therefore, we also focus our attention on their role in differentiation and development. SncRNAs and lncRNAs play critical roles in defining DNA methylation patterns, as well as chromatin remodeling thus having a substantial effect in epigenetics. The identification of some overlaps in their biogenesis pathways and functional roles raises the hypothesis that these molecules play concerted functions *in vivo*, creating complex regulatory networks where cooperation with regulatory proteins is necessary. We also highlighted the implications of biogenesis and gene expression deregulation of sncRNAs and lncRNAs in human diseases like cancer.

Keywords: sncRNAs; lncRNAs; miRNAs; siRNAs; piRNAs; gene expression regulation; epigenetic regulation

1. Introduction

1.1. The Incredible RNA Molecules

RNA has been known since the late 1800s, but its importance in cell functioning has long been in the shadow of DNA and proteins. In the 1950s, with the establishment of the molecular structure of DNA, it was proposed that RNA would be an intermediate molecule in the information flux between DNA and proteins. Later, this was experimentally demonstrated revealing that during gene expression, DNA is copied in a molecule of messenger RNA (mRNA) that is then translated into proteins with the help of other RNA molecules like transfer RNA (tRNA) and ribosomal RNAs (rRNAs). The idea that RNAs are much more than molecules involved in storage/transfer of information emerged with the discovery of ribozymes, RNA molecules that have, like proteins, active roles as catalysts of chemical reactions in cells. The two ribozymes identified first have RNAs as substrates and were the *Tetrahymena* intron of the 26S rRNA that is a self-sufficient catalytic unit capable of autoexcision and autocyclization [1], and the ribonucleoprotein, RNase P, an enzyme containing an RNA subunit essential for the catalysis required for the synthesis of tRNAs [2]. These discoveries clearly encouraged a variety of studies to search for potential new roles of RNA molecules *in vivo*, and led to the re-evaluation of RNAs as crucial molecules in the evolution of life. In view of the ability of RNAs to catalyze biological reactions, it is conceivable that the first organisms could rely only on RNA molecules and that only later an evolution of a more complex system based on proteins was established. This hypothesis gave support to the model of a primordial "RNA World" (for review [3,4]).

Progressively, the participation of RNAs in other critical molecular processes in eukaryotic cells was revealed, as in the case of DNA replication (RNA primers allow DNA polymerases to start the process), protein translation and RNA transcript maturation. For example, several ribosome functions required for protein synthesis were shown to be, at least in part, RNA-mediated, including peptidyl transferase activity [5], decoding functions [6], and the tRNA acceptor site interaction with 23S rRNA [7]. On the other hand, many small non-coding RNA molecules were isolated and characterized as being associated with proteins originating from ribonucleoprotein complexes (RNP), later identified as the components of the splicesome, including U1, U2, U4, U5 and U6 small nuclear RNA (snRNA) [8].

Furthermore, the information content of tRNA, rRNA and mRNA molecules can be biochemically altered after transcription by different molecular mechanisms that are generally designated by RNA editing [9]. These include sequence changes such as nucleoside modifications from C to U and A to I deaminations, as well as non-templated nucleotide additions and insertions. In general, RNA editing mechanisms are based on protein or protein-RNA complexes responsible for the RNA editing reaction and require a "guide RNA" molecule, which, through base-pairing with the target RNA molecule, determines the editing site. By this mechanism an mRNA sequence may be post-transcriptionally altered and consequently the amino acid sequence of the protein will then differ from that predicted by the genomic DNA sequence. Moreover, post-transcriptional processing and modifications of rRNAs are important for the production of efficient and accurate ribosomes which is directed by two large guide families of small nucleolar RNAs (snoRNA) [10].

In the mid-1980s Blackburn and Greider, demonstrated the existence of an enzymatic activity within cell extracts that added tandem hexanucleotides to chromosome ends and led to the discovery of telomerase [11]. Today it is well established that telomerase is a specialized reverse transcriptase that uses an internal RNA template sequence that is responsible for the synthesis of telomeric repeats [12]. More recently, Quiao and Cech [13] have described that the non-template RNA part of telomerase works together with the protein reverse-transcriptase motifs to facilitate catalysis, using a mechanism resembling that of pure ribozymes [13]. According to the "RNA first" model it was speculated that telomerase arose by the association of an ancient ribozyme with the reverse-transcriptase subunit. In light of this hypothesis the telomerase RNA may be a molecular fossil and telomerase a missing link in the evolution from RNA enzymes to protein enzymes [13]. This type of close functional collaboration is also observable in snoRNPs.

At this point the growing descriptions of the importance of RNA molecules for cell function started to push them to the limelight, but the complexity of their roles and the wide variety of molecular mechanisms where RNA molecules are critical players was still far from clear. In recent years, the use of genome wide approaches and the large output of genome sequencing technologies have revealed that the mammalian transcriptome is much more complex than previously thought since it includes a large number of small non-coding RNAs (sncRNAs) and long noncoding RNAs (lncRNAs) [14,15]. In most cases, these molecules present complex and precise patterns of expression during differentiation and development, tissue specificity, and some have been related to different pathophysiological states [16]. For example, it became clear that snoRNA guide families are widely diverse, which seems to be related to variant snoRNA structures and multiple cellular RNA targets, and consequently to cellular functions beyond ribosome biogenesis [10,17]. Indeed, snoRNAs have been recently implicated in alternative splicing and in cell transformation, tumorigenesis, and metastasis (for review, see [18]) showing that we are far from having a complete picture of their roles *in vivo*. Importantly, the observations that exogenously introduced double stranded RNA (dsRNA) molecules and plasmids expressing short hair-pin RNA (shRNA) specifically base-pairing with target mRNA molecules were able to trigger mRNA degradation (RNA interference -RNAi) [19,20] revealed, for the first time, that specific silencing pathways based on sncRNAs operate in eukaryotic cells. Moreover, these observations led to the development of the powerful RNA interference (RNAi) technique that has been extensively used in the study of gene function.

The aim of this review is to give a summarized overview of the biogenesis pathways of distinct classes of sncRNAs, including miRNA, piRNA, and siRNA, as well as lncRNAs, focusing on the miRNA and lncRNAs gene regulatory roles in distinct cellular functions and developmental regulatory programs. We will highlight the implications of the deregulation of miRNA and lncRNAs biogenesis pathways further illustrating the role of these molecules in the establishment of human diseases such as cancer. Finally, we will bring to discussion the fact that the pathways where distinct family members of sncRNAs and lncRNA funtion are probably interconnected, establishing a complex network of interactions and actions required for rapid and fine-tuned gene expression regulation at multiple levels.

1.2. The Small Non-Coding RNAs

Three classes of sncRNAs, namely short interfering RNAs (siRNAs), microRNAs (miRNAs) and PIWI-interacting RNAs (piRNAs), have been extensively studied in the last decade and have been associated with pathways that lead to silencing of specific genes and to the protection of the cell/genome against viruses, mobile repetitive DNA sequences, retro-elements and transposons [16].

1.2.1. siRNAs and miRNAs

The siRNAs and miRNAs (~20–30 nucleotides long) originate from double-stranded RNA (dsRNA) precursors that are introduced into, or produced endogenously by gene transcription of both sense and anti-sense DNA strands and of pseudogenes and inverted repeats. These molecules are critical in pathways involved in mRNA degradation, translational repression, or both, therefore regulating gene expression.

In the case of siRNAs, they are small RNA duplex molecules produced by the action of Dicer, a ribonuclease III (RNaseIII) enzyme that creates RNA duplexes with 2-nt overhangs at their 3' ends and phosphate groups at their 5' ends [21].

The miRNAs are mostly transcribed by RNA polymerase II as primary-miRNA (pri-miRNA) molecule precursors that possess a characteristic stem loop structure and are subsequently subjected to processing mechanisms [22]. In animals, the first step occurs in the nucleus where the RNaseIII Drosha acts over pri-mRNAs generating a pre-miRNA, a small RNA duplex of ~65–70 nucleotides containing the hair pin. This action can be facilitated by RNA processing proteins such as hnRNP A1 [23]. The pre-miRNAs are then exported to the cytoplasm by a nuclear transport receptor complex, exportin-5–RanGTP [24] where they are processed by Dicer into ~22-nt mature miRNAs (miRNA–miRNA* duplexes, where miRNA is the antisense, or guide/mature strand, and miRNA* is the sense, or passenger strand).

An alternative nuclear pathway for miRNA biogenesis was described in invertebrates [25] where the pre-miRNA is processed via splicing/spliceosome, instead of Drosha. Accordingly, spliced lariats linearized by the lariat debranching enzyme accept monophosphates and 3' hydroxyls, the same moieties found in pre-miRNAs, that were designated by-miRNAs/introns, “mirtrons” (for review [24]). These mirtrons are subsequently exported to the cytoplasm and processed by a Dicer protein.

The next step, for both siRNA and miRNA production, is the subsequent association with members of the Argonaute protein family that have diverged into specialized clades (or subfamilies), each recognizing different sncRNA types and conferring the specific features of the various silencing pathways operating in cells [26]. Argonaute loading occurs in the RNA-induced silencing complex (RISC)-loading complex, a ternary complex that consists of an Argonaute protein, Dicer and a dsRNA-binding protein (known as TRBP in humans). During loading, the non-guide strand is cleaved by an Argonaute protein [22].

The selection of the different Argonaute proteins seems to be based on the small interfering RNA duplex structure. For example, siRNAs that are perfect duplexes in terms of base pairing are loaded into Argonaute 2 (Ago2), whereas duplexes presenting mismatches, as in the case of miRNAs, are generally driven to Argonaute 1 (Ago1) [27,28]. When the complementarity between the miRNA bound to Ago1 and the target RNA is high, this causes miRNA tailing and 3'- to 5'-trimming. The discrimination between Ago1 and Ago2 seems to depend on the action of Hen1 an enzyme that adds the 2'-*O*-methyl group at the 3' ends of small RNAs bound to Ago2, but not those bound to Ago1 [29]. This methyl group is known to block tailing and trimming of the miRNA. The maturation and function of certain miRNAs can be also associated to enzymatic post-transcriptional modifications, like mono-uridylation [30]. These modifications will increase the variety of miRNAs and their precursor pools allowing more complex schemes of regulation in different backgrounds.

In the small RNA duplex of the siRNA the guide strand seems to be the one whose 5' end is less tightly paired to its complement [31]. In both siRNAs and miRNAs the guide strands drives the RISCs to the target mRNAs that contain complementary sequences thereby causing their degradation or translation inhibition (for review [16,32]). Recently, it has been shown that the target choice can also depend on accessory factors that interact with Dicer. For example, the *Drosophila* Loqs-PB Dicer-partner cleaves pre-miR-307a, generating a longer miRNA isoform with a distinct seed sequence and target specificity [33]. The mammalian TRBP homologue also acts together with Dicer to cleave pre-miR-132 generating a longer miRNA and consequently targets different mRNA molecules [34].

In fission yeast a specialized nuclear complex, known as the RNA-induced transcriptional silencing complex (RITS), mediates transcriptional gene silencing by inducing heterochromatin formation [35]. The RITS complex consists of Chp1 (H3K9me binding protein), Ago1, a poorly characterized protein Tas3, and siRNAs derived from centromeric repeat sequences [36]. These studies also showed the existence of a tight coupling of both siRNA and H3K9 methylation that appears to be important for the recruitment of RITS for heterochromatin assembly [37]. Therefore, there seems to exist a complex interplay between the RNAi pathway and the chromatin modifying machinery [37].

1.2.2. The piRNAs

The piRNAs are the least characterized class of sncRNAs and, contrary to the siRNAs and miRNAs that are widely expressed in different tissues and cell types, the piRNAs have been essentially detected in the germline cells of mammals, fish and *Drosophila melanogaster* [38,39] where they are important for germ line development and to suppress transposon activity. Mutations that disrupt the piRNA biogenesis pathway in mouse and fish cause germline-specific cell death and sterility, and are also associated with increased transposon expression [40].

The piRNAs (~24–31 nucleotides) got their name from the fact that they only associate to the PIWI subfamily of the Argonaute protein family (Piwi proteins). These sncRNAs usually have a uridine at the 5' end, hold a 5' monophosphate, and present a 2'-*O*-methyl (2'-*O*-Me) modification on the nucleotide at the 3' end (for review, see [32]).

Although not much is known concerning the intervening factors involved in piRNAs biogenesis pathways and transcription regulation, it is now well documented that they diverge from siRNAs and miRNAs by being generated by RNaseIII-independent pathways that do not involve dsRNA precursors. These sncRNAs are generated from long single-stranded precursors [41,42] that are preferentially cleaved at U residues and loaded onto Piwi proteins.

In *Drosophila*, as in mammals, the majority of piRNAs are transcribed from discrete genomic *loci* that are clustered in large pericentromeric or subtelomeric domains, generally spanning from 50–100 kb, and that comprise mainly various transposable DNA elements and their remnants [41].

Other piRNAs are derived from 3' UTRs of protein coding genes and dispersed euchromatin copies of transposable elements [41,43]. Most of these clusters are active specifically in germ cells, while only a single major cluster (*flamenco*) impels transposon silencing in the soma. Interestingly, if new transposons are introduced into piRNA clusters, and if they are heritable by the progeny, novel piRNAs will be produced that can lead to the control of the new transposons indicating that the mechanisms that drive adaptation to transposon invasion might be mediated by the piRNA pathway [44].

The critical role of piRNAs on transposon silencing was demonstrated by loss-of function mutations in *Drosophila* piRNAs and genes coding for the proteins involved in their biogenesis. In the germline these mutations cause a retro-transposition up-regulation causing the loss of germ cells and a variety of defects due to alterations in microtubule cytoskeleton polarization, with consequences to the polarized localization of specific proteins and mRNAs required for normal oogenesis [45]. However, it was found that the derepression of transposons activates the Chk2 DNA damage checkpoint [44] suggesting that the described phenotypes are probably an indirect consequence of transposon overexpression and DNA damage signaling (for review [44]).

Other studies also reveal that besides being involved in keeping genome integrity, a subset of piRNA genes have been implicated in the assembly of the telomere protection complex [46].

Detailed analysis of the small RNAs associated with the Piwi sub-family (PIWI, Aubergine and Argonaute 3) [41,47] in the *Drosophila* female germline showed that these sncRNAs have in their structure and sequences, signatures that give clues about their biogenesis. The most abundant piRNAs are mainly generated from the antisense strand of retro-transposons and these preferentially associate with Piwi and Aubergine proteins [41,47]. Those present in the single major somatic cluster are mainly originated from the sense strand and are associated with Argonaute 3 (Ago3).

The piRNAs from the germ cells seem to be generated by a self-amplifying loop designated by ping-pong cycle. Specifically, PIWI and/or Aubergine form complexes with antisense piRNAs that direct the slicing of sense strand transposon transcripts [41,47]. The sliced sense strands are then bound by Ago3, and this complex directs the slicing of antisense transposon transcripts [41]. A similar mechanism seems to operate in other animal genomes [42,48]. The piRNAs derived from genomic regions depleted of transposons, seem to be generated by a different pathway not completely understood called "primary processing" that operates in somatic cells, and may have a role in the regulation of target mRNAs (for review [32]).

Recent studies have shown that, in addition to their role in germ line transposon regulation and genome stability, piRNAs have a broader function in heterochromatin formation and developmental gene regulation. The analysis of a high-throughput small RNA sequencing data in *Drosophila*, mouse and rhesus macaque samples demonstrated that piRNAs are widespread and are abundant in other tissues as much as in the germline [49]. In fact, their involvement in the regulation of gene expression was demonstrated in *Drosophila*, where the degradation of a subset of maternal RNAs, *i.e.*, embryonic posterior morphogen Nanos (Nos), at the maternal-to-zygotic transition, was shown to require the zygotic expression of a piRNA cluster [50]. When this expression is inhibited, the Nos mRNA is stabilized which was accompanied by a reduced deadenylation and translational derepression, resulting in head development defects. Because the piRNAs involved in this regulation are produced from transposable elements, the authors suggested the existence of a direct developmental function for transposable elements in the regulation of gene expression through piRNAs [50].

The importance of the piRNAs pathway in the nervous system and in epigenetic regulation has also been gaining support. In the hippocampus, the inhibition of piRNAs causes a decrease of the dendrite spine area suggesting that these sncRNAs are required for spine morphogenesis [51]. More recently, in Aplysia sensory neurons, a Piwi/piRNA complex was described to facilitate the methylation of a conserved CpG island in the promoter of the transcriptional repressor of memory, CREB2, in a serotonin-dependent manner [52]. Consequently, this Piwi/piRNA complex is at the cross-roads between a transient external stimuli and alterations in the gene-expression of neurons involved in long term memory storage. Sienski *et al.* [53] have also shown that in *Drosophila* ovarian somatic cells piRNAs mediate the silencing of hundreds of transposon copies at the transcriptional level by establishing heterochromatic methylation of H3K9 on transposons and their genomic surroundings. The involvement of the piRNA pathway in *de novo* methylation of the differentially methylated region of the imprinted mouse *Rasgrf1 locus* [54] shows that the role of this pathway in methylation is also extendable to mammalian genomes.

There is also growing evidence that piRNA-pathway dependent mechanisms may have been critical during evolution, in the establishment of developmental robustness. In fact, the piRNA-pathway seems to be required for preventing phenotypic variation despite genotypic variation and environmental influences (canalization) [55]. The Hsp90 protein was previously described as a capacitator [56] being able to prevent phenotypic variation by suppressing the mutagenic activity of transposons [57]. Interestingly, it was shown in *Drosophila* that a protein complex composed of Hsp90, Piwi and Hop, is involved in canalization, probably through phosphorylation regulation of the Piwi protein by Hsp90 and Hop [55]. Therefore, it is possible that the Piwi-piRNA pathway will mediate canalization by both suppressing the generation of new genotypes and epigenetically silencing the expression of existing genetic variants [55].

The piRNAs, contrary to miRNAs, are less conserved through the eukaryotic lineage. This difference has been explained by the possible co-evolvement of miRNA with their RNA targets which have created sequence divergence constraints. There are increasing examples that piRNAs play roles in somatic cells regulating protein encoding genes. It is possible that piRNAs are more likely to be involved in epigenetic regulation rather than post-transcriptional regulation [58]. These puzzling facts suggest that our knowledge of the mechanistic relationships between piRNAs and the regulatory mechanisms based on regulatory proteins is far from being understood. On the other hand, the initial evidence that piRNAs may be involved in epigenetic regulation in tumorigenesis [59,60] requires additional attention.

The role of piRNAs in protecting genomes against parasitic nucleic acids seems to have developed early in evolution since ciliates present a mechanism that resembles that of piRNAs. Ciliates are single celled organisms that present a polyploid macronucleus that guarantees the vegetative growth of cells (the somatic nucleus) and the diploid micronucleus that is only active during sexual conjugation and constitutes the germline [61]. After conjugation the zygotic macronucleus differentiates from the micronucleus by undergoing an extensive developmentally programmed genome reorganization [62]. This reorganization involves chromosome fragmentation and elimination of germline limited sequences (internal eliminated sequences (IES), transposons and other repeated sequences) according to the pre-existing rearrangements of the maternal somatic genome. This seems to rely on a global comparison of the germline and somatic genomes and a genomic subtraction between meiosis-specific, germline scnRNAs (small RNAs that resemble piRNAs) and longer non-coding transcripts from the somatic genome (for review, see [63]). This mechanism that parallels the patterns of heterochromatin formation in other eukaryotes allows the maintenance of an epigenetic memory of rearrangement patterns across sexual generations and establishes, in an ancestral unicellular organism, a relationship between piRNAs and development. The ancestrality of the piRNAs and the fact that they have been placed in developmental frameworks being protagonists in the establishment of developmental robustness strongly supports the view that they have been critical factors in eukaryotic evolution.

2. Long Noncoding RNAs

It is now clear that the mammalian genome produces a large transcriptome of long noncoding RNA (lncRNA, defined as RNA >100 nucleotides in length). The number of gene members integrating this class of ncRNAs is still under debate and ranges from 10,000 to >200,000 [64].

The lncRNAs can be transcribed from intergenic regions, promoter regions or be interleaved, overlapping or antisense to annotated protein-coding genes [44]. There is also growing evidence that lncRNAs molecules might be produced by transcriptional active pseudogenes [65]. Although the majority of lncRNAs are transcribed from the nuclear genome, recently it was found that some can be generated from mitochondrial genomes [66]. Like coding genes, lncRNAs undergo post-transcriptional processing, including 5'capping, alternative splicing, RNA editing, and polyadenylation [67,68].

The referred transcriptional origins have been used to establish classification classes for lncRNA, as for example promoter-associated long RNAs (lpaRNAs) [68], natural antisense transcripts (NATs) or opposite-strand transcripts [69], large intervening noncoding RNA (lincRNA) [70], and enhancer associated RNAs (eRNA) [71,72]. However, other criteria should probably be used since frequently one lncRNA molecule can be associated with more than one class.

Mammalian genomes encode a large number of natural antisense transcripts (NATs) [64,73]. For instance, the FANTOM-3 mouse transcriptome sequencing consortium identified natural antisense transcripts for more than 70% of the transcription units, the majority of which represent non-protein-coding RNAs [73].

NATs have been defined as endogenous RNA molecules at least partially complementary to transcripts of known function [74]. NATs can be transcribed from the opposite strand at the same genomic locus of their sense counterparts and will present perfect sequence complementarity being designated by *cis*-NATs. On the other hand those transcribed from different genomic loci may have imperfect sequence complementarity and are named *trans*-NATs [75]. Sense and antisense RNA pairs can present different relative orientations and variable overlapping regions. For example, they can overlap by their 5' regions (5' to 5'), by their 3' regions (3' to 3'), or fully-overlap (one gene included within the region of the other) [76]. Antisense RNAs have a tendency to have lost introns and typically show lower abundance compared with sense transcripts [77].

Studies performed in various organisms have suggested that NATs can participate in a broad range of regulatory events that will be discussed later.

3. The Emerging Roles of lncRNAS and miRNAs

3.1. LncRNAs: Implications in Different Levels of Gene Expression Regulation and Differentiation

LncRNAs have emerged as pivotal molecules for the regulation of gene expression [76]. These transcripts are biologically relevant as supported by their cell-specific expression pattern [78], subcellular distribution [79], developmental regulation and possible association with human diseases.

LncRNAs encompass a wide variety of functions which include almost all levels of gene expression regulation, ranging from epigenetic to translational regulation, including transcriptional and post-transcriptional control. The main functions of lncRNAs are summarized below.

3.1.1. Epigenetic Regulation

lncRNAs modulate chromatin through the specific recruitment of histone and chromatin modifying complexes on one hand and by the recruitment of transcription factors on the other hand. X chromosome inactivation (XCI) is the classic example of the former type of regulation and is caused by the lncRNA "Xist" which physically associates with the Polycomb repressive complex 2 (PRC2) recruiting it to the X chromosome ultimately leading to its inactivation [80]. More precisely, it is a 1.6-kb ncRNA (RepA) within Xist that targets PRC2. Depletion of RepA abolishes full-length Xist induction and trimethylation on lysine 27 of histone H3 of the X (thus abolishing X inactivation). In addition it was demonstrated that PRC2 deficiency compromises Xist up-regulation [80]. A similar process to XCI is genomic imprinting, an epigenetic event in which genes are expressed from the allele of only one parent. One of the first lncRNAs to be identified was H19, which is reciprocally imprinted with insulin-like growth factor 2 (Igf2). Even though this lncRNA is highly expressed, its deletion has no phenotype and, in fact, recently it has been proposed to function as a microRNA precursor [81].

Other lncRNAs (*i.e.*, Air, Kcnq1ot1, HOTAIR) can control chromatin states in *cis* and/or in *trans*, thereby regulating gene expression through the association with chromatin-modifying complexes [82,83]. Specifically, HOTAIR is a *trans*-acting lncRNA that serves as a scaffold for two histone modification complexes: it binds both to polycomb repressive complex 2 (PRC2) and to LSD1 (in complex with CoREST/REST). This coordinates targeting of PCR2 and LSD1 to chromatin for coupled histone H3 at lysine 27 methylation and lysine 4 demethylation leading to subsequent gene silencing [84]. Also, in the plant *Arabidopsis* it was demonstrated that environmental conditions, such as cold, are able to induce the transcription of related NATs (*i.e.*, COOLAIR) that are involved in the silencing of a flower repressor *locus* designated by flowering locus c (*FLC*) [85]. More recently it was discovered that a lncRNA, named COLDAIR, that differs from COOLAIR by the fact that it is transcribed in the sense direction relative to *FLC* mRNA transcription, interacts directly with PRC2 and targets it to *FLC*, establishing an epigenetic memory [86]. Interestingly, winter cold triggers the methylation of H3 at *FLC* and it was shown that COLDAIR is induced by cold, demonstrating that lncRNas participate in the integration of signals from the environment to cell signaling pathways.

Other *trans*-acting lncRNAs have different functions some of which remain incompletely defined. For example, the p21-associated ncRNA DNA damage-activated (PANDA) lncRNA is induced upon DNA damage in a p53-dependent manner and it interacts with the transcription factor NF-YA to limit expression of pro-apoptotic genes [87]. Mistral is another example of an lncRNA that acts on the recruitment of the transcription factor MLL1 thereby activating Hoxa6 and Hoxa7 expression and subsequent stem cell differentiation [88].

Another group of lncRNAs that play a role in mammalian genomes are the long intergenic non-coding RNAs (lincRNAs) that range in size from ~300 nucleotides to several thousands and that, in humans, have been estimated to be around 3300, although a more correct number may be closer to 4500 [89]. This group of transcripts is heterogeneous but show significant evolutionary conservation relative to neutral sequences [70], which support the idea that they have important functions. In fact, it has been described that some groups of lincRNAs present expression patterns that correlate with those observed for protein-coding genes involved in cellular processes as diverse as cell-cycle regulation, innate immunity responses, and stem cell pluripotency [70,90]. In agreement, a reference catalog of 8195 human lincRNAs based on integratingRNA-seq data from 24 tissues and cell types showed that lincRNAs are expressed in a more tissue-specific manner than protein-coding genes [91]. By using co-immunoprecipitation and RNAi approaches it was also demonstrated that lincRNAs are associated with chromatin-modifying complexes to specific genomic loci to regulate gene expression [89]. The capacity to bind chromatin-modifying proteins or transcription factors, as exemplified, in combination with the abundance of lncRNAs suggests that lncRNAs may be part of a broad epigenetic regulatory network (reviewed in [92,93]).

3.1.2. Transcriptional Regulation

The discovery and characterization of several ncRNAs that are able to associate with promoters (promoter associated RNAs—paRNA) is also changing the traditional view of how genes encoding proteins are regulated at the transcriptional level. Promoter associated RNAs paRNAs are transcribed approximately from the start of or within the promoter, and include long, short and tiny RNA molecules (for review [49]). The long paRNAs were found at a single-gene level and were also associated with the modification of DNA methylation and demethylation patterns [94], inhibition of transposition expression in *Saccharomyces cerevisiae* [95] and gene expression in humans [96].

Interestingly, long (antisense) paRNAs have the potential to form double stranded molecules that can be processed into endo-siRNAs, and that, due to their sequence complementarity to that of a promoter, are able to induce transcriptional gene silencing [97–99] or activation [100–102] in a similar way to short paRNAs [49]. This picture is far from being complete since an increasing amount of experimental data are supporting the idea that enhancers can be transcribed and the resulting enhancer-non coding transcripts (eRNAs) may, in some cases, have functional roles, rather than represent mere transcriptional noise (for review see [103,104]).

On the other hand, lncRNAs can modulate the function of transcription factors by acting as co-regulators, modulators of transcription factors activity or by regulating the association and activity of co-regulators, among others. The ncRNA Evf-2, for example, functions as a co-activator for the homeobox transcription factor Dlx2, which plays important roles in forebrain development and neurogenesis [105]. Local ncRNAs can also recruit transcriptional factors and co-activating molecules to regulate adjacent protein-coding gene expression. The RNA binding protein TLS, binds to and inhibits the CREB binding protein (CCND1) and p300 histone acetyltransferase activities on a repressed gene target, cyclin D1. The recruitment of TLS to the promoter of cyclin D1 is directed by single stranded, low copy number lncRNA transcripts tethered to 5' regulatory regions of CCND1 in response to DNA damage signals [106].

Finally, lncRNAs also regulate the basal transcription machinery by targeting transcription factors required for the RNAP II transcription of all genes [107]. These general factors include components of the initiation complex that assemble on promoters or are involved in transcription elongation. An example of lncRNA-mediated regulation of basal transcription is the formation of a stable RNA-DNA triplex within the major promoter of the dihydrofolate reductase (DHFR) by an lncRNA that is transcribed from an upstream minor promoter of the DHFR gene. This complex prevents the binding of the transcriptional co-factor TFIIB [96].

3.1.3. Post-Transcriptional Regulation

LncRNAs can act on splicing, on mRNA stability and translation. It has been shown that lnc antisense RNA may bind to the sense RNA, masking the splice sites and thereby changing the balances between splice variants. Thyroid hormone receptor alpha gene (*TRα*) is an example where the antisense transcript *RevErbAα* influences splicing of *TRα1* and *TRα2* mRNAs [108]. Recently, it was discovered that a new class of sno-lncRNAs, whose ends correspond to positions of intronic snoRNA, are able to interact with the splicing factor Fox2 and alter splicing patterns [109]. The authors also showed that some of these sno-lncRNAs map to a genomic region that is deleted in the patients presenting Prader-Willi syndrome, strongly suggesting an association of these sno-lncRNAs with the disease. LncRNAs can also recruit proteins to mRNA to promote its degradation or stabilization. There's evidence for lncRNA binding to sequences present in the 3' UTR of specific mRNAs, thus creating a recognition site for Staufen, a protein that binds double-stranded mRNA and induces its decay [110]. By contrast, the lncRNA TINCR (terminal differentiation-induced ncRNA) also interacts with Staufen 1 but the complex between TINCR-STAU1 seems to mediate stabilization of mRNAs encoding differentiation factors such as Keratin 80 [111]. TINCR-mRNA interaction occurs through a motif of 25 nt that is abundantly present in target interacting mRNAs [111]. Another example is that of the mRNA of BACE1, a β-secretase responsible for β-amyloid production, that is stabilized and protected from RNase cleavage by base pairing of its antisense (BACE1-AS) [112]. Therefore, different lncRNAs are able to differentially regulate factors involved in mRNA stability regulation. Translational regulation is yet another proposed function for lncRNAs. Such is the case of the antisense for *PU.1* mRNA. *PU.1* mRNA translation is inhibited by a noncoding antisense transcript, which is a polyadenylated RNA with a lower concentration but a half-life longer than the sense *PU.1* transcript [113]. On the other hand the lncRNA Uchl1, shuttles from the nucleus to the cytoplasm under the control of the mTOR pathway and is involved in the translation up-regulation of the ubiquitin carboxy terminal hydrolase L1 (UCHL1) mRNA by promoting its association with polysomes [114]. Interestingly, the UCHL1 is a specific neuronal protein involved in rampamycin neuroprotective function and more generally in cellular stress response, that has been associated with neurodegenerative diseases. The various referred examples clearly show that lncRNAs present a vast repertoire of strategies to post-transcriptionally regulate protein encoding genes and different molecules are able to differentially modulate a specific regulatory molecule or pathway.

3.1.4. Modulation of mRNA Nuclear Trafficking and Control of Nuclear Compartmentalization

NRON is a non-coding repressor of nuclear factor of the activated T cells (NFAT), which interacts with multiple proteins including members of the importin-beta superfamily and likely functions as a specific regulator of NFAT nuclear trafficking [115].

The lncRNA nuclear-enriched autosomal transcript 1 (NEAT1), and abundant 4 kb ncRNA, is retained in nuclei *foci* that are coincident with "paraspeckles" [79]. It has been demonstrated that it contributes to the formation of these dynamic structures of the interchromatin space that are implicated in mRNA retention [79].

3.1.5. Formation of Endogenous siRNA

It has been described that NATs can originate siRNAs that will be involved in mRNA down-regulation. This mechanism requires the formation of a sense:anti-sense pair of transcripts that are then processed into siRNAs. This pair can be originated directly from the same *loci* (*cis*-NATs) or from different *loci* (*trans*-NATs). Interestingly, it was observed that certain *trans*-NATs are produced from pseudogene transcription. For example, in rice, a small number of pseudogenes are transcribed and processed into siRNAs, after pairing with the coding gene or a paralogous pseudogene transcript [116]. A similar observation was reported in mammals where pseudogene transcripts can be processed into small interfering RNAs (siRNA) with the ability to repress gene expression in mouse oocytes [117,118]. Therefore, NATs play their gene expression regulatory role through a mechanism equivalent to that of miRNAs and siRNAs (see Figure 1). Until recently, pseudogenes were envisaged only as copies of protein-coding genes that have lost the ability to produce functional proteins therefore constituting junk DNA in genomes [119]. Pseudogenes can be created by diverse processes, including: (1) spontaneous mutations, preventing transcription of the gene, or translation of the protein [120]; (2) duplication, in which pseudogenes are originated via tandem duplication or uneven crossing-over leading to the loss of promoters or enhancers or the appearance of crippling mutations such as frame shifts or premature stop codons [119]; and (3) retro-transposition, the mRNA transcript being reverse-transcribed and integrated into the genome at a new location originating retro-transposed or processed pseudogenes [121,122]. Therefore, their origin directly makes them prone to participate in post-transcriptional regulatory mechanisms promoted by lncRNAs. These observations started to change the vision that pseudogenes are mere junk in the genomes of organisms, and suggested that they can play important biological roles. In agreement with this hypothesis is the fact that the transcription of NATs is generally regulated in a tissue-specific manner and varying sense/antisense ratios are found [123].

It is evident that the critical roles of lncRNAs at different levels of gene expression regulation will largely contribute to establish differential profiles of gene expression required for development [124–127]. This is supported by the observation that lncRNAs such as Xist [128], TUG1 [129], PINC [130], and HOTAIR [131] have important roles in development. Moreover, Dinger *et al.* [90] using a microarray to examine the expression profiles of mouse embryonic stem cells differentiating as embryoid bodies over a 16 day time course have identified 945 ncRNAs, of which 174 were differentially expressed, many correlating with pluripotency or specific differentiation events [90]. Accordingly, it was also observed that the expression of some lincRNAs is increased in induced pluripotent stem cells (iPSCs) in comparison to those found in stem cells. This suggests that their activation may promote the emergence of iPSCs. It was also demonstrated that one of this lincRNA (lincRNARoR) modulates the reprogramming process leading to pluripotent stem cells [132].

Figure 1. Gene silencing: mRNA post-transcriptional regulation by lncRNA and miRNA. lncRNAs can be transcribed as natural antisense transcripts, from the same loci (*cis*-NAT, the same gene is transcribed in both directions) or from a different loci (*trans*-NAT, for example from a pseudogene). These NATs transcripts can pair with the coding transcripts, originating dsRNA molecules that will activate the siRNA machinery leading to mRNA degradation. miRNAs are also complementary of coding mRNAs and can pair with a perfect match leading to the activation of the siRNA machinery or they can pair with gaps leading to translation interference.

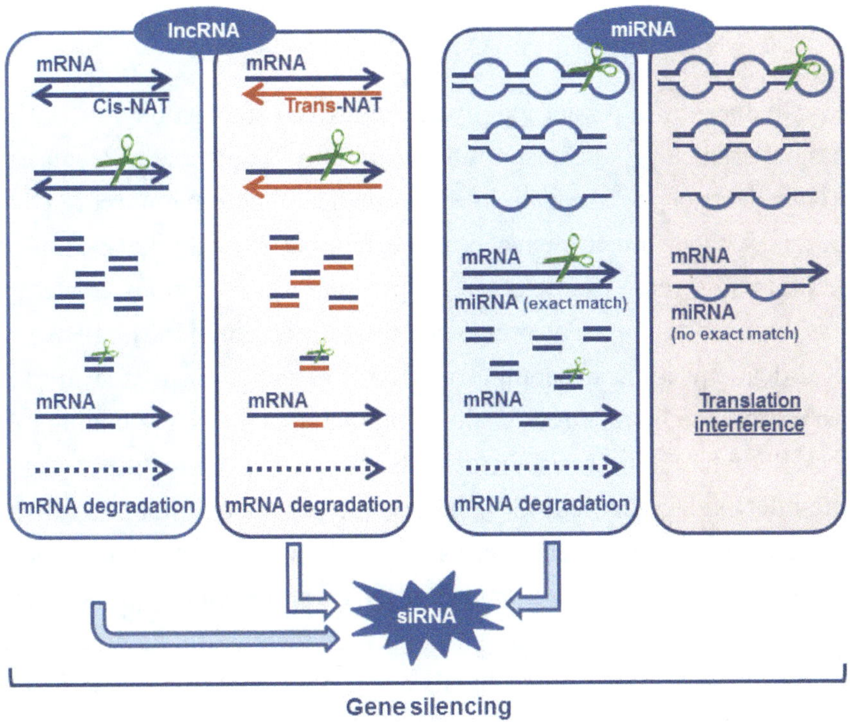

Spermatogenesis is a very complex developmental process that requires precise microtubule cytoskeleton remodeling, creating complex microtubule structures such as the manchette and the flagellum of the sperm [133]. During this process it was observed that the gene encoding TBCA, a protein that interacts with β-tubulin and is involved in the folding and dimerization of new tubulin heterodimers (the building blocks of microtubules) is regulated by a *Tbca* pseudogene that is transcribed in both directions [134]. The *Tbca* pseudogene is down-regulated leading to the increase of the *Tbca* mRNA, during testis maturation suggesting that this *Tbca* lncRNA is required for the undifferentiated state of spermatids. Similarly, the gene encoding the nitric oxide synthase protein (NOS2A) is transcribed into a noncoding RNA containing a region of significant antisense homology with the NOS2A mRNA. As in the case of *Tbca* lncRNA, the expression patterns of the anti-NOS2A RNA and the NOS2A mRNA exhibit opposite changes in undifferentiated human embryonic stem cells (hESCs) and in hESCs induced to differentiate into neurogenic precursors [74]. In conclusion, lncRNAs are clearly required to regulate programs of differentiation during development and seem to be generally associated with the undifferentiated states, repressing critical target genes whose expression is crucial for the cells to reach their fate.

3.2. miRNAs as Critical Regulators of Target Degradation and Translation

miRNAs act as sequence-specificity guides for the RNAi machinery to mediate repression of target gene expression. First identified as regulators of larval development in nematodes [19], miRNAs are now known to serve key roles in the regulation of almost every important cellular process in all multicellular eukaryotes. These include cell development, proliferation, differentiation, apoptosis and oncogenic transformation [135]. The genome of human cells encodes over 1000 miRNA species that regulate 60% of all protein-coding genes [32]. Most mRNA targets contain multiple miRNA binding sites, and each miRNA can regulate multiple genes. Therefore, the deregulation of miRNA levels might perturb the expression of many genes, thereby playing a key role in the occurrence of diseases (see below).

It is still unclear whether miRNAs act mainly at the mRNA translational or transcriptional levels. The miRNA repression, at the level of transcriptional inhibition, can occur as a consequence of mRNA decay, direct mRNA cleavage or through miRNA-mediated chromatin reorganization. Decay of targeted mRNA occurs without direct cleavage at the binding site. Unlike in translational inhibition where only a slight protein decrease can be obtained, protein level reductions greater than 33% indicate that mRNA decay is the major component of miRNA-driven silencing [136]. miRNA-mediated mRNA decay can occur via deadenylation, decapping or 5' to 3' degradation of the mRNA [137]. Dicer1, Ago1 and Ago2 were shown to be required for the rapid decay of mRNA containing AU-rich elements (AREs) in the 3' UTR of tumor necrosis factor-alpha suggesting that miRNA targeting of ARE is essential to mediate mRNA degradation [137]. It was also shown that upon GW182 interaction with AGO1, there is recruitment of deadenylases and decapping enzymes, leading to mRNA degradation [138]. The mRNA cleavage, another miRNA transcription repressive mechanism that is rare in animals, but frequent in plants, normally occurs when there is full complementarity between the miRNA and its mRNA target [139]. miRNAs also have the capacity to reorganize chromatin by increasing methylation of the targeted mRNA promoters thereby inhibiting their expression [140].

Finally, the repressed mRNAs, Ago proteins and miRNAs are frequently accumulated in processing bodies (P-bodies), which are cytoplasmic structures enriched in the mRNA degradation machinery but where the translational machinery is normally absent [141].

The second major mechanism of miRNAs activity includes repression of translation initiation and/or elongation, premature termination and nascent polypeptide degradation. Inhibition of translation initiation can occur at the level of cap-40S association or via 40S-AUG-60S association. Endogenous let-7 micro-ribonucleoproteins (miRNPs) or the tethering of Ago proteins to reporter mRNAs in human cells inhibit m(7)G-cap-dependent translation initiation, suggesting that miRNPs interfere with the recognition of the cap [142]. The cap-binding protein eukaryotic initiation factor 4E has in fact been proposed as a molecular target of miRNA function [143]. Ago2 represses the initiation of mRNA translation by directly binding to the m(7)G-cap of mRNA targets, thus likely precluding the recruitment of eIF4E [144]. Another Ago1 was shown to interact with GW182, this interaction being essential for miRNA-mediated inhibition of translation [145]. It was also shown that miRNA-repressed mRNAs contain 40S but not 60S components suggesting that miRNAs repress translation initiation by preventing the 60S subunit from joining to miRNA-targeted mRNAs [146]. It has also been reported that some miRNAs can inhibit translation initiation by inducing the formation of dense miRNPs (pseudo-polysomes) [147]. The fact that various studies showed repressed mRNA targets to be associated with polyribosomes seems to indicate that miRNAs can also repress translation at the elongation step [148–150]. Silencing by miRNAs can also occur before completion of the nascent polypeptide chain causing a decrease in translational read through at a stop codon, with ribosomes on repressed mRNAs dissociating more rapidly after a block of initiation of translation, than those of control mRNAs [149]. These observations pinpoint a role for miRNAs in ribosome drop-off-mediated repression.

Intriguingly, there is also evidence for transcriptional [100] and translational [151] activation by miRNAs. The miRNA-373 was shown to induce expression of genes with complementary promoter sequences [100]. miRNA-10a can bind to the 5' UTR of ribosomal protein mRNAs and enhances their translation [152]. Further, a growing series of studies has demonstrated that miRNAs and their associated complexes (microRNPs) elicit alternate functions that enable stimulation of gene expression in addition to their assigned repressive roles [151,153].

While the global importance of miRNAs is clearly illustrated by the developmental failure of Dicer-deficient embryonic stem cells (*in vitro*) and embryos (*in vivo*) [154], unique spatial and temporal expression patterns in distinct hematopoietic and neuronal lineages are clearly suggestive of multiple roles for miRNA in hematopoiesis, immune responses and neurological differentiation. The specific profiling of hPSCs by microarray and sequencing methods has allowed the identification of miRNAS that have potential roles in differentiation and development (reviewed in [155]). Several miRNA families, including the human (hsa)-miR-302, hsa-miR-106, hsa-miR-372, hsa-miR-17, hsa-miR-520, hsa-miR-195 and hsa-miR-200 families [155] were up-regulated specifically in hPSCs compared to mature differentiated cell types. Interestingly, the "seed" sequences (short sequence at nucleotides 2-8 on the 5' end of the miRNA that binds to the 3' UTRs of their target mRNAs) for most of these miRNAs are closely related, suggesting that these miRNA families may share mRNA targets. Thus, their regulatory functions might help maintain the unique characteristics of PSCs. Contrary to the miRNA families the hsa let-7 family [155] is expressed at significantly lower levels in hPSCs than in differentiated cells. The miRNA-dependent post-transcriptional gene regulation is also crucial for neural and immune cell development. Early evidence for miRNA function in the nervous system development came partly from knockout mutations of the miRNA processing genes present in the miRNA pathway. Pioneering studies of nervous system development using maternal-zygotic mutants of zebrafish dicer revealed gross morphological defects specifically in early brain patterning and morphogenesis [156]. Detailed studies of later stages in neural development have begun to suggest a more extensive contribution of miRNAs in the formation of synaptic connections, circuit maturation, and the activity-driven plasticity of these connections. For example, the mRNA processing enzyme DGCR8 mutant mice exhibited abnormalities in synaptic connectivity due to a reduction in the number and size of dendritic spines, reduced synaptic complexity, impaired synaptic transmission, and altered short-term plasticity [157].

In the immune system, miRNAs mediate the regulation of T cell development and function, as confirmed by the observation of defective thymic and peripheric T cell subsets in Dicer deficient mice [158,159].

Individual miRNAs play different roles at distinct developmental stages. For example, miR-125b and miR-132 regulate dendritic spine development. More specifically, miR-125b and miR-132 (as well as several other miRNA) are associated with fragile X mental retardation protein (FMRP) in mouse brain. The miR-125b overexpression results in longer, thinner processes of hippocampal neurons. FMRP knockdown is shown to ameliorate the effect of overexpressed miR-125b and miR-132 on spine morphology. It has been proposed that miR-125b negatively regulates its target, NR2A, along with FMRP and AGO1 [160].

Focusing on T cells, miRNA expression patterns vary among stages of development and T cell subsets, which indicate that these molecules may contribute to the identity of the cell subsets or their functional state [161]. Consistent with this, recent reports have demonstrated that various miRNAs, namely miR-101, miR-150, miR-155, miR181a, miR-29a, miR-146a and miR-326, are expressed in particular T cell subsets and regulate several aspects of their differentiation and function [162–164] .

Like lncRNAs, miRNAs are required to regulate differentiation programs during development. However, they are associated with both undifferentiated and differentiated states repressing target genes involved in maintaining those programs.

4. ncRNAs Active Players in Cancer and Other Human Diseases

The deregulation of gene expression networks, responsible for normal cellular identity, growth and differentiation leads to cancer. The large majority of genome-wide association studies (GWAS) identify cancer risk loci outside of protein-coding regions. Of 301 single-nucleotide polymorphisms (SNPs) currently linked to cancer, only 12 (3.3%) change the protein amino-acid sequence. Most are located in the introns of protein-coding genes (40%) or intergenic regions (44%), raising the question of the function of these noncoding loci and their role in cancer development [165]. These facts, associated with the observations that miRNA and lncRNAs are involved in programs of differentiation and development soon raise the hypothesis that alterations in their profiles of expression could be correlated with cancer development. In the last years, numerous evidences have confirmed this hypothesis since miRNas and long ncRNAs, that present tissue-specific expression, were found to be deregulated in distinct types of cancers. For example, data coming from microarray expression from a wide range of distinct cancers showed that alterations in miRNAs are almost always present in the analyzed tumors [166]. More specifically, overexpression of miR-155 was reported in hematopoietic cancers, breast, lung and colon cancer [167], whereas miR-21 was found to be overexpressed in glioblastoma and to have antiapoptotic properties [168,169]. Also, transgenic mice overexpressing miR-17-92 developed lymphoproliferative disorders [170] and retroviral overexpression of the cluster accelerated lymphoma formation. The miR-17-92 cluster was also found to be overexpressed in lung, colon and gastric cancer [171]. Like miRNAs, lncRNAs have also been associated with cancer development. For example, the lncRNA MALAT1 is up-regulated in several cancer types and its overexpression has been linked to an increase in cell proliferation and migration in lung and colorectal cancer cells [165]. These phenotypes are probably related to the role of MALAT1 in controlling alternative splicing of pre-mRNAs [172]. However, this relationship is probably too simplistic since a more recent study indicates that MALAT1 may also have a role in the regulation of gene expression, different from alternative splicing, in lung metastasis [173].

Many studies have also shown that miRNA and LncRNAs themselves can function as tumor suppressor genes or oncogenes, [174–176]. Several studies found that the tumor suppressor p53 transcriptionally regulates the three gene members of the miR-34 family. On the other hand, the miR-34 activation resembles p53 activity, including induction of cell-cycle arrest and promotion of apoptosis, and loss of miR-34 can impair p53-mediated apoptosis [177]. However, the interaction between p53 and miR-34 is much more complex since mice possessing the combined loss of all three miR-34 members are viable and fertile, do not display morphological defects and are not prone to spontaneous tumor formation [178].

Similarly to miRNAs, some lincRNAs are transcriptional targets of p53 like the lincRNA-p21 that plays a role as a transcriptional repressor in the p53 pathway by triggering apoptosis. The lincRNA-p21 binds to hnRNP-K that allows for the correct localization of hnRNP-K, probably by influencing their target preference, and therefore the transcriptional repression of p53-regulated genes [179]. The precise mechanism by which lincRNA-p21 contributes to repression at specific loci remains to be defined.

Although most of the mechanisms that implicate lncRNA in cancer biology are uncovered, the growing available data show that they are probably linked for example to chromatin remodeling. For example lncRNAs that are known to be involved in the recruitment of epigenetic modifiers to specific loci such as ANRIL, XIST, HOTAIR and KCNQ1OT1 were observed to have modified expression in a variety of cancers [176]. Also the lncRNA named TERRA, which binds telomerase, inhibiting its activity *in vitro* [180] is downregulated in many cancer cells which may be related to the longevity of cancer cells.

The broad functional classes of genes and regulatory pathways that involve ncRNA participation clearly justifies that the deregulation of their biogenesis and roles could are not restricted to cancer development (Figure 2). Perturbations in the biogenesis and actions of ncRNAs have also been associated with diverse neurodegenerative diseases such as Huntington's disease [181], Alzheimer [182] and Parkinson [183]. Moreover, recent studies have shown that miRNAs have unique expression profiles in cells of the innate and adaptive immune systems, suggesting that these molecules are important regulators of immune cell functions (reviewed in [184]). In fact, the role of miRNAs have been linked to autoimmune disorders (e.g., systemic lupus erythematosus, rheumatoid arthritis, multiple sclerosis, inflammatory bowel disease, psoriasis) and inflammatory pathologies of distinct organ (e.g., atherosclerosis, osteoarthritis, atopic eczema) and/or systemic locations like allergy. Chromatin remodeling by lncRNA is not exclusively related with cancer but is also linked to other diseases like facioscapulohumeral muscular dystrophy (FSHD) [185], lethal lung developmental disorder [186] and the HELLP syndrome, a pregnancy-associated disease [187].

The presented examples directly implicate long ncRNA and miRNAs in cancer biology and other human diseases and indicate that a complex interplay between their biogenesis pathways, their regulatory mechanisms and their targets should be seriously taken into consideration not only in cancer research but in other human pathologies and also in the definition of future strategies of diagnostics and therapeutics.

Figure 2. Diagram of functional relationships among lncRNAs, siRNAs, miRNas and piRNAs. This "venn diagram" depicts the specific function of each RNA molecule (inside each circle) as well as the shared functions (overlapping areas). Some of the disorders caused by deregulation in the expression patterns of these RNA molecules are indicated outside the circles.

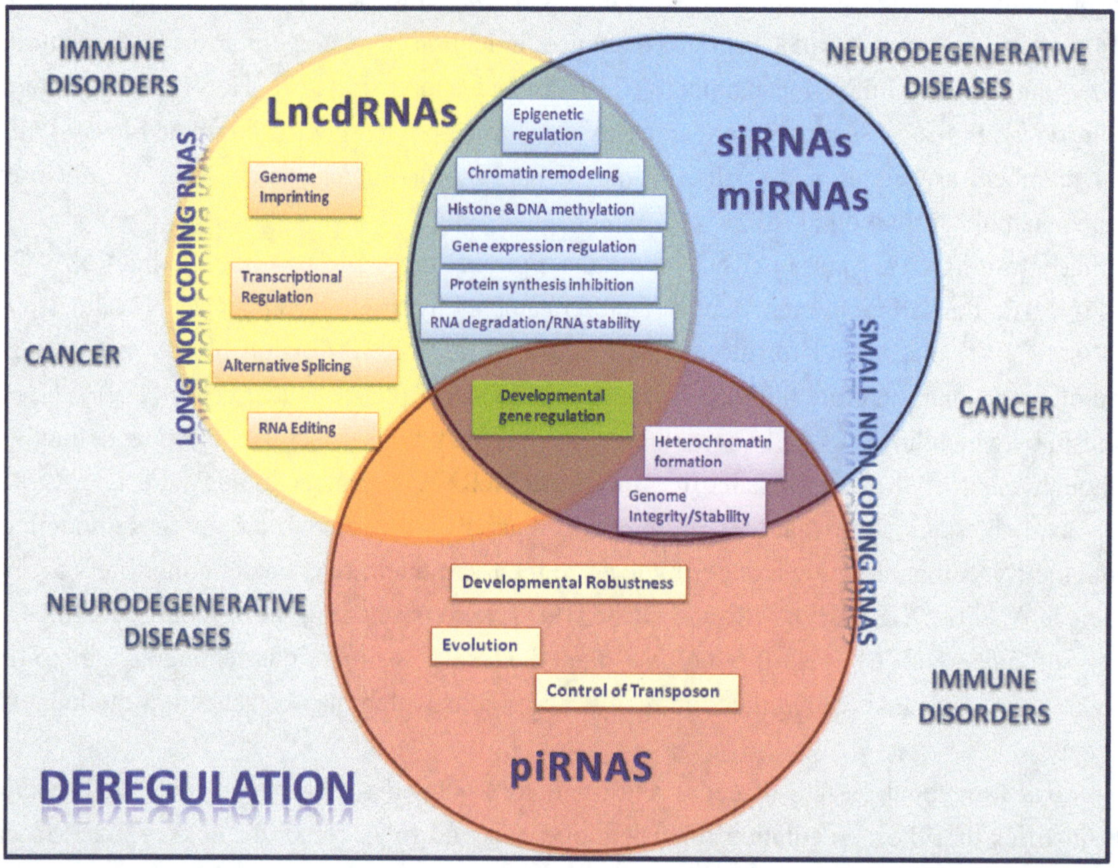

5. Concluding Remarks

In the last years we have witnessed an unprecedented discovery of numerous functions of non-coding RNAs in eukaryotic cells ranging from gene expression regulation to genome imprinting roles that were previously attributed to proteins.

This means that proteins are likely to cooperate with ncRNAs to control gene expression at different levels of regulation. They cooperate in the regulation of the transcription of genes encoding proteins, to process and maturate their transcripts and finally to regulate their mRNA stability and translation. Moreover, upstream of these regulatory steps, cooperation will also be required for altering DNA methylation profiles and the remodeling of chromatin contributing to epigenetic regulation. This means that complex networks between proteins and RNAs have been established during the course of evolution. We can envisage and speculate that due to their biochemical nature and biogenesis, ncRNAs will contribute to speed-up, make more flexible, transform and ultimately make more accurate the regulatory pathways conducted by regulatory proteins, pushing gene expression regulation to a new level. It is predictable that this complex regulatory web will have several hubs that will be composed of ncRNAs and proteins or alternatively only proteins or ncRNAs, which will also allow a rapid and better integration of different environmental signals. Although, the field of ncRNAs has been growing fast we are still far from understanding the complexity and the mechanisms underlying the establishment of the regulatory networks between RNAs and proteins.

From the evolutionary point of view it seems that the "invention" of proteins like telomerase was a critical step in the establishment of accurate spatial and temporal regulatory processes which probably allowed the evolution of eukaryotic complexity and later on, the appearance of multicellularity. In the view of an "RNA World hypothesis" it is tempting to speculate that the first RNA activities were to maintain the viability and integrity of "cell precursors" defending them from destructive invader molecules; these ancestral defense functions of ncRNAs that are still present and operate in "modern cells" seem to havebeen extended to mechanisms of gene regulation.

It should also be pointed out that the close analysis of different classes of ncRNAs (sncRNA and lncRNA), and the fact that we can detect biogenesis (see Figure 1) and functional overlaps between them (see Figure 2), strongly supports the idea that they could also have close interactions, not only at the level of their biogenesis pathways, but also at the functional level. This has been probably missed to a certain extent by the fact that they have been essentially separately studied.

From what has been compiled, the deregulation of biogenesis and functional roles of ncRNAs were, as expected at the crossroads of different human pathologies ranging from cancer to neurodegenerative and immune diseases. Finally the continued understanding of the molecular mechanisms and signaling pathways where ncRNAs participate should offer new insights to define new diagnostic strategies and open new avenues for therapies.

Acknowledgments

We are deeply in debt to João Gonçalves (Mount Sinai Hospital, Samuel Lunefeld Institute, Canada) for critical reviewing the manuscript. This work was supported by Fundação para a Ciência e a Tecnologia (FCT), PEst-OE/QUI/UI0612/2013.

References

1. Cech, T.R.; Zaug, A.J.; Grabowski, P.J. *In vitro* splicing of the ribosomal RNA precursor of Tetrahymena: Involvement of a guanosine nucleotide in the excision of the intervening sequence. *Cell* **1981**, *27*, 487–496.
2. Guerrier-Takada, C.; Gardiner, K.; Marsh, T.; Pace, N.; Altman, S. The RNA moiety of ribonuclease P is the catalytic subunit of the enzyme. *Cell* **1983**, *35*, 849–857.
3. Dworkin, J.P.; Lazcano, A.; Miller, S.L. The roads to and from the RNA world. *J. Theor. Boil.* **2003**, *222*, 127–134.
4. Lehman, N. RNA in evolution. *Wiley Interdiscip. Rev.* **2010**, *1*, 202–213.
5. Noller, H.F.; Hoffarth, V.; Zimniak, L. Unusual resistance of peptidyl transferase to protein extraction procedures. *Science* **1992**, *256*, 1416–1419.
6. Von Ahsen, U.; Noller, H.F. Identification of bases in 16S rRNA essential for tRNA binding at the 30S ribosomal P site. *Science* **1995**, *267*, 234–237.
7. Moazed, D.; Noller, H.F. Sites of interaction of the CCA end of peptidyl-tRNA with 23S rRNA. *Proc. Natl. Acad. Sci.USA* **1991**, *88*, 3725–3728.
8. Wassarman, D.A.; Steitz, J.A. Interactions of small nuclear RNA's with precursor messenger RNA during *in vitro* splicing. *Science* **1992**, *257*, 1918–1925.

9. Brennicke, A.; Marchfelder, A.; Binder, S. RNA editing. *FEMS Microbial. Rev* **1999**, *23*, 297–316.
10. Bachellerie, J.P.; Cavaille, J.; Huttenhofer, A. The expanding snoRNA world. *Biochimie* **2002**, *84*, 775–790.
11. Greider, C.W.; Blackburn, E.H. Identification of a specific telomere terminal transferase activity in *Tetrahymena* extracts. *Cell* **1985**, *43*, 405–413.
12. Greider, C.W.; Blackburn, E.H. A telomeric sequence in the RNA of *Tetrahymena* telomerase required for telomere repeat synthesis. *Nature* **1989**, *337*, 331–337.
13. Qiao, F.; Cech, T.R. Triple-helix structure in telomerase RNA contributes to catalysis. *Nat. Struct. Mol. Boil.* **2008**, *15*, 634–640.
14. Birney, E.; Stamatoyannopoulos, J.A.; Dutta, A.; Guigo, R.; Gingeras, T.R.; Margulies, E.H.; Weng, Z.; Snyder, M.; Dermitzakis, E.T.; Thurman, R.E.; *et al.* Identification and analysis of functional elements in 1% of the human genome by the ENCODE pilot project. *Nature* **2007**, *447*, 799–816.
15. Mattick, J.S.; Makunin, I.V. Non-coding RNA. *Hum. Mol. Genet.* **2006**, *15*, R17–R29.
16. Moazed, D. Small RNAs in transcriptional gene silencing and genome defence. *Nature* **2009**, *457*, 413–420.
17. Eliceiri, G.L. Small nucleolar RNAs. *Cell. Mol. Life Sci.* **1999**, *56*, 22–31.
18. Mannoor, K.; Liao, J.; Jiang, F. Small nucleolar RNAs in cancer. *Biochim. Biophys. Acta* **2012**, *1826*, 121–128.
19. Lee, R.C.; Feinbaum, R.L.; Ambros, V. The C. elegans heterochronic gene lin-4 encodes small RNAs with antisense complementarity to lin-14. *Cell* **1993**, *75*, 843–854.
20. Reinhart, B.J.; Slack, F.J.; Basson, M.; Pasquinelli, A.E.; Bettinger, J.C.; Rougvie, A.E.; Horvitz, H.R.; Ruvkun, G. The 21-nucleotide let-7 RNA regulates developmental timing in Caenorhabditis elegans. *Nature* **2000**, *403*, 901–906.
21. Meister, G.; Tuschl, T. Mechanisms of gene silencing by double-stranded RNA. *Nature* **2004**, *431*, 343–349.
22. Jinek, M.; Doudna, J.A. A three-dimensional view of the molecular machinery of RNA interference. *Nature* **2009**, *457*, 405–412.
23. Michlewski, G.; Guil, S.; Semple, C.A.; Cáceres, J.F. Posttranscriptional regulation of miRNAs harboring conserved terminal loops. *Mol. Cell* **2008**, *7*, 383–393.
24. Zeng, Y.; Cullen, B.R. Structural requirements for pre-microRNA binding and nuclear export by Exportin 5. *Nucleic Acids Res.* **2004**, *32*, 4776–4785.
25. Berezikov, E.; Chung, W.J.; Willis, J.; Cuppen, E.; Lai, E.C. Mammalian mirtron genes. *Mol. Cell* **2007**, 28, 328–336.
26. Farazi, T.A.; Juranek, S.A.; Tuschl, T. The growing catalog of small RNAs and their association with distinct Argonaute/Piwi family members. *Development* **2008**, *135*, 1201–1214.
27. Forstemann, K.; Horwich, M.D.; Wee, L.; Tomari, Y.; Zamore, P.D. Drosophila microRNAs are sorted into functionally distinct argonaute complexes after production by dicer-1. *Cell* **2007**, *130*, 287–297.
28. Tomari, Y.; Du, T.; Zamore, P.D. Sorting of *Drosophila* small silencing RNAs. *Cell* **2007**, *130*, 299–308.

29. Ameres, S.L.; Horwich, M.D.; Hung, J.H.; Xu, J.; Ghildiyal, M.; Weng, Z.; Zamore, P.D. Target RNA-directed trimming and tailing of small silencing RNAs. *Science* **2010**, *328*, 1534–1539.
30. Heo, I.; Ha, M.; Lim, J.; Yoon, M.J.; Park, J.E.; Kwon, S.C.; Chang, H.; Kim, V.N. Mono-uridylation of pre-microRNA as a key step in the biogenesis of group II let-7 microRNAs. *Cell* **2012**, *151*, 521–532.
31. Khvorova, A.; Reynolds, A.; Jayasena, S.D. Functional siRNAs and miRNAs exhibit strand bias. *Cell* **2003**, *115*, 209–216.
32. Siomi, H.; Siomi, M.C. On the road to reading the RNA-interference code. *Nature* **2009**, *457*, 396–404.
33. Fukunaga, R.; Han, B.W.; Hung, J.H.; Xu, J.; Weng, Z.; Zamore, P.D. Dicer partner proteins tune the length of mature miRNAs in flies and mammals. *Cell* **2012**, *151*, 533–546.
34. Lee, H.Y.; Doudna, J.A. TRBP alters human precursor microRNA processing *in vitro*. *RNA* **2012**, *18*, 2012–2019.
35. Grewal, S.I.; Elgin, S.C. Transcription and RNA interference in the formation of heterochromatin. *Nature* **2007**, *447*, 399–406.
36. Verdel, A.; Jia, S.; Gerber, S.; Sugiyama, T.; Gygi, S.; Grewal, S.I.; Moazed, D. RNAi-mediated targeting of heterochromatin by the RITS complex. *Science* **2004**, *303*, 672–676.
37. Creamer, K.M.; Partridge, J.F. RITS-connecting transcription, RNA interference, and heterochromatin assembly in fission yeast. *Wiley Interdiscip. Rev.* **2011**, *2*, 632–646.
38. Aravin, A.A.; Lagos-Quintana, M.; Yalcin, A.; Zavolan, M.; Marks, D.; Snyder, B.; Gaasterland, T.; Meyer, J.; Tuschl, T. The small RNA profile during *Drosophila melanogaster* development. *Dev. cell* **2003**, *5*, 337–350.
39. Lau, N.C.; Seto, A.G.; Kim, J.; Kuramochi-Miyagawa, S.; Nakano, T.; Bartel, D.P.; Kingston, R.E. Characterization of the piRNA complex from rat testes. *Science* **2006**, *313*, 363–367.
40. Aravin, A.A.; Hannon, G.J.; Brennecke, J. The Piwi-piRNA pathway provides an adaptive defense in the transposon arms race. *Science* **2007**, *318*, 761–764.
41. Brennecke, J.; Aravin, A.A.; Stark, A.; Dus, M.; Kellis, M.; Sachidanandam, R.; Hannon, G.J., Discrete small RNA-generating loci as master regulators of transposon activity in *Drosophila*. *Cell* **2007**, *128*, 1089–1103.
42. Houwing, S.; Kamminga, L.M.; Berezikov, E.; Cronembold, D.; Girard, A.; van den Elst, H.; Filippov, D.V.; Blaser, H.; Raz, E.; Moens, C.B.; *et al.* A role for Piwi and piRNAs in germ cell maintenance and transposon silencing in Zebrafish. *Cell* **2007**, *129*, 69–82.
43. Robine, N.; Lau, N.C.; Balla, S.; Jin, Z.; Okamura, K.; Kuramochi-Miyagawa, S.; Blower, M.D.; Lai, E.C. A broadly conserved pathway generates 3' UTR-directed primary piRNAs. *Curr. Biol.* **2009**, *19*, 2066–2076.
44. Khurana, J.S.; Theurkauf, W. piRNAs, transposon silencing, and *Drosophila* germline development. *J. Cell Biol.* **2010**, *191*, 905–913.
45. Chen, Y.; Pane, A.; Schupbach, T. Cutoff and aubergine mutations result in retrotransposon upregulation and checkpoint activation in *Drosophila*. *Curr. Biol.* **2007**, *17*, 637–642.
46. Khurana, J.S.; Xu, J.; Weng, Z.; Theurkauf, W.E. Distinct functions for the *Drosophila* piRNA pathway in genome maintenance and telomere protection. *PLoS Genet.* **2010**, *6*, e1001246.

47. Gunawardane, L.S.; Saito, K.; Nishida, K.M.; Miyoshi, K.; Kawamura, Y.; Nagami, T.; Siomi, H.; Siomi, M.C. A slicer-mediated mechanism for repeat-associated siRNA 5' end formation in Drosophila. *Science* **2007**, *315*, 1587–1590.
48. Ishizu, H.; Siomi, H.; Siomi, M.C. Biology of PIWI-interacting RNAs: New insights into biogenesis and function inside and outside of germlines. *Genes Dev.* **2012**, *26*, 2361–2373.
49. Yan, B.X.; Ma, J.X. Promoter-associated RNAs and promoter-targeted RNAs. *Cell. Mol. Life Sci.* **2012**, *69*, 2833–2842.
50. Rouget, C.; Papin, C.; Boureux, A.; Meunier, A.C.; Franco, B.; Robine, N.; Lai, E.C.; Pelisson, A.; Simonelig, M. Maternal mRNA deadenylation and decay by the piRNA pathway in the early Drosophila embryo. *Nature* **2010**, *467*, 1128–1132.
51. Lee, E.J.; Banerjee, S.; Zhou, H.; Jammalamadaka, A.; Arcila, M.; Manjunath, B.S.; Kosik, K.S. Identification of piRNAs in the central nervous system. *RNA* **2011**, *17*, 1090–1099.
52. Rajasethupathy, P.; Antonov, I.; Sheridan, R.; Frey, S.; Sander, C.; Tuschl, T.; Kandel, E.R. A role for neuronal piRNAs in the epigenetic control of memory-related synaptic plasticity. *Cell* **2012**, *149*, 693–707.
53. Sienski, G.; Donertas, D.; Brennecke, J. Transcriptional silencing of transposons by Piwi and maelstrom and its impact on chromatin state and gene expression. *Cell* **2012**, *151*, 964–980.
54. Watanabe, T.; Tomizawa, S.; Mitsuya, K.; Totoki, Y.; Yamamoto, Y.; Kuramochi-Miyagawa, S.; Iida, N.; Hoki, Y.; Murphy, P.J.; Toyoda, A.; *et al.* Role for piRNAs and noncoding RNA in *de novo* DNA methylation of the imprinted mouse Rasgrf1 locus. *Science* **2011**, *332*, 848–852.
55. Gangaraju, V.K.; Yin, H.; Weiner, M.M.; Wang, J.; Huang, X.A.; Lin, H. *Drosophila* Piwi functions in Hsp90-mediated suppression of phenotypic variation. *Nat. genet.* **2011**, *43*, 153–158.
56. Rutherford, S.L.; Lindquist, S. Hsp90 as a capacitor for morphological evolution. *Nature* **1998**, *396*, 336–342.
57. Specchia, V.; Piacentini, L.; Tritto, P.; Fanti, L.; D'Alessandro, R.; Palumbo, G.; Pimpinelli, S.; Bozzetti, M.P. Hsp90 prevents phenotypic variation by suppressing the mutagenic activity of transposons. *Nature* **2010**, *463*, 662–665.
58. Kim, V.N. Small RNAs just got bigger: Piwi-interacting RNAs (piRNAs) in mammalian testes. *Genes Dev.* **2006**, *20*, 1993–1997.
59. Siddiqi, S.; Matushansky, I. Piwis and piwi-interacting RNAs in the epigenetics of cancer. *J. Cell. Biochem.* **2012**, *113*, 373–380.
60. Siddiqi, S.; Terry, M.; Matushansky, I. Hiwi mediated tumorigenesis is associated with DNA hypermethylation. *PLoS One* **2012**, *7*, e33711.
61. Prescott, D.M., The DNA of ciliated protozoa. *Microbiol. Rev.* **1994**, *58*, 233–267.
62. Chalker, D.L. Dynamic nuclear reorganization during genome remodeling of *Tetrahymena*. *Biochim. Biophys. Acta* **2008**, *1783*, 2130–2136.
63. Duharcourt, S.; Lepere, G.; Meyer, E. Developmental genome rearrangements in ciliates: A natural genomic subtraction mediated by non-coding transcripts. *Trends Genet.* **2009**, *25*, 344–350.
64. Lee, J.T. Epigenetic regulation by long noncoding RNAs. *Science* **2012**, *338*, 1435–1439.
65. Pink, R.C.; Wicks, K.; Caley, D.P.; Punch, E.K.; Jacobs, L.; Carter, D.R. Pseudogenes: Pseudo-functional or key regulators in health and disease? *RNA* **2011**, *17*, 792–798.

66. Rackham, O.; Shearwood, A.M.; Mercer, T.R.; Davies, S.M.; Mattick, J.S.; Filipovska, A. Long noncoding RNAs are generated from the mitochondrial genome and regulated by nuclear-encoded proteins. *RNA* **2011**, *17*, 2085–2093.
67. Carninci, P.; Kasukawa, T.; Katayama, S.; Gough, J.; Frith, M.C.; Maeda, N.; Oyama, R.; Ravasi, T.; Lenhard, B.; Wells, C.; *et al.* The transcriptional landscape of the mammalian genome. *Science* **2005**, *309*, 1559–1563.
68. Kapranov, P.; Cheng, J.; Dike, S.; Nix, D.A.; Duttagupta, R.; Willingham, A.T.; Stadler, P.F.; Hertel, J.; Hackermuller, J.; Hofacker, I.L.; *et al.* RNA maps reveal new RNA classes and a possible function for pervasive transcription. *Science* **2007**, *316*, 1484–1488.
69. Magistri, M.; Faghihi, M.A.; St Laurent, G., 3rd; Wahlestedt, C. Regulation of chromatin structure by long noncoding RNAs: Focus on natural antisense transcripts. *Trends Genet.* **2012**, *28*, 389–396.
70. Guttman, M.; Amit, I.; Garber, M.; French, C.; Lin, M.F.; Feldser, D.; Huarte, M.; Zuk, O.; Carey, B.W.; Cassady, J.P.; *et al.* Chromatin signature reveals over a thousand highly conserved large non-coding RNAs in mammals. *Nature* **2009**, *458*, 223–227.
71. Orom, U.A.; Derrien, T.; Beringer, M.; Gumireddy, K.; Gardini, A.; Bussotti, G.; Lai, F.; Zytnicki, M.; Notredame, C.; Huang, Q.; *et al.* Long noncoding RNAs with enhancer-like function in human cells. *Cell* **2010**, *143*, 46–58.
72. Kim, T.K.; Hemberg, M.; Gray, J.M.; Costa, A.M.; Bear, D.M.; Wu, J.; Harmin, D.A.; Laptewicz, M.; Barbara-Haley, K.; Kuersten, S.; *et al.* Widespread transcription at neuronal activity-regulated enhancers. *Nature* **2010**, *465*, 182–187.
73. Katayama, S.; Tomaru, Y.; Kasukawa, T.; Waki, K.; Nakanishi, M.; Nakamura, M.; Nishida, H.; Yap, C.C.; Suzuki, M.; Kawai, J.; *et al.* Antisense transcription in the mammalian transcriptome. *Science* **2005**, *309*, 1564–1566.
74. Korneev, S.A.; Korneeva, E.I.; Lagarkova, M.A.; Kiselev, S.L.; Critchley, G.; O'Shea, M. Novel noncoding antisense RNA transcribed from human anti-NOS2A locus is differentially regulated during neuronal differentiation of embryonic stem cells. *RNA* **2008**, *14*, 2030–2037.
75. Lavorgna, G.; Dahary, D.; Lehner, B.; Sorek, R.; Sanderson, C.M.; Casari, G. In search of antisense. *Trends Biochem. Sci.* **2004**, *29*, 88–94.
76. Werner, A.; Carlile, M.; Swan, D. What do natural antisense transcripts regulate? *RNA Biol.* **2009**, *6*, 43–48.
77. He, Y.; Vogelstein, B.; Velculescu, V.E.; Papadopoulos, N.; Kinzler, K.W. The antisense transcriptomes of human cells. *Science* **2008**, *322*, 1855–1857.
78. Ravasi, T.; Suzuki, H.; Pang, K.C.; Katayama, S.; Furuno, M.; Okunishi, R.; Fukuda, S.; Ru, K.; Frith, M.C.; Gongora, M.M.; *et al.* Experimental validation of the regulated expression of large numbers of non-coding RNAs from the mouse genome. *Genome Res.* **2006**, *16*, 11–19.
79. Clemson, C.M.; Hutchinson, J.N.; Sara, S.A.; Ensminger, A.W.; Fox, A.H.; Chess, A.; Lawrence, J.B. An architectural role for a nuclear noncoding RNA: NEAT1 RNA is essential for the structure of paraspeckles. *Mol. Cell* **2009**, *33*, 717–726.
80. Zhao, J.; Sun, B.K.; Erwin, J.A.; Song, J.J.; Lee, J.T. Polycomb proteins targeted by a short repeat RNA to the mouse X chromosome. *Science* **2008**, *322*, 750–756.

81. Cai, X.; Cullen, B.R. The imprinted H19 noncoding RNA is a primary microRNA precursor. *RNA* **2007**, *13*, 313–316.
82. Hung, T.; Chang, H.Y. Long noncoding RNA in genome regulation: Prospects and mechanisms. *RNA Biol.* **2010**, *7*, 582–585.
83. Mohammad, F.; Mondal, T.; Guseva, N.; Pandey, G.K.; Kanduri, C. Kcnq1ot1 noncoding RNA mediates transcriptional gene silencing by interacting with Dnmt1. *Development* **2010**, *137*, 2493–2499.
84. Tsai, M.C.; Manor, O.; Wan, Y.; Mosammaparast, N.; Wang, J.K.; Lan, F.; Shi, Y.; Segal, E.; Chang, H.Y. Long noncoding RNA as modular scaffold of histone modification complexes. *Science* **2010**, *329*, 689–693.
85. Swiezewski, S.; Liu, F.; Magusin, A.; Dean, C. Cold-induced silencing by long antisense transcripts of an *Arabidopsis* polycomb target. *Nature* **2009**, *462*, 799–802.
86. Heo, J.B.; Sung, S. Vernalization-mediated epigenetic silencing by a long intronic noncoding RNA. *Science* **2011**, *331*, 76–79.
87. Hung, T.; Wang, Y.; Lin, M.F.; Koegel, A.K.; Kotake, Y.; Grant, G.D.; Horlings, H.M.; Shah, N.; Umbricht, C.; Wang, P.; *et al.* Extensive and coordinated transcription of noncoding RNAs within cell-cycle promoters. *Nat. Genet.* **2011**, *43*, 621–629.
88. Bertani, S.; Sauer, S.; Bolotin, E.; Sauer, F. The noncoding RNA Mistral activates Hoxa6 and Hoxa7 expression and stem cell differentiation by recruiting MLL1 to chromatin. *Mol. Cell* **2011**, *43*, 1040–1046.
89. Khalil, A.M.; Guttman, M.; Huarte, M.; Garber, M.; Raj, A.; Rivea Morales, D.; Thomas, K.; Presser, A.; Bernstein, B.E.; van Oudenaarden, A.; *et al.* Many human large intergenic noncoding RNAs associate with chromatin-modifying complexes and affect gene expression. *Proc. Natl. Acad. Sci. USA* **2009**, *106*, 11667–11672.
90. Dinger, M.E.; Amaral, P.P.; Mercer, T.R.; Pang, K.C.; Bruce, S.J.; Gardiner, B.B.; Askarian-Amiri, M.E.; Ru, K.; Solda, G.; Simons, C.; *et al.* Long noncoding RNAs in mouse embryonic stem cell pluripotency and differentiation. *Genome Res.* **2008**, *18*, 1433–1445.
91. Cabili, M.N.; Trapnell, C.; Goff, L.; Koziol, M.; Tazon-Vega, B.; Regev, A.; Rinn, J.L. Integrative annotation of human large intergenic noncoding RNAs reveals global properties and specific subclasses. *Genes Dev.* **2011**, *25*, 1915–1927.
92. Mercer, T.R.; Mattick, J.S. Structure and function of long noncoding RNAs in epigenetic regulation. *Nat. Struct. Mol. Biol.* **2013**, *20*, 300–307.
93. Su, W.Y.; Xiong, H.; Fang, J.Y. Natural antisense transcripts regulate gene expression in an epigenetic manner. *Biochem. Biophys. Res. Commun.* **2010**, *396*, 177–181.
94. Imamura, T.; Yamamoto, S.; Ohgane, J.; Hattori, N.; Tanaka, S.; Shiota, K. Non-coding RNA directed DNA demethylation of Sphk1 CpG island. *Biochem. Biophys. Res. Commun.* **2004**, *322*, 593–600.
95. Berretta, J.; Pinskaya, M.; Morillon, A. A cryptic unstable transcript mediates transcriptional *trans*-silencing of the Ty1 retrotransposon in *S. cerevisiae*. *Genes Dev.* **2008**, *22*, 615–626.
96. Martianov, I.; Ramadass, A.; Serra Barros, A.; Chow, N.; Akoulitchev, A. Repression of the human dihydrofolate reductase gene by a non-coding interfering transcript. *Nature* **2007**, *445*, 666–670.

97. Morris, K.V.; Chan, S.W.; Jacobsen, S.E.; Looney, D.J. Small interfering RNA-induced transcriptional gene silencing in human cells. *Science* **2004**, *305*, 1289–1292.
98. Napoli, S.; Pastori, C.; Magistri, M.; Carbone, G.M.; Catapano, C.V. Promoter-specific transcriptional interference and c-myc gene silencing by siRNAs in human cells. *EMBO J.* **2009**, *28*, 1708–1719.
99. Hawkins, P.G.; Santoso, S.; Adams, C.; Anest, V.; Morris, K.V. Promoter targeted small RNAs induce long-term transcriptional gene silencing in human cells. *Nucleic Acids Res.* **2009**, *37*, 2984–2995.
100. Place, R.F.; Li, L.C.; Pookot, D.; Noonan, E.J.; Dahiya, R. microRNA-373 induces expression of genes with complementary promoter sequences. *Proc. Natl. Acad. Sci. USA* **2008**, *105*, 1608–1613.
101. Li, L.C.; Okino, S.T.; Zhao, H.; Pookot, D.; Place, R.F.; Urakami, S.; Enokida, H.; Dahiya, R. Small dsRNAs induce transcriptional activation in human cells. *Proc. Natl. Acad. Sci. USA* **2006**, *103*, 17337–17342.
102. Janowski, B.A.; Younger, S.T.; Hardy, D.B.; Ram, R.; Huffman, K.E.; Corey, D.R. Activating gene expression in mammalian cells with promoter-targeted duplex RNAs. *Nat. Chem. Biol.* **2007**, *3*, 166–173.
103. Natoli, G.; Andrau, J.C. Noncoding transcription at enhancers: General principles and functional models. *Annu. Rev. Genet.* **2012**, *46*, 1–19.
104. Orom, U.A.; Shiekhattar, R. Long non-coding RNAs and enhancers. *Curr. Opin. Genet. Dev.* **2011**, *21*, 194–198.
105. Panganiban, G.; Rubenstein, J.L. Developmental functions of the Distal-less/Dlx homeobox genes. *Development* **2002**, *129*, 4371–4386.
106. Wang, X.; Arai, S.; Song, X.; Reichart, D.; Du, K.; Pascual, G.; Tempst, P.; Rosenfeld, M.G.; Glass, C.K.; Kurokawa, R. Induced ncRNAs allosterically modify RNA-binding proteins in *cis* to inhibit transcription. *Nature* **2008**, *454*, 126–130.
107. Goodrich, J.A.; Kugel, J.F. Non-coding-RNA regulators of RNA polymerase II transcription. *Nat. Rev.* **2006**, *7*, 612–616.
108. Hastings, M.L.; Ingle, H.A.; Lazar, M.A.; Munroe, S.H. Post-transcriptional regulation of thyroid hormone receptor expression by *cis*-acting sequences and a naturally occurring antisense RNA. *J. Boil. Chem.* **2000**, *275*, 11507–11513.
109. Yin, Q.F.; Yang, L.; Zhang, Y.; Xiang, J.F.; Wu, Y.W.; Carmichael, G.G.; Chen, L.L. Long noncoding RNAs with snoRNA ends. *Mol. Cell* **2012**, *48*, 219–230.
110. Gong, C.; Maquat, L.E. lncRNAs transactivate STAU1-mediated mRNA decay by duplexing with 3' UTRs via Alu elements. *Nature* **2011**, *470*, 284–288.
111. Kretz, M.; Siprashvili, Z.; Chu, C.; Webster, D.E.; Zehnder, A.; Qu, K.; Lee, C.S.; Flockhart, R.J.; Groff; A.F.; Chow, J.; *et al.* Control of somatic tissue differentiation by the long non-coding RNA TINCR. *Nature* **2013**, *493*, 231–235.
112. Faghihi, M.A.; Modarresi, F.; Khalil, A.M.; Wood, D.E.; Sahagan, B.G.; Morgan, T.E.; Finch, C.E.; St Laurent, G., 3rd; Kenny, P.J.; Wahlestedt, C. Expression of a noncoding RNA is elevated in Alzheimer's disease and drives rapid feed-forward regulation of beta-secretase. *Nat. Med.* **2008**, *14*, 723–730.

113. Ebralidze, A.K.; Guibal, F.C.; Steidl, U.; Zhang, P.; Lee, S.; Bartholdy, B.; Jorda, M.A.; Petkova, V.; Rosenbauer, F.; Huang, G.; *et al.* PU.1 expression is modulated by the balance of functional sense and antisense RNAs regulated by a shared *cis*-regulatory element. *Genes Dev.* **2008**, *22*, 2085–2092.
114. Carrieri, C.; Cimatti, L.; Biagioli, M.; Beugnet, A.; Zucchelli, S.; Fedele, S.; Pesce, E.; Ferrer, I.; Collavin, L.; Santoro, C.; *et al.* Long non-coding antisense RNA controls Uchl1 translation through an embedded SINEB2 repeat. *Nature* **2012**, *491*, 454–457.
115. Willingham, A.T.; Orth, A.P.; Batalov, S.; Peters, E.C.; Wen, B.G.; Aza-Blanc, P.; Hogenesch, J.B.; Schultz, P.G. A strategy for probing the function of noncoding RNAs finds a repressor of NFAT. *Science* **2005**, *309*, 1570–1573.
116. Guo, X.; Zhang, Z.; Gerstein, M.B.; Zheng, D. Small RNAs originated from pseudogenes: *cis*- or *trans*-acting? *PLoS Comput. Biol.* **2009**, *5*, e1000449.
117. Tam, O.H.; Aravin, A.A.; Stein, P.; Girard, A.; Murchison, E.P.; Cheloufi, S.; Hodges, E.; Anger, M.; Sachidanandam, R.; Schultz, R.M.; *et al.* Pseudogene-derived small interfering RNAs regulate gene expression in mouse oocytes. *Nature* **2008**, *453*, 534–538.
118. Watanabe, T.; Totoki, Y.; Toyoda, A.; Kaneda, M.; Kuramochi-Miyagawa, S.; Obata, Y.; Chiba, H.; Kohara, Y.; Kono, T.; Nakano, T.; *et al.* Endogenous siRNAs from naturally formed dsRNAs regulate transcripts in mouse oocytes. *Nature* **2008**, *453*, 539–543.
119. Mighell, A.J.; Smith, N.R.; Robinson, P.A.; Markham, A.F. Vertebrate pseudogenes. *FEBS Lett.* **2000**, *468*, 109–114.
120. Zhang, Z.D.; Frankish, A.; Hunt, T.; Harrow, J.; Gerstein, M. Identification and analysis of unitary pseudogenes: Historic and contemporary gene losses in humans and other primates. *Genome Biol.* **2010**, *11*, R26.
121. Maestre, J.; Tchenio, T.; Dhellin, O.; Heidmann, T. mRNA retroposition in human cells: Processed pseudogene formation. *EMBO J.* **1995**, *14*, 6333–6338.
122. D'Errico, I.; Gadaleta, G.; Saccone, C. Pseudogenes in metazoa: Origin and features. *Briefings Funct. Genomics Proteomics* **2004**, *3*, 157–167.
123. Werner, A.; Schmutzler, G.; Carlile, M.; Miles, C.G.; Peters, H. Expression profiling of antisense transcripts on DNA arrays. *Physiol. Genomics* **2007**, *28*, 294–300.
124. Fanarraga, M.L.; Parraga, M.; Aloria, K.; del Mazo, J.; Avila, J.; Zabala, J.C. Regulated expression of p14 (cofactor A) during spermatogenesis. *Cell Motil. Cytoskeleton* **1999**, *43*, 243–254.
125. Pavlicek, A.; Gentles, A.J.; Paces, J.; Paces, V.; Jurka, J. Retroposition of processed pseudogenes: The impact of RNA stability and translational control. *Trends Genet.* **2006**, *22*, 69–73.
126. Borsani, O.; Zhu, J.; Verslues, P.E.; Sunkar, R.; Zhu, J.K. Endogenous siRNAs derived from a pair of natural *cis*-antisense transcripts regulate salt tolerance in *Arabidopsis*. *Cell* **2005**, *123*, 1279–1291.
127. Okada, Y.; Tashiro, C.; Numata, K.; Watanabe, K.; Nakaoka, H.; Yamamoto, N.; Okubo, K.; Ikeda, R.; Saito, R.; Kanai, A.; *et al.* Comparative expression analysis uncovers novel features of endogenous antisense transcription. *Hum. Mol. Genet.* **2008**, *17*, 1631–1640.
128. Okamoto, I.; Arnaud, D.; le Baccon, P.; Otte, A.P.; Disteche, C.M.; Avner, P.; Heard, E. Evidence for *de novo* imprinted X-chromosome inactivation independent of meiotic inactivation in mice. *Nature* **2005**, *438*, 369–373.

129. Young, T.L.; Matsuda, T.; Cepko, C.L. The noncoding RNA taurine upregulated gene 1 is required for differentiation of the murine retina. *Curr. Biol.* **2005**, *15*, 501–512.
130. Ginger, M.R.; Shore, A.N.; Contreras, A.; Rijnkels, M.; Miller, J.; Gonzalez-Rimbau, M.F.; Rosen, J.M. A noncoding RNA is a potential marker of cell fate during mammary gland development. *Proc. Natl. Acad. Sci. USA* **2006**, *103*, 5781–5786.
131. Rinn, J.L.; Kertesz, M.; Wang, J.K.; Squazzo, S.L.; Xu, X.; Brugmann, S.A.; Goodnough, L.H.; Helms, J.A.; Farnham, P.J.; Segal, E.; *et al.* Functional demarcation of active and silent chromatin domains in human HOX loci by noncoding RNAs. *Cell* **2007**, *129*, 1311–1323.
132. Loewer, S.; Cabili, M.N.; Guttman, M.; Loh, Y.H.; Thomas, K.; Park, I.H.; Garber, M.; Curran, M.; Onder, T.; Agarwal, S.; *et al.* Large intergenic non-coding RNA-RoR modulates reprogramming of human induced pluripotent stem cells. *Nat. Genet.* **2010**, *42*, 1113–1117.
133. Kierszenbaum, A.L.; Tres, L.L. The acrosome-acroplaxome-manchette complex and the shaping of the spermatid head. *Arch. Histol. Cytol.* **2004**, *67*, 271–284.
134. Nolasco, S.; Bellido, J.; Goncalves, J.; Tavares, A.; Zabala, J.C.; Soares, H. The expression of tubulin cofactor A (TBCA) is regulated by a noncoding antisense Tbca RNA during testis maturation. *PLoS One* **2012**, *7*, e42536.
135. Bartel, D.P. microRNAs: Target recognition and regulatory functions. *Cell* **2009**, *136*, 215–233.
136. Lim, L.P.; Lau, N.C.; Garrett-Engele, P.; Grimson, A.; Schelter, J.M.; Castle, J.; Bartel, D.P.; Linsley, P.S.; Johnson, J.M. Microarray analysis shows that some microRNAs downregulate large numbers of target mRNAs. *Nature* **2005**, *433*, 769–773.
137. Jing, Q.; Huang, S.; Guth, S.; Zarubin, T.; Motoyama, A.; Chen, J.; di Padova, F.; Lin, S.C.; Gram, H.; Han, J. Involvement of microRNA in AU-rich element-mediated mRNA instability. *Cell* **2005**, *120*, 623–634.
138. Behm-Ansmant, I.; Rehwinkel, J.; Doerks, T.; Stark, A.; Bork, P.; Izaurralde, E. mRNA degradation by miRNAs and GW182 requires both CCR4: NOT deadenylase and DCP1:DCP2 decapping complexes. *Genes Dev.* **2006**, *20*, 1885–1898.
139. Llave, C.; Xie, Z.; Kasschau, K.D.; Carrington, J.C. Cleavage of Scarecrow-like mRNA targets directed by a class of *Arabidopsis* miRNA. *Science* **2002**, *297*, 2053–2056.
140. Kim, D.H.; Saetrom, P.; Snove, O., Jr.; Rossi, J.J. microRNA-directed transcriptional gene silencing in mammalian cells. *Proc. Natl. Acad. Sci. USA* **2008**, *105*, 16230–16235.
141. Liu, J.; Valencia-Sanchez, M.A.; Hannon, G.J.; Parker, R. microRNA-dependent localization of targeted mRNAs to mammalian P-bodies. *Nat. Cell Biol.* **2005**, *7*, 719–723.
142. Pillai, R.S.; Bhattacharyya, S.N.; Artus, C.G.; Zoller, T.; Cougot, N.; Basyuk, E.; Bertrand, E.; Filipowicz, W. Inhibition of translational initiation by Let-7 microRNA in human cells. *Science* **2005**, *309*, 1573–1576.
143. Humphreys, D.T.; Westman, B.J.; Martin, D.I.; Preiss, T. microRNAs control translation initiation by inhibiting eukaryotic initiation factor 4E/cap and poly(A) tail function. *Proc. Natl. Acad. Sci. USA* **2005**, *102*, 16961–16966.
144. Kiriakidou, M.; Tan, G.S.; Lamprinaki, S.; de Planell-Saguer, M.; Nelson, P.T.; Mourelatos, Z. An mRNA m7G cap binding-like motif within human Ago2 represses translation. *Cell* **2007**, *129*, 1141–1151.

145. Eulalio, A.; Huntzinger, E.; Izaurralde, E. GW182 interaction with Argonaute is essential for miRNA-mediated translational repression and mRNA decay. *Nat. Struct. Mol. Biol.* **2008**, *15*, 346–353.
146. Wang, B.; Yanez, A.; Novina, C.D. microRNA-repressed mRNAs contain 40S but not 60S components. *Proc. Natl. Acad. Sci. USA* **2008**, *105*, 5343–5348.
147. Thermann, R.; Hentze, M.W. *Drosophila* miR2 induces pseudo-polysomes and inhibits translation initiation. *Nature* **2007**, *447*, 875–878.
148. Maroney, P.A.; Yu, Y.; Fisher, J.; Nilsen, T.W. Evidence that microRNAs are associated with translating messenger RNAs in human cells. *Nat. Struct. Mol. Biol.* **2006**, *13*, 1102–1107.
149. Petersen, C.P.; Bordeleau, M.E.; Pelletier, J.; Sharp, P.A. Short RNAs repress translation after initiation in mammalian cells. *Mol. Cell* **2006**, *21*, 533–542.
150. Lytle, J.R.; Yario, T.A.; Steitz, J.A. Target mRNAs are repressed as efficiently by microRNA-binding sites in the 5' UTR as in the 3' UTR. *Proc. Natl. Acad. Sci. USA* **2007**, *104*, 9667–9672.
151. Vasudevan, S.; Steitz, J.A. AU-rich-element-mediated upregulation of translation by FXR1 and Argonaute 2. *Cell* **2007**, *128*, 1105–1118.
152. Orom, U.A.; Nielsen, F.C.; Lund, A.H. microRNA-10a binds the 5' UTR of ribosomal protein mRNAs and enhances their translation. *Mol. Cell* **2008**, *30*, 460–471.
153. Lee, S.; Vasudevan, S. Post-transcriptional stimulation of gene expression by microRNAs. *Adv. Exper. Med. Biol.* **2013**, *768*, 97–126.
154. Bernstein, E.; Kim, S.Y.; Carmell, M.A.; Murchison, E.P.; Alcorn, H.; Li, M.Z.; Mills, A.A.; Elledge, S.J.; Anderson, K.V.; Hannon, G.J. Dicer is essential for mouse development. *Nat. Genet.* **2003**, *35*, 215–217.
155. Leonardo, T.R.; Schultheisz, H.L.; Loring, J.F.; Laurent, L.C. The functions of microRNAs in pluripotency and reprogramming. *Nat. Cell Biol.* **2012**, *14*, 1114–1121.
156. Giraldez, A.J.; Cinalli, R.M.; Glasner, M.E.; Enright, A.J.; Thomson, J.M.; Baskerville, S.; Hammond, S.M.; Bartel, D.P.; Schier, A.F. microRNAs regulate brain morphogenesis in zebrafish. *Science* **2005**, *308*, 833–838.
157. Stark, K.L.; Xu, B.; Bagchi, A.; Lai, W.S.; Liu, H.; Hsu, R.; Wan, X.; Pavlidis, P.; Mills, A.A.; Karayiorgou, M.; *et al.* Altered brain microRNA biogenesis contributes to phenotypic deficits in a 22q11-deletion mouse model. *Nat. Genet.* **2008**, *40*, 751–760.
158. Cobb, B.S.; Nesterova, T.B.; Thompson, E.; Hertweck, A.; O'Connor, E.; Godwin, J.; Wilson, C.B.; Brockdorff, N.; Fisher, A.G.; Smale, S.T.; *et al.* T cell lineage choice and differentiation in the absence of the RNase III enzyme Dicer. *J. Exp. Med.* **2005**, *201*, 1367–1373.
159. Muljo, S.A.; Ansel, K.M.; Kanellopoulou, C.; Livingston, D.M.; Rao, A.; Rajewsky, K. Aberrant T cell differentiation in the absence of Dicer. *J. Exp. Med.* **2005**, *202*, 261–269.
160. Edbauer, D.; Neilson, J.R.; Foster, K.A.; Wang, C.F.; Seeburg, D.P.; Batterton, M.N.; Tada, T.; Dolan, B.M.; Sharp, P.A.; Sheng, M. Regulation of synaptic structure and function by FMRP-associated microRNAs miR-125b and miR-132. *Neuron* **2010**, *65*, 373–384.
161. Monticelli, S.; Ansel, K.M.; Xiao, C.; Socci, N.D.; Krichevsky, A.M.; Thai, T.H.; Rajewsky, N.; Marks, D.S.; Sander, C.; Rajewsky, K.; *et al.* microRNA profiling of the murine hematopoietic system. *Genome Biol.* **2005**, *6*, R71.

162. Rodriguez, A.; Vigorito, E.; Clare, S.; Warren, M.V.; Couttet, P.; Soond, D.R.; van Dongen, S.; Grocock, R.J.; Das, P.P.; Miska, E.A.; *et al.* Requirement of bic/microRNA-155 for normal immune function. *Science* **2007**, *316*, 608–611.
163. Boldin, M.P.; Taganov, K.D.; Rao, D.S.; Yang, L.; Zhao, J.L.; Kalwani, M.; Garcia-Flores, Y.; Luong, M.; Devrekanli, A.; Xu, J.; *et al.* miR-146a is a significant brake on autoimmunity, myeloproliferation, and cancer in mice. *J. Exp. Med.* **2011**, *208*, 1189–1201.
164. Steiner, D.F.; Thomas, M.F.; Hu, J.K.; Yang, Z.; Babiarz, J.E.; Allen, C.D.; Matloubian, M.; Blelloch, R.; Ansel, K.M. microRNA-29 regulates T-box transcription factors and interferon-gamma production in helper T cells. *Immunity* **2011**, *35*, 169–181.
165. Cheetham, S.W.; Gruhl, F.; Mattick, J.S.; Dinger, M.E. Long noncoding RNAs and the genetics of cancer. *Br. J. Cancer* **2013**, *108*, 2419–2425.
166. Munker, R.; Calin, G.A. microRNA profiling in cancer. *Clin. Sci.* **2011**, *121*, 141–158.
167. Faraoni, I.; Antonetti, F.R.; Cardone, J.; Bonmassar, E. miR-155 gene: A typical multifunctional microRNA. *Biochim. Biophys. Acta* **2009**, *1792*, 497–505.
168. Chan, J.A.; Krichevsky, A.M.; Kosik, K.S. microRNA-21 is an antiapoptotic factor in human glioblastoma cells. *Cancer Res.* **2005**, *65*, 6029–6033.
169. Volinia, S.; Calin, G.A.; Liu, C.G.; Ambs, S.; Cimmino, A.; Petrocca, F.; Visone, R.; Iorio, M.; Roldo, C.; Ferracin, M.; *et al.* A microRNA expression signature of human solid tumors defines cancer gene targets. *Proc. Natl. Acad. Sci. USA* **2006**, *103*, 2257–2261.
170. Xiao, C.; Srinivasan, L.; Calado, D.P.; Patterson, H.C.; Zhang, B.; Wang, J.; Henderson, J.M.; Kutok, J.L.; Rajewsky, K. Lymphoproliferative disease and autoimmunity in mice with increased miR-17-92 expression in lymphocytes. *Nat. Immunol.* **2008**, *9*, 405–414.
171. Concepcion, C.P.; Bonetti, C.; Ventura, A. The microRNA-17–92 family of microRNA clusters in development and disease. *Cancer J.* **2012**, *18*, 262–267.
172. Tripathi, V.; Ellis, J.D.; Shen, Z.; Song, D.Y.; Pan, Q.; Watt, A.T.; Freier, S.M.; Bennett, C.F.; Sharma, A.; Bubulya, P.A.; *et al.* The nuclear-retained noncoding RNA MALAT1 regulates alternative splicing by modulating SR splicing factor phosphorylation. *Mol. Cell* **2010**, *39*, 925–938.
173. Gutschner, T.; Hammerle, M.; Eissmann, M.; Hsu, J.; Kim, Y.; Hung, G.; Revenko, A.; Arun, G.; Stentrup, M.; Gross, M.; *et al.* The noncoding RNA MALAT1 is a critical regulator of the metastasis phenotype of lung cancer cells. *Cancer Res.* **2013**, *73*, 1180–1189.
174. Drakaki, A.; Iliopoulos, D. microRNA gene networks in oncogenesis. *Curr. Genomics* **2009**, *10*, 35–41.
175. Dalmay, T.; Edwards, D.R. microRNAs and the hallmarks of cancer. *Oncogene* **2006**, *25*, 6170–6175.
176. Gutschner, T.; Diederichs, S. The hallmarks of cancer: A long non-coding RNA point of view. *RNA Biol.* **2012**, *9*, 703–719.
177. He, L.; He, X.; Lim, L.P.; de Stanchina, E.; Xuan, Z.; Liang, Y.; Xue, W.; Zender, L.; Magnus, J.; Ridzon, D.; *et al.* A microRNA component of the p53 tumour suppressor network. *Nature* **2007**, *447*, 1130–1134.
178. Concepcion, C.P.; Han, Y.C.; Mu, P.; Bonetti, C.; Yao, E.; D'Andrea, A.; Vidigal, J.A.; Maughan, W.P.; Ogrodowski, P.; Ventura, A. Intact p53-dependent responses in miR-34-deficient mice. *PLoS Genet.* **2012**, *8*, e1002797.

179. Huarte, M.; Guttman, M.; Feldser, D.; Garber, M.; Koziol, M.J.; Kenzelmann-Broz, D.; Khalil, A.M.; Zuk, O.; Amit, I.; Rabani, M.; *et al.* A large intergenic noncoding RNA induced by p53 mediates global gene repression in the p53 response. *Cell* **2010**, *142*, 409–419.
180. Redon, S.; Reichenbach, P.; Lingner, J. The non-coding RNA TERRA is a natural ligand and direct inhibitor of human telomerase. *Nucleic Acids Res.* **2010**, *38*, 5797–5806.
181. Packer, A.N.; Xing, Y.; Harper, S.Q.; Jones, L.; Davidson, B.L. The bifunctional microRNA miR-9/miR-9* regulates REST and CoREST and is downregulated in Huntington's disease. *J. Neurosci.* **2008**, *28*, 14341–14346.
182. Cogswell, J.P.; Ward, J.; Taylor, I.A.; Waters, M.; Shi, Y.; Cannon, B.; Kelnar, K.; Kemppainen, J.; Brown, D.; Chen, C.; *et al.* Identification of miRNA changes in Alzheimer's disease brain and CSF yields putative biomarkers and insights into disease pathways. *J. Alzheimer's Disease* **2008**, *14*, 27–41.
183. Kim, J.; Inoue, K.; Ishii, J.; Vanti, W.B.; Voronov, S.V.; Murchison, E.; Hannon, G.; Abeliovich, A. A microRNA feedback circuit in midbrain dopamine neurons. *Science* **2007**, *317*, 1220–1224.
184. Contreras, J.; Rao, D.S. microRNAs in inflammation and immune responses. *Leukemia* **2012**, *26*, 404–413.
185. Cabianca, D.S.; Casa, V.; Bodega, B.; Xynos, A.; Ginelli, E.; Tanaka, Y.; Gabellini, D. A long ncRNA links copy number variation to a polycomb/trithorax epigenetic switch in FSHD muscular dystrophy. *Cell* **2012**, *149*, 819–831.
186. Szafranski, P.; Dharmadhikari, A.V.; Brosens, E.; Gurha, P.; Kolodziejska, K.E.; Zhishuo, O.; Dittwald, P.; Majewski, T.; Mohan, K.N.; Chen, B.; *et al.* Small noncoding differentially methylated copy-number variants, including lncRNA genes, cause a lethal lung developmental disorder. *Genome Res.* **2013**, *23*, 23–33.
187. Van Dijk, M.; Thulluru, H.K.; Mulders, J.; Michel, O.J.; Poutsma, A.; Windhorst, S.; Kleiverda, G.; Sie, D.; Lachmeijer, A.M.; Oudejans, C.B. HELLP babies link a novel lincRNA to the trophoblast cell cycle. *J. Clin. Invest.* **2012**, *122*, 4003–4011.

Permissions

All chapters in this book were first published by MDPI; hereby published with permission under the Creative Commons Attribution License or equivalent. Every chapter published in this book has been scrutinized by our experts. Their significance has been extensively debated. The topics covered herein carry significant findings which will fuel the growth of the discipline. They may even be implemented as practical applications or may be referred to as a beginning point for another development.

The contributors of this book come from diverse backgrounds, making this book a truly international effort. This book will bring forth new frontiers with its revolutionizing research information and detailed analysis of the nascent developments around the world.

We would like to thank all the contributing authors for lending their expertise to make the book truly unique. They have played a crucial role in the development of this book. Without their invaluable contributions this book wouldn't have been possible. They have made vital efforts to compile up to date information on the varied aspects of this subject to make this book a valuable addition to the collection of many professionals and students.

This book was conceptualized with the vision of imparting up-to-date information and advanced data in this field. To ensure the same, a matchless editorial board was set up. Every individual on the board went through rigorous rounds of assessment to prove their worth. After which they invested a large part of their time researching and compiling the most relevant data for our readers.

The editorial board has been involved in producing this book since its inception. They have spent rigorous hours researching and exploring the diverse topics which have resulted in the successful publishing of this book. They have passed on their knowledge of decades through this book. To expedite this challenging task, the publisher supported the team at every step. A small team of assistant editors was also appointed to further simplify the editing procedure and attain best results for the readers.

Apart from the editorial board, the designing team has also invested a significant amount of their time in understanding the subject and creating the most relevant covers. They scrutinized every image to scout for the most suitable representation of the subject and create an appropriate cover for the book.

The publishing team has been an ardent support to the editorial, designing and production team. Their endless efforts to recruit the best for this project, has resulted in the accomplishment of this book. They are a veteran in the field of academics and their pool of knowledge is as vast as their experience in printing. Their expertise and guidance has proved useful at every step. Their uncompromising quality standards have made this book an exceptional effort. Their encouragement from time to time has been an inspiration for everyone.

The publisher and the editorial board hope that this book will prove to be a valuable piece of knowledge for researchers, students, practitioners and scholars across the globe.

List of Contributors

Anna Musto, Angelica Navarra, Giuseppina Minopoli, Tommaso Russo and Silvia Parisi
Department of Molecular Medicine and Medical Biotechnology, University of Naples, "Federico II", Via Sergio Pansini 5, 80131 Naples, Italy
Ceinge Biotecnologie Avanzate, Via Gaetano Salvatore 486, 80145 Naples, Italy

Marica Battista
Department of Molecular Medicine and Medical Biotechnology, University of Naples, "Federico II", Via Sergio Pansini 5, 80131 Naples, Italy
Ceinge Biotecnologie Avanzate, Via Gaetano Salvatore 486, 80145 Naples, Italy
European School of Molecular Medicine (SEMM), 80145 Naples, Italy

Franck Morceau, Sébastien Chateauvieux, Anthoula Gaigneaux and Mario Dicato
Laboratoire de Biologie Moléculaire et Cellulaire du Cancer, Kirchberg Hospital, 9 Rue Edward Steichen, 2540, Luxembourg

Marc Diederich
Department of Pharmacy, College of Pharmacy, Seoul National University, Seoul 151-742, Korea

Maren Thomas, Kerstin Lange-Grünweller, Dorothee Hartmann, Lara Golde, Julia Schlereth, Dennis Streng, Arnold Grünweller and Roland K. Hartmann
Institut für Pharmazeutische Chemie, Philipps-Universität Marburg, 35032 Marburg, Germany

Achim Aigner
Medizinische Fakultät, Rudolf-Boehm-Institut für Pharmakologie und Toxikologie, Klinische Pharmakologie, Universität Leipzig, 04107 Leipzig, Germany

Raymond Wai-Ming Lung and Joanna Hung-Man Tong
Department of Anatomical and Cellular Pathology, State Key Laboratory in Oncology in South China, The Chinese University of Hong Kong, Hong Kong, China

Ka-Fai To
Department of Anatomical and Cellular Pathology, State Key Laboratory in Oncology in South China, The Chinese University of Hong Kong, Hong Kong, China
Institute of Digestive Diseases, Li Ka-Shing Institute of Health Sciences, The Chinese University of Hong Kong, Hong Kong, China

Hidekazu Suzuki
Division of Gastroenterology and Hepatology, Department of Internal Medicine, Keio University School of Medicine, 35 Shinanomachi, Shinjuku-ku, Tokyo 160-8582, Japan

Toshihiro Nishizawa
Division of Gastroenterology and Hepatology, Department of Internal Medicine, Keio University School of Medicine, 35 Shinanomachi, Shinjuku-ku, Tokyo 160-8582, Japan
Division of Gastroenterology, National Hospital Organization Tokyo Medical Center, 2-5-1 Higashigaoka, Meguro-ku, Tokyo 152-8902, Japan

Michael Hadjiargyrou
Department of Life Sciences, Theobald Science Center, Room 420, New York Institute of Technology, Old Westbury, NY 11568, USA

Nicholas Delihas
Department of Molecular Genetics and Microbiology, School of Medicine, Stony Brook University, Stony Brook, NY 11794, USA

Michelle L. Stoller, Henry C. Chang and Donna M. Fekete
Department of Biological Sciences, Purdue University, 915 W State St, West Lafayette, IN 47907-1392, USA
Purdue University Center for Cancer Research, Purdue University, 201 S University Dr, West Lafayette, IN 47907-2064, USA

Pernille B. L. Hansen and Peter Bie
Department of Cardiovascular and Renal Research, Institute of Molecular Medicine, University of Southern Denmark, DK-5000 Odense, Denmark

Tilde V. Eskildsen, Pia L. Jeppesen, Charlotte H. Jensen, Ditte C. Andersen and Søren P. Sheikh
Department of Cardiovascular and Renal Research, Institute of Molecular Medicine, University of Southern Denmark, DK-5000 Odense, Denmark
Department of Clinical Biochemistry and Pharmacology, Odense University Hospital, Sdr. Boulevard 29, DK-5000 Odense, Denmark

Mikael Schneider, Maria B. Sandberg and Lars M. Rasmussen
Department of Clinical Biochemistry and Pharmacology, Odense University Hospital, Sdr. Boulevard 29, DK-5000 Odense, Denmark

Anne Y. Nossent
Department of Cardiovascular and Renal Research, Institute of Molecular Medicine, University of Southern Denmark, DK-5000 Odense, Denmark
Department of Clinical Biochemistry and Pharmacology, Odense University Hospital, Sdr. Boulevard 29, DK-5000 Odense, Denmark
Department of Vascular Surgery, Einthoven Laboratory for Experimental Vascular Medicine, Leiden University Medical Center, 2333 ZA Leiden, The Netherlands

Maria L. Hansen
Department of Clinical Biochemistry and Pharmacology, Odense University Hospital, Sdr. Boulevard 29, DK-5000 Odense, Denmark
Department of Cardiac and Thoracic Surgery, Odense University Hospital, Sdr. Boulevard 29, DK-5000 Odense, Denmark

Niels Marcussen
Department of Pathology, University of Southern Denmark, DK-5000 Odense, Denmark

Chia-Hsien Lee, Chen-Ching Lin and Yen-Jen Oyang
Graduate Institute of Biomedical Electronics and Bioinformatics, National Taiwan University, Taipei 106, Taiwan

Wen-Hong Kuo
Department of Physiology, College of Medicine, National Taiwan University, Taipei 100, Taiwan

Hsuan-Cheng Huang
Institute of Biomedical Informatics and Center for Systems and Synthetic Biology, National Yang-Ming University, Taipei 112, Taiwan

Hsueh-Fen Juan
Graduate Institute of Biomedical Electronics and Bioinformatics, National Taiwan University, Taipei 106, Taiwan
Institute of Molecular and Cellular Biology and Department of Life Science, National Taiwan University, Taipei 106, Taiwan

Anita Quintal Gomes
Health Technology College of Lisbon – Polytechnic Institute of Lisbon, 1990-096 Lisbon, Portugal
Institute of Molecular Medicine, Faculty of Medicine, University of Lisbon, 1649-028 Lisbon, Portugal

Sofia Nolasco
Health Technology College of Lisbon – Polytechnic Institute of Lisbon, 1990-096 Lisbon, Portugal
Gulbenkian Science Institute, 2780-256 Oeiras, Portugal
Interdisciplinary Centre of Research in Animal Health (CIISA), Faculty of Veterinary Medicine, 1300-666 Lisbon, Portugal

Helena Soares
Health Technology College of Lisbon – Polytechnic Institute of Lisbon, 1990-096 Lisbon, Portugal
Gulbenkian Science Institute, 2780-256 Oeiras, Portugal
Center for Chemistry and Biochemistry, Department of Chemistry and Biochemistry, Faculty of Sciences, University of Lisbon, 1749-016 Lisbon, Portugal

Index

www.ingramcontent.com/pod-product-compliance
Lightning Source LLC
LaVergne TN
LVHW080949210926
848485LV00004B/809